Bulk Solids Handling

Bulk Solids Handling

Equipment Selection and Operation

Edited by

Don McGlinchey

Reader

Centre for Industrial Bulk Solids Handling

Glasgow Caledonian University

UK

Blackwell Publishing

Blackwell Publishing editorial offices:
Blackwell Publishing Ltd, 9600 Garsington Road, Oxford OX4 2DQ, UK
Tel: +44 (0)1865 776868
Blackwell Publishing Professional, 2121 State Avenue, Ames, Iowa 50014-8300, USA
Tel: +1 515 292 0140
Blackwell Publishing Asia Pty Ltd, 550 Swanston Street, Carlton, Victoria 3053, Australia
Tel: +61 (0)3 8359 1011

First published 2008 by Blackwell Publishing Ltd

ISBN: 978-1-4051-5825-1

Library of Congress Cataloging-in-Publication Data
Bulk solids handling : equipment selection and operation / edited by Don McGlinchey. – 1st ed.
p. cm.
Includes bibliographical references and index.
ISBN-13: 978-1-4051-5825-1 (hardback : alk. paper)
ISBN-10: 1-4051-5825-5 (hardback : alk. paper) 1. Bulk solids handling–Equipment and supplies.
I. McGlinchey, Don.

TS180.8.B8B853 2008
621.8′6–dc22

A catalogue record for this title is available from the British Library

Set in 10/12 pt Times by Aptara Inc., New Delhi, India

For further information on Blackwell Publishing, visit our website:
www.blackwellpublishing.com

Contents

Contributors

L. Bates	Ajax Equipment Ltd, Milton Works, Mule Street, Bolton, BL2 2AR, UK
J.W. Carson	Jenike & Johanson, Inc, 400 Business Park Dr., Tyngsboro MA 01879, USA
A. Cowell	School of Engineering and Computing, Glasgow Caledonian University, Cowcaddens Road, Glasgow, G4 0BA, UK
M. Jones	Division of Mechanical Engineering, The University of Newcastle, Callaghan Campus, Callaghan, NSW 2308, Australia
D. McGlinchey	School of Engineering and Design, Glasgow Caledonian University, Cowcaddens Road, Glasgow, G4 0BA, UK
D. Mills	Pneumatic Conveying Consultant, 9 Cherry Orchard, Old Wives Lees, Canterbury, Kent, CT4 8BQ, UK
J.M. Rotter	Institute for Infrastructure and Environment, School of Engineering and Electronics, University of Edinburgh, Edinburgh, EH9 3JL, Scotland, UK
N. Stanley-Wood	Selwood, Brock Road, St Peter Port, Guernsey, GY1 1RB, UK

Aims and scope

The readership is expected to be engineers, scientists and technologist, most likely process operators and chemical engineers or mechanical engineers or physicists. They may be young engineers or 'role changers' with little formal education in bulk solids handling who have responsibility for handling, storage, processing or production involving particulate materials.

The topics covered by the book will include the characterisation of individual particles and bulk particulate materials, silo design for strength and flow, pneumatic conveying systems, mechanical conveying, small-scale operations. Belt conveying will not be covered in this volume.

The material in the book will provide an overview and discuss the limitations and applications of the technology. Guidance will be given on making appropriate equipment choices. Operational issues will be discussed and will include example calculations and case studies.

The book aims to provide the reader with the breadth of knowledge to give a good general understanding of the major technologies involved in the storage and handling of particulate materials from large grains to fine cohesive materials. This will place the reader in a better position of being able to diagnose solids handling and storage problems in industry and to deal with experts and equipment suppliers from an informed standpoint.

The material contained in the book is designed to be equally applicable to engineers and scientists working in a broad spectrum of industries, for example agriculture, agrochemicals, cement, construction, food, bulk and fine chemicals industries, minerals and metals, petro-chemicals, pharmaceuticals, plastics, pigments, power generation and waste handling.

Acknowledgements

I wish to thank all the authors who made such significant contributions to this book. I thank all at Wiley-Blackwell for their advice and help in bringing the book together. I would also like to thank colleagues at Glasgow Caledonian University for their support and time.

1 Bulk powder properties: instrumentation and techniques

NAYLAND STANLEY-WOOD

1.1 Introduction

The success of powdered material process engineering and manufacturing is dependent upon the measurement and control of micro-, nano- and bio-scale particulate products as well as the behaviour of bulk powders on a macro-scale. Great strides have been accomplished with on-line, in-line and in-process characterisation of micro-scale particles, but in order to optimise the multivariate products generated nowadays, fundamental understanding must be gained in bulk processing and product behaviour at all levels of the industrial scale.

Powders can be regarded as being either two- (solid–liquid) or three-phasic (solid–liquid–gas) systems which, at times, can be treated as a single continuum. Because there is a great variety in the bulk powder properties and characteristics of such assemblies there is a need to understand the mechanico-physical properties of discrete solid particles as well as the physico-chemical interaction of adsorbed solid–fluid boundaries on particles.

Thus, there is an awareness to regard the continuous fluid phase (generally a gas) not only as a supportive phase, but also as a major contributing addition to the solid particulate phase and to treat powders as a solid–fluid system.

Powders, in reality and practice, are not solely solid units of matter, which can withstand shear and deformation under mild stress, but may, at times, also be regarded as liquids. With a degree of applied stress powders can be encouraged to flow from hoppers and bins and may, with excessive aeration, flood-like liquids. If desired these particulate systems can also be compacted into coherent shapes such as granules, pellets or compacts. Although powders can be packed tightly in packed beds and compressed into discrete forms they do not behave as gases and have no gaseous-like properties.

There are many of empirical instruments which may measure bulk powder characteristics, which affects the behaviour of powders in processing plants. And some of these empirical laboratory instruments have now been commercialised. However, the sophistication of instrumentation available for the measurement of bulk powder properties in real time is not as advanced as the instruments currently used to measure particle size on a micro- or nano-scale. The phenomenon of flow of a powder out of any orifice or from a specified piece of process plant equipment should be unambiguously termed 'rate of powder flow' and not confused with the bulk powder property of flowability which can be used discriminate the 'handability' of bulk powders and may eventually contribute to enhance the rate of powder flow.

The common definition of flowability is the ability of a powder to flow from a specific item of plant equipment at a desired degree of flowability. Flowability is generally quantified by a range of mobility, from 'free flow' to 'non-flow' which can, at times, be measured in a laboratory with specific flow property instrumentation. A bulk powder may be described as being free flowing, cohesive or non-flowing, and thus powder flowability tends to become an integral part of the description of bulk powder property measurement. Correlations between powder bulk properties in terms of both micro-scaled particulate factors as well as macro-

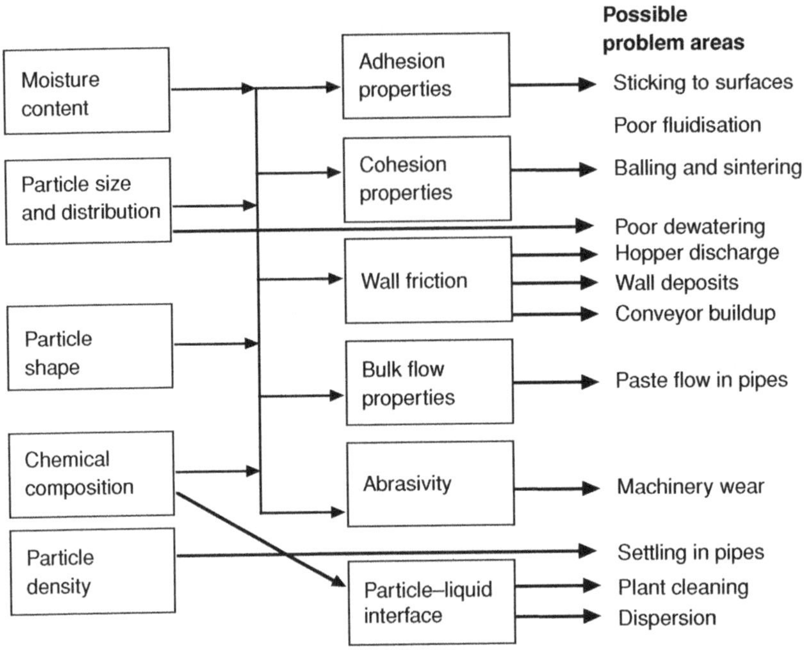

Figure 1.1 Micro- and macro-scale properties of bulk powders allied to possible production problems.

scale bulk properties can therefore enhance and possibly optimise process engineering and the handling of powders.

'Bulk powder properties' refer to the results from tests performed either under specified laboratory conditions or with pilot plant equipment under known processing conditions. With knowledge of the 'bulk properties of a powder', under specified conditions, problems in the handling and flow of the multitude of powders currently produced may be highlighted and hopefully corrected or alleviated. All too often powder handling problems experienced in modern industrial chemical and pharmaceutical plants is due mainly to inadequate information associated with macro-scale powder bulk properties of real industrial powders as opposed to idealised laboratory micro-scale particulate systems. This section will therefore itemise some of the available micro-particle and macro-powder properties that may influence, and possible prevent, some inherent problems seen in the processing of industrial powders (Figure 1.1).

To aid the process plant engineer, information must be available in terms of bulk powder behaviour under specific conditions. Powders should therefore be treated as a biphasic assembly and handled with knowledge of the interactions between a fluid and the solid external and internal surfaces. In this way, powder behaviour can be visualised as behaving in two separate manners either in terms of *fluid-like behaviour* or *compacted solid-like behaviour*. Both of these conceptual viewpoints will be considered in this section, but the emphasis will be on the fluid-like (flow) behaviour of powders.

Since in the physical world there are only three dimensions: mass, length and time; bulk powder properties can only be described in terms of the inter-relationship between these three dimensions.

1.1.1 *Density*

The relationship between mass and volume is density and the determination of this basic property is essential in the subsequent understanding of bulk powder behaviour.

1.1.2 *Particle–particle bonding*

In the relationship of mass to mass the phenomena of particle–particle and particle–fluid–particle interaction, bonding is of great importance in terms of either cohesive, or free-flowing powders. Whilst at the opposite scale of flow, the discharge from plant orifices is reliant upon the interaction between particles and plant equipment walls as seen with cohesive, non-flowing, 'sticky' or difficult to flow powders.

1.1.3 *Particle packing*

The relationship of mass or volume of the solid contents of a powder to the total volume of the bulk powder shows its influence in terms of the intensity of particle packing, because of the various degrees to which powders may be compressed. Manufactured powders may be subjected to a range of compressions which can range from that experienced in the process of fluidisation to that of high compaction, as seen in the pharmaceutical and ceramic industries when a coherent-shaped pellet, by die compaction, is at times the ultimate end product.

1.1.4 *Flowability*

The largest category of bulk powder behaviour, expressed in the terms of the three-dimensional parameters, is either a mass–time or a volume–time relationship. This category measures the rate of discharge from orifices, as opposed to the bulk powder properties of flowability and/or floodability of powders (Table 1.1).

1.1.5 *Variables contributing to bulk powder properties*

The micro- and macro-properties of powders are illustrated in Figure 1.2, which shows the influence that both particle characteristics and macro-particle (bulk powder) characteristics have upon powder flow or powder rheology and powder density or powder packing. Thus both particle and powder properties, in terms of either particle properties of size, shape and surface, or mechanico-physical powder factors together with the particle–continuum (gas or liquid) properties can enormously influence possible problems in powder processing (Figures 1.1 and 1.2).

It must be realised, however, that it is not easy, with any degree of certainty, to predict the rate of flow of powder assemblies and agglomerates from orifices solely by static individual conditions.

In order to aid the characterisation of an assembly of particles or bulk mass, knowledge of the distribution of the physical particle properties is required. Although distribution of particle size and recently particle shape can be readily obtained by laboratory instrumentation,

Table 1.1 Categories of particle assemblies in terms of mass, length and time.

Dimensional groups	*Bulk powder properties*	
Volume:volume	Packing Densification	
Mass:volume (density)	Single particle	Porous Non-porous Intra-particle space
	Assembly of particles	Compressed Aerated Tap Poured
Mass:mass	Particle–particle	Cohesiveness Tensile strength Densification Agglomeration Comminution Flowability Sedimentation
	Particle–fluid	Pneumatic transport Hydraulic slurries Fluidisation Catalysis Granulation Floodability
Mass:time/volume:time	Transportation (particle–particle fluid)	Flow rate Conveying Hopper discharge Extrusion

the distribution of other powder properties, as opposed to the physical size, shape and, at times, the physico-chemical properties of particles, is difficult to achieve and analyse.

The properties and phenomena associated with an assembly of particles and powder behaviour in processing are due to the combination of many individual particle and powder properties, some of which are listed below:

1. Particle size distribution and specific surface area
2. Particle shape distribution
3. Packing property (bulk density, voidage)
4. Rate and compressibility of packing
5. Flowability
6. Failure properties and angle of internal friction
7. Cohesion, strength and adhesion

This chapter will highlight the factors of packing (bulk density, voidage), rate of flow, compressibility of packing, flowability, failure properties and angle of internal friction, cohesion, strength and adhesion. The factors of particle size distribution, specific surface area and particle shape distribution have already been dealt with elsewhere (Stanley-Wood 2000, 2005).

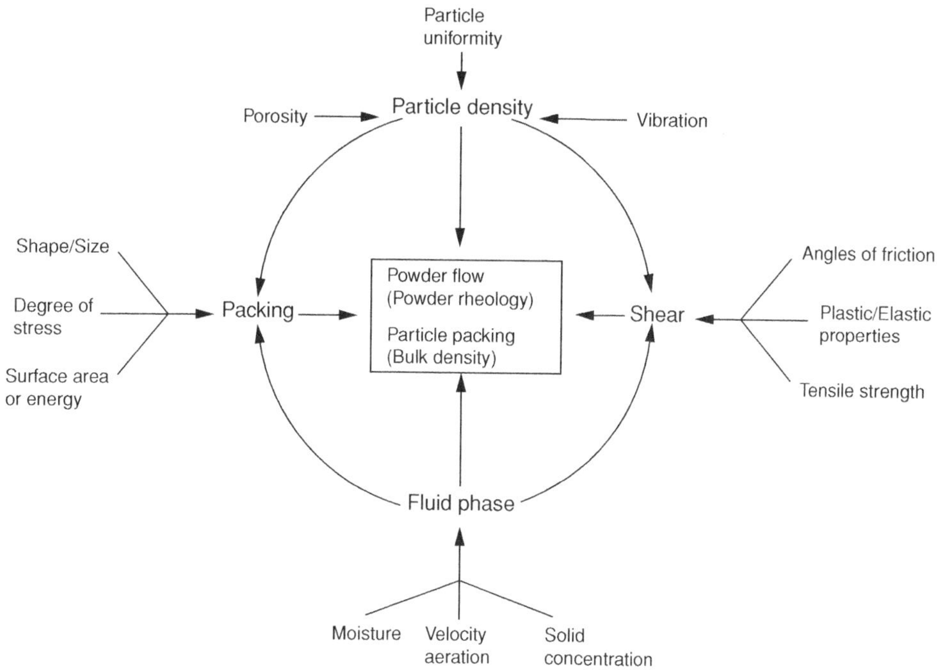

Figure 1.2 Micro-scale particle and macro-scale powder parameters contribution to bulk powder behaviour.

1.2 Particle and powder densities

Penetration into the internal structure of porous material with a fluid to determine particle density can result in ambiguous volume measurements. Thus, immersion methods have minimal use unless specific conditions are defined.

In practice:

- Large pieces of porous material must be used
- The liquid immersion method is dependent upon penetration of liquid into pores
- Agglomeration of fine particles can occur with liquids
- The degree of penetration into particle pores varies with the liquid used, the surface tension of the liquid as well as the pressure exerted
- Gas density displacement methods are limited to non-absorbable gases
- The impregnation or the coating of porous material to prevent liquid penetration is limited to large particles and not applicable to fine material

The bulk density of a powder is its mass divided by the bulk volume it occupies. This volume includes the inter-particle space between the particles as well as the particle volume and should not be confused with the term particle volume, which does not include the inter-particle space.

Bulk packed and poured densities are known to be dependent upon the rate of powder feed when material is directed into a specific volume containment. Thus the rate of powder flow,

Table 1.2 A selection of densities from National and International Standards to show the multitude of defined densities used in the process industries.

Bulk densities			
Poured	Aerated	Cap loaded	Unpacked
Tapped	Sack loaded	Bagged	Vibrated
Weight/length ratio	Pipe	Weight per unit length	
Geometric densities	Particle	Powder	Compact
	Ergun	Mercury	Pill
	Bulk	Pellet	Feed
Skeletal or internal porous structure	Adsorption	Mercury	
	Surface area	Permeability	

height of fall, impact pressure and powder yield strength together with other physical powder properties, such as particle size, particle shape, particle density and powder cohesiveness contribute influentially to powder bulk density.

The range of terms used to describe the general property of the bulk density of a powder is legion but can readily be subdivided into:

- Densities related to the bulk of the powder
- Densities related to the external geometry of the particles within a powder or
- Densities related to the internal geometry of particles within powders

Thus powder density as well as particle density may have many descriptions and definitions (BS EN 1097-3: 1998; BS EN 543: 2003; BS EN 725-8: 1997; BS EN 993-17: 1999; ISO 6770: 1982) (Table 1.2).

The bulk density of a material is not therefore a single definite number like a true solid density. Initially both particle and bulk densities were regarded as being easy to measure. Particle density is now, however, regarded as a complex material property due to the exclusion or inclusion of open and/or closed pore space within a solid mass or particle (Table 1.3). The variety of particle densities is due to the variation in penetration of the displacement fluid into a particle needed to record the volume of the porous particle when undergoing measurement. Whilst the reason why the bulk density of a powder has no unique value is because bulk densities vary as a function of the consolidation pressure applied to the powder within the containing space and volume. It is thus essential to specify the degree of compaction on an assembly of particles contained in a specified volume before a measurement value can be recorded (BS 2955: 1993; BS 4140-23: 1987; BS EN 23923-2: 1993; BS EN ISO 7837: 2001; ASTM B329-98: 2003; ISO 3923-1:1979).

Although there is no direct linear relationship between the flow of a powder and its bulk density, the bulk density, with its numerous variations in definition (Table 1.3), is important because of bin or hopper capacity:

- Indication of the size of particles, which will either flood or pack together
- Determination of the compression strength of powders or compaction strength of powders
- Classification of particles into fluidisable and non-fluidisable powders

Table 1.3 Glossary of terms used to define particle and powder densities.

Particle densities	
Apparent particle density	Mass of a particle divided by the volume of the particle excluding open pores but still including the closed pores The mass of a particle divided by its apparent volume
Apparent particle volume	The total volume of the particle, excluding open pores, but including closed pores
Effective particle density	Mass of a particle divided by the volume of the particle including open pores and closed pores The density of particles as determined by a given fluid displacement method
Effective particle volume	The mass of the particles divided by the effective particle density
Envelope volume	The external volume of a particle, powder or monolith such as would be obtained by tightly shrinking a film to contain it
True particle density	Mass of a particle divided by the volume of the particle excluding open pores and closed pores
Powder densities	
Absolute powder density	Mass of powder divided by the absolute powder volume
Absolute powder volume	Space occupied by a powder excluding pores, interstices and voids
Absolute volume	The volume of the solid matter after exclusion of all the spaces (pores and voids)
Apparent powder density	Mass of powder divided by the apparent volume
Apparent powder volume	The total volume of solid matter, open pores and closed pores and interstices
Bulk powder density	The apparent powder density under defined conditions
Effective solid density	The density of powdered particles by a specified liquid displacement method
Green density	Apparent density of a compact before sintering or other heat treatment
Pressed density	Apparent density of a compact
Immersed density	Mass of a powder expressed as a percentage volume of suspension media displaced
Tap density	Apparent powder density of a powder bed formed in a container of stated and specified dimensions when subjected to vibration or tapping under known and stated conditions
Aerated density	Apparent powder density of a powder bed formed in a container of stated and specified dimensions when subjected to minimum compression to ensure an open loosely packed volume
Poured density	Aerated bulk density; the minimum bulk density of a motionless powder
Pressed density	The apparent density of a compact
Voids	The space between particles in a bed of powder

Since powders can be regarded as a mixture of solid and fluid (usual air) the density for a bulk powder can vary between that of the continuum (gas/fluid) and the solid phase. The value of bulk density is thus dependent upon the packed state of the powder; and there may be a range of densities for one powder:

(i) An aerated bulk density
(ii) Poured bulk density
(iii) Tap (or tapped) density
(iv) Fluidised bulk density
(v) Compressed or compacted density

all of which are dependent upon the history or treatment which the powdered sample has undergone prior to bulk density measurement. Although each method can offer reproducibility of measurement, these laboratory results may not mimic or represent the actual compaction or stress which powders made undergo in process plant.

Results of the variation of bulk density with the degree of compaction, typically shown as a log–log plot, can be beneficial in the determination of the type of plant equipment required for transfer, storage and processing (Carson & Marinelli 1994) (see Section 1.3.5).

1.3 Determination and protocol for bulk density

1.3.1 *Aerated bulk density*

Confusion exists as to when a powder is aerated or not aerated. An aerated bulk density must be determined under specific conditions (BS EN 1097-3: 1998).

Particles should be separated from one another by a film of fluid either air (or a specified gas) and should not be in direct contact with each other.

Equipment for an aerated bulk density generally consists of a screen cover, a screen sieve, a spacer ring and a chute attached to a vibrator. A stationary non-vibrated chute is aligned with the centre of the pre-weighed cup. Powder, when poured through the vibrating sieve, falls a fixed height (generally 25 cm), via the guiding stationary chute, into a standard 100 mL cylindrical cup (Figure 1.3).

The vibration of the chute, which does not impart any consolidation force to the powder, is governed to ensure that the standard cup is filled within 30 seconds. Excess powder is removed without causing any compaction of the loosely vibrated powder and the cup and powder weighed.

1.3.1.1 *Hosokawa micron powder tester* A practical way in which reproducibility of aeration or the attainment of conditions of low stress packing can be achieved is to use a Hosokawa Powder Tester (Hosokawa Micron Ltd) (Figure 1.3). This powder tester deposits an aerated powder into a test cup, which thus gives an aerated volume. From the mass of this aerated bulk powder the density, with a very open structure, can be calculated.

Fluidisation may also be used to measure aerated bulk densities, but this technique is normally used to determine the Ergun particle density (see Section 1.4.4).

1.3.1.2 *Carr's aerated and packed bulk density box* A square box with a bottomless extension containing a coarse screen (10 mesh or 2 mm) positioned slightly up from the bottom of the extension was constructed and assembled (Figure 1.4). Material is screened into the known dimensioned bottom box. After removal of the extension and levelling of the powder in the bottom box, the box is weighed. This weight times an appropriate numerical factor gives the aerated bulk density (Carr 1965b).

The packed density is determined by replacement of the extension box onto the powder filled aerated bottom box. Placement of more powder into the extension box, clamping the two together and vibration for 5 minutes packs the powder into the bottom box. Extension removal, levelling of excess powder and weighing gives, with the appropriate factor, the packed bulk density. This technique has now been incorporated into a standard test

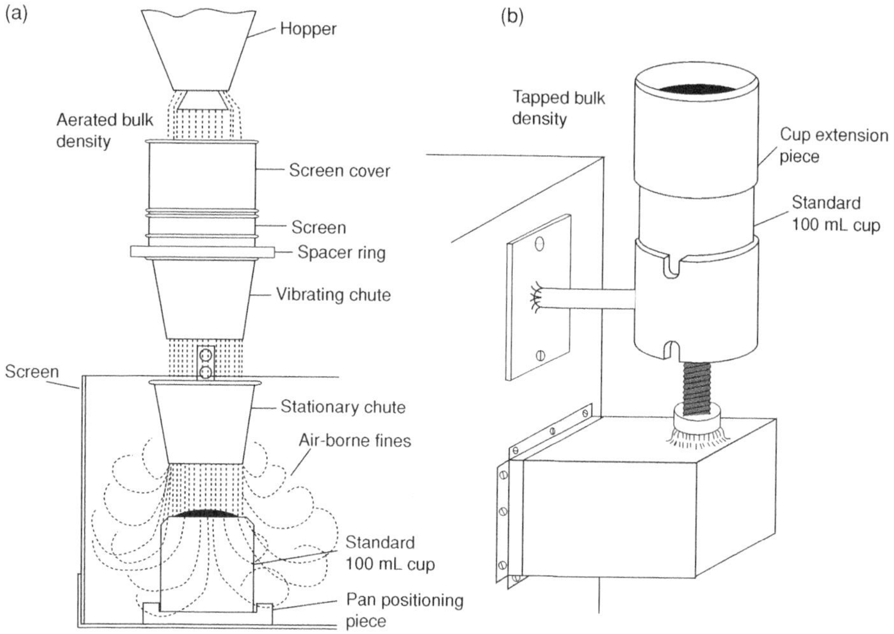

Figure 1.3 Hosokawa apparatus for aerated (a) and tapped (b) density.

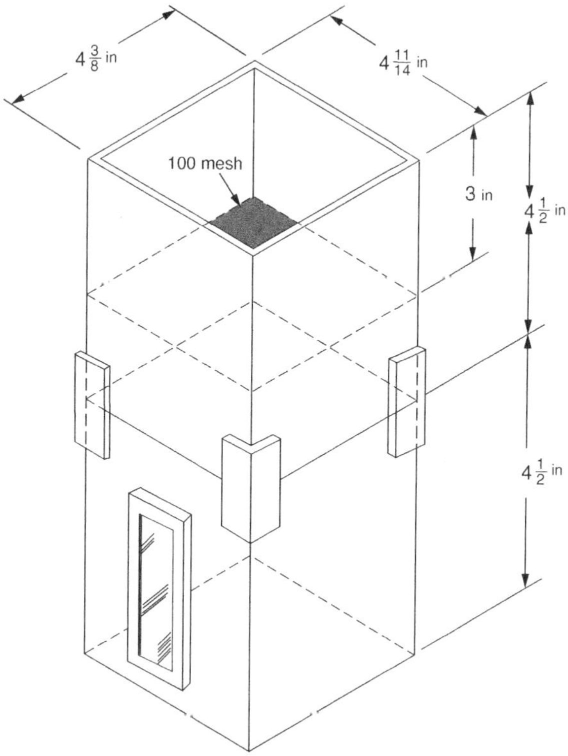

Figure 1.4 Carr's aerated and tap density box (Carr 1965c; ASTM D 6393-99: 1999).

method (ASTM D 6393-99: 1999) for bulk solid characterisation by Carr's indices, and has been applied to improve the flowability, handling and storage of distiller's grains (Ileleji 2005).

1.3.2 *Poured bulk density*

This is probably the most widely used bulk powder property: although different industries have different protocols in its measurement (ISO 3923-1: 1979; ASTM B329-98: 2003).

Generally, since few industries have a standard procedure, the methodology consists of a funnel, with a short down comer containing a tap or stop, which when opened pours powder into a known volume or measurement container. The container usually has a 2:1 length to diameter ratio. Standardisation of the height of fall is essential; as height of fall or gravitational impaction force affects the degree of particle packing and compaction. This type of stress can change values of any poured bulk density measured.

1.3.2.1 *Scott volumeter* Since the definition of bulk density is related to the apparent powder volume under defined conditions, the methodology outlined in ISO 3923-2 (1981) and EN 23923 (1992) may be used to determine the bulk density of metals and other powders using the Scott volumeter to give an 'as poured' bulk density.

1.3.3 *Tap density*

Application of external energy into the system by vibration or tapping causes the particles to rearrange and give denser packing.

Tap density is the bulk density of a powder compacted by tapping or being subjected to vibration. A specified mass or volume of powder should be poured into a container or measuring cylinder and then tapped against a hard surface from a standard height (usually 15 mm) for a specified number of times (usually 10).

Various methods recommend a fixed mass (10–100 g) or a fixed volume (usually 50 mL) with a special density cup coupled with an extension to accommodate the extra powder that will eventually flow into the tapped volume and fill the fixed volume cup.

The end point can be determined either by a specified number of taps, their frequency or the number of taps to reach constant volume.

Use of a mechanical reproducible tapping device is better than manual tapping in order to achieve standardisation and more reproducible tap densities. The Hosokawa Powder Tester (Figure 1.3) has a cam-operated tapping unit which displaces a standard cup of 100 mL upwards and thus allows it to fall downwards under gravity at a rate of once every 1.2 seconds. A cup extension is fitted to the top of the standard cup and powder is continuously added during the tapping process. This ensures that powder is never below the top of the standard cup. Usually after 180 taps (216 s) excess powder is lightly scraped off from the top of the standard cup. Cup and powder are then weighed to calculate tap bulk density.

There is, however, an enormous variety of commercial instruments available for the measurement of tap density.

1.3.3.1 *Hall test cup method* This standard test cup has a 25 mL volume with a 25 mL extension volume. After the cup and extension have been filled and excess powder removed – similar to the Hosokawa method – the sample is tapped for a given number of cycles using a sieve shaker. The cup extension is removed after tapping has been completed, and excess powder scraped off the upper surface of the cup. The standard cup, filled with a known volume of powder, is removed and weighed to eventually give the tap bulk density. Vibration of the cup and extension is *not recommended* because of low non-reproducible results, although in one American Standard (ASTM D-4180-03) a vibrated funnel is used.

1.3.3.2 *Kostelnik and Beddow Ro-Tap method* A Ro-Tap method was used by Kostelnik and Beddow (1970) for tap bulk densities and they proposed that when the aerated bulk density was determined for subsequent use in the Hausner ratio (Section 1.4.1) the same standard test cup should be used. Visual determination of the volume end point is difficult and thus the weight end point methodology is preferable.

For particles in the range 0.8–4.8 mm, a 250 mL cylinder is recommended with mechanical tapping. The tapping unit is marketed as a Tap-Pak Volumeter (J. Engelsmann Company). Powder poured into this measuring cylinder is tapped 1000 times and the volume measured. From the initial weight of powder a tapped bulk density can be determined.

1.3.4 *Fluidised bulk density*

The faster a fluidisable phase (liquid or gas) is moved through a bed of powder the greater the volume the fluidised powdered material will have. Thus, the weight per unit volume of a fluidised bed or pneumatically conveyed powder may be less than the aerated bulk density. This bulk density is sometimes referred to as the weight per unit length density, pipe density or weight/length ratio (Table 1.2).

1.3.5 *Compressed and compact bulk density*

The density of a powder compressed or compacted to a degree greater than that achieved by tap densities is, at times, required as a ceramic, metallurgical or pharmaceutical end product.

A civil engineering uniaxial oedometer or consolidometer (BS 1377-7: 1990; ISO/TS 17892-5: 2004) consists of a cylindrical ring container which compresses, between two porous plates, a powdered sample at known normal forces in the range 10–3200 MPa. The porous plates allow fluids (gases/liquids) to flow through the compressed media.

A coefficient of volume compressibility or coefficient of consolidation can be calculated from the thickness, volume or void ratio versus logarithmic applied stress relationship.

The compaction characteristic of powders obtained by this uniaxial technique may also be used to determine the elasticity, plasticity, deformation, fragmentation and coherency of cohesive and non-cohesive powders (Stanley-Wood 1983, 1987).

1.4 Flow properties from powder bulk densities

1.4.1 *Hausner ratio*

The Hausner ratio is defined as the ratio of tap density to aerated bulk density. The Hausner ratio (1966, 1967) was initially used in the characterisation and evaluation of the bulkiness, particle shape and degree of friction required to prevent the flow of metal powders. The

Hausner ratio, in shape analysis, the behavioural effects in fluidisation, mixing and handling, is a simple factor which is relatively easy to measure but which has been found to have worthwhile industrial powder qualitative assurance potential.

It has been subsequently used to characterise the flow, sieving and compaction of powders and many other aspects of powder technology for a diversity of industrial powders (Adler 1969; Kostelnik *et al.* 1968; Roberts & Beddow 1968; Woodcock & Mason 1987; Wong 2000).

The most complicated technique for the determination of powder flowability is that proposed by Jenike (Section 1.6.1), in which the ratio of the maximum compaction stress, σ, to an unconfined stress, f_c, at a specified shear stress of 3 kPa – the flow function, *ff* – is calculated to predict the ease of powder flow. This Jenike flow function, $ff = \sigma_1/f_c$, has remained the main consensual parameter designated for mass flow hopper design and the flowability of powders.

Using the Jenike concept of the ratio of maximum compaction stress to an unconfined stress – flow function *ff* – Stanley-Wood and Abdelkarim (1982) and Abdelkarim (1982) superimposed the experimental parameters of unconfined and maximum compaction stresses, for a range of industrial powders (liquorice, chalk 2, fine talc, chalk 1 and di-calcium phosphate (DCP)) classed as either free flowing/non-cohesive, intermediate flowability or cohesive, onto the various degrees of cohesiveness postulated by the Jenike flow function, as outlined in Bulletin 123 (Figure 1.5).

Subsequently, this collection of powders was re-investigated in terms of a simple modified Hausner ratio or flow ratio (FR) (Stanley-Wood *et al.* 1993) to discriminate between those powders which could readily flow and those which had some degree of cohesiveness. This

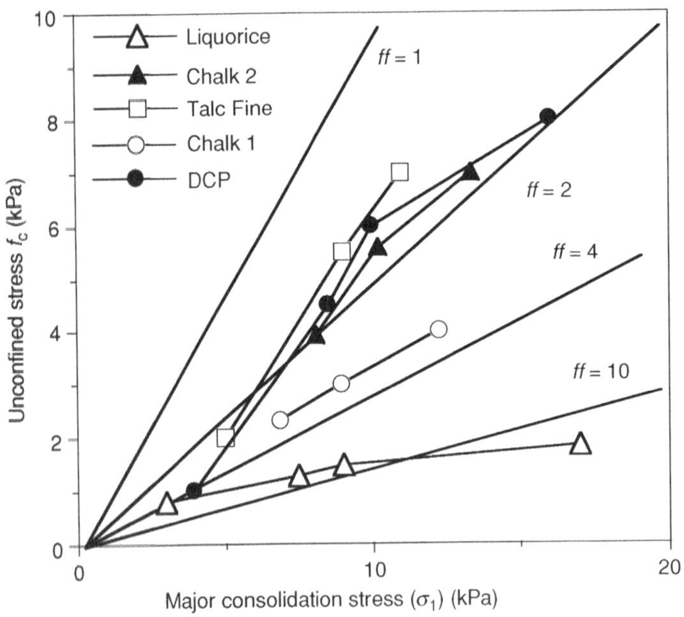

Figure 1.5 Jenike flow function *ff* (degree of flowability), for a selection of industrial powders.

Table 1.4 Calculation of flow ratios, for a variety of industrial powders, from the tapped and aerated densities.

Powder cgs units	Tapped density (gm/cm^3) TD	Aerated density (gm/cm^3) AD	Flow ratio (TD/AD) (−) FR	Comments FR < 1.25 free-flowing >1.55 cohesive
Liquorice-HSC (Fraction 153.5 μm)	0.461	0.400	1.15	Free flowing
Microcoat D	0.4627	0.3688	1.25	Free flowing
Microcoat H	0.4950	0.3869	1.28	Intermediate
Liquorice-BMC (Fraction 50.5 μm)	0.732	0.473	1.28	Intermediate
Liquorice-BMC (Fraction 153.5 μm)	0.530	0.457	1.28	Intermediate
PVdC (ICI)	0.5328	0.3964	1.34	Intermediate
Liquorice-HSC (Fraction 76.5 μm)	0.526	0.390	1.35	
Liquorice-HSC (Fraction 50.5 μm)	0.598	0.409	1.46	Intermediate
Liquorice-BMC (Fraction 50.5 μm)	0.732	0.473	1.55	Intermediate/cohesive
HBA sample 1	0.8476	0.5364	1.58	Cohesive
TiO_2 pigment	1.2709	0.789	1.61	Cohesive
HBA sample 2	0.79117	0.4883	1.62	Cohesive
Aerated acidulant	1.3197	0.7714	1.71	Cohesive
TiO_2 MS	1.1704	0.5904	1.98	Cohesive
Powdered sugar	1.0378	0.5141	2.02	Cohesive
Calcium carbonate (Chalk 1)	0.326	0.155	2.10	Cohesive
Di calcium phosphate	1.1126	0.4900	2.27	Cohesive
Chalk 2	0.980	0.363	2.70	Cohesive
Steamic talc	0.5401	0.1754	3.08	Cohesive
Fine talc	0.3470	0.0982	3.53	Cohesive

Note: HSC, high speed cutter; BMC, ball mill comminuted.

simple FR was then compared with other, more elaborate and time-consuming measurement techniques for a range of industrial powders (Table 1.4).

Figure 1.6 shows the correlation between the complex Jenike failure function *ff* with the simple FR (Table 1.4). These results indicate that for values of FR less than 1.25, the industrial powders tested were free flowing. Above an FR value of 1.55, the powders could be classified as cohesive with intermediate degrees of flowability between FR values of 1.25 and 1.55 as illustrated in Table 1.5.

1.4.2 *Carr's percentage compressibility from bulk densities*

Although the Hausner ratio or FR is a reliable indicator for the prediction of the flowability of powders the ratio may not represent the actual compaction behaviour a powder may

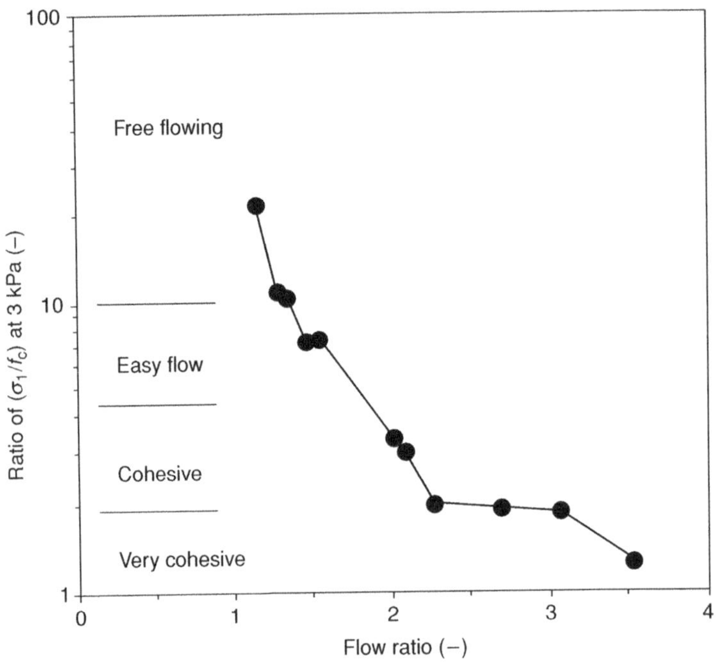

Figure 1.6 Relationship between Jenike failure function ($ff = \sigma_1/f_c$) and a simple flow ratio (FR) for various industrial powders.

undergo when subjected to industrial powder processing. Powder compressibility can be expressed in alternative ways such as the difference between tap and aerated densities as defined by Carr (1965b,c), rather than as a ratio:

$$\text{Carr's compressibility} = 100 \times \frac{\rho_t - \rho_A}{\rho_t} \tag{1.1}$$

where ρ_t is the tapped and ρ_A is the aerated density of the bulk powder.

Compressibilities in excess of 20% tend to create particle bridges in hoppers while powders with a compressibility percentage greater than 40% are difficult to discharge (Table 1.6). At present the Hausner ratio is used more widely than compressibility, although Carr (1965c) used his parameter extensively to classify various types of powder into categories of excellent or very poor flow. The Carr approach to flow has now been subsumed into a standard test method for bulk solid flowability (ASTM D 6393-99: 1999).

The Carr's compressibility percentages have been effectively correlated to measure:

- Uniformity of size
- Uniformity of shape

Table 1.5 Simple modified flow ratio (FR) of free flowing and cohesive powders.

Flow ratio < 1.28:	powder tends to be free flowing
1.28 < flow ratio < 1.57:	powder has an intermediate flowability
Flow ratio > 1.57:	powder is cohesive

Table 1.6 Carr's percentage compressibility of materials.

5–15%	Free flowing – excellent flow granules
12–16%	Free flowing – good flow powders
18–21%	Fair to passable powdered granule flow
23–28%	Easy fluidisable powders – poor flow
33–38%	Cohesive powders – very poor flow
28–35%	Cohesive powders – poor flow
>40%	Cohesive powder – very very poor flow

- Hardness of particles
- Bulk density
- Cohesiveness of particles
- Moisture content
- Relative flow
- Potential strength of arches in bin/hoppers

An alternative and possibly a more comprehensive parameter may be one where the degree of powder compaction is related to the degree of applied pressure in a log–log plot of bulk density versus consolidation pressure. This degree of compressibility and degree of relaxation is beneficial for the control of paste consistency or grindability of plastic-like materials as well as the determination of capacities required for processing and storage. Jenike (1964) and Carson and Marinelli (1994) plotted log bulk density versus log consolidating stress and determined the slope from the relationship:

$$\rho = \rho_0 \left(\frac{\sigma}{\sigma_0} \right) B \qquad (1.2)$$

where ρ and ρ_0 are the bulk densities at stress σ and σ_0 respectively, σ_0 being an arbitrary stress chosen to be less than σ. For bulk solids the slope B was found to vary between zero and unity. Most industrial and pharmaceutical powders, however, fell in the range 0.1–0.01 when compressed at high uniaxial stress (Prescott & Barnum 2000).

1.4.3 *Compacted bulk density*

The density of powders compacted with greater forces than those applied by the mechanism of tapping can be used to ascertain the hardness, deformability and mechanico-physical properties of powders.

When porous material is broken into small pieces, particle density tends to increase because of the elimination of internal pores within the particles. This increase in density with particle size decrease will ultimately progress towards the true density of the solid, because internal pores will have been slowly eliminated leaving smaller and smaller pores in smaller and smaller particles.

With catalysts, sintered products and porous materials the Ergun gas flow method can determine a particle density of porous material regardless of particle size and porosity because the external gas flows around the porous particles and does not penetrate the porous interstructure of the particle (see Section 1.4.4).

The problem of determining particle density of porous material is essentially dependent upon the determination of the enveloped (apparent) volume of the material, since the determination of the weight of the material presents no problem.

The magnitude of the void space within a complex matrix of a highly densified material becomes difficult to measure because of the need to distinguish between intra-particle and inter-particle space. Bockstiegel (1966) and Stanley-Wood and Johansson (1980) found that, for metals and inorganic powders, the inter-particle space in powder matrices disappeared in an orderly fashion; smaller and smaller spaces disappearing with increase in densification or compaction stress. It is, therefore, beneficial to measure the void volume in terms of void sizes rather than just total volume within a solid containing a range of void sizes. Void sizes in the range from 10 nm to 100 μm may be measured by the use of two techniques. One method is the adsorption of a fluid onto the external and into the internal surfaces of a compact, whilst the second method relies upon the penetration of mercury under pressure into the voids of the compact (Stanley-Wood 2000, 2005). The influence of voidage or volume of space contained either in intra- and inter-particulate systems has a marked effect upon the mechanical strength of granulated material and uniaxially compacted powder (Table 1.7) (Stanley-Wood 1987; Stanley-Wood & Abdelkarim 1983a,b; Stanley-Wood & Sarrafi 1988; Stanley-Wood et al. 1990).

There is no specialised national or international standard procedure to test all the mechanical and physical tests which may be applied to highly dense and compacted materials. A selection of national or international standards allied to bulk powder properties is, however, shown in Table 1.7.

1.4.4 *Ergun particle density*

In fluidised and fixed beds, knowledge of particle density is necessary for calculation of the effective void bed volume (Ergun equation: Ergun 1952) and specific surface areas of solids by permeametry using the Kozeny–Carman equation (BS 4359: Pt 2: 1982; Carman 1937).

Kozeny–Carman equation

$$S_k = \left[\frac{A\varepsilon^3\Delta p}{K(1-\varepsilon)^2 L\eta q}\right]^{1/2} \tag{1.3}$$

where S_k is the effective permeability volume-specific surface of a powder assuming that only viscous flow occurs in the determination of permeability; A the cross-sectional area of a bed of powder perpendicular to the direction of the flow of gas, generally air; ε the porosity of the bed; Δp the pressure difference across the bed of powder; K the Kozeny constant (usually taken as 5); L the linear dimension of a bed of powder parallel to the direction of the flow of air (commonly known as the height of the powder bed); η the viscosity of air at its temperature at the time of the determination, q the rate of flow of an incompressible fluid through a bed of powder.

The Ergun density may be determined from a well-mixed bed of porous, non-cohesive powder of known weight by fluidisation at a known pressure drop, Δp. After the gas flow has been stopped the height of the settled bed is measured. An Ergun density may then be calculated from the Kozeny–Carman equation. Use of the Kozeny–Carman equation for the determination of the Ergun particle density becomes, however, invalid with coarse powders

Table 1.7 A selection of National and International Standards allied to bulk powder properties.

Compression/compaction	
Uniaxial	ASTM B331-95 (2002): Standard Test Method for Compressibility of Metal Powders in Uniaxial Compaction ASTM E9-89a (2000): Standard Test Methods of Compression Testing of Metallic Materials at Room Temperature
Moisture effects	ASTM D1075-96 (2000): Standard Test Method for Effect of Water on Compressive Strength of Compacted Bituminous Mixtures
Hardness	
Micro hardness	ASTM E1268-01: Standard Practice for Assessing the Degree of Banding or Orientation of Microstructures
Brinell	ASTM E10-01: Standard Test Method for Brinell Hardness of Metallic Materials
Indentation	ASTM E103-84: (2002): Standard Test Method for Rapid Indentation Hardness Testing of Metallic Materials
Strength	
Fracture	ASTM E1290-02: Standard Test Method for Crack-Tip Opening Displacement (CTOD) Fracture Toughness Measurement ASTM B646-04: Standard Practice for Fracture Toughness Testing of Aluminium Alloys BS ISO 12108 :2002 Method for Determination of the Rate of Fatigue Crack Growth in Metallic Materials. Fatigue crack growth rates of above 10^{-8} m per cycle.
Tumble test	ASTM C421-05: Standard Test Method for Tumbling Friability of Preformed Block-Type Thermal Insulation ASTM C1421-01b: Standard Test Methods for Determination of Fracture Toughness of Advanced Ceramics at Ambient Temperatures
Crack displacement	ASTM E1290-02: Standard Test Method for Crack-Tip Opening Displacement (CTOD) Fracture Toughness Measurement
Compact tensile	
Strength	BS 1881-102: 1983 (withdrawn) now BS EN 12350-2: 2000 Method for the Determination of Slump BS 1881-103: 1993 Method for the Determination of Compacting Factor BS 1881-119: Testing Concrete. Method for determination of compressive strength using portions of beams broken in flexure BS 1881-207: 1992 Testing Concrete. Recommendations for determination of strain in concrete
Toughness	
Fracture	ASTM B646-04: Standard Practice for Fracture Toughness Testing of Aluminium Alloys ASTM B909-00: Standard Guide for Plane Strain Fracture Toughness Testing of Non-Stress Relieved Aluminium Products ASTM E399-05: Standard Test Method for Linear–Elastic Plane–Strain Fracture Toughness K_{Ic} of Metallic Materials BS 5447: 1987 (withdrawn) BS 7448-1: 1991 Methods of Test for Plane Strain Fracture Toughness (K_{Ic}) of Metallic Materials BS 7448-2: 1997 Fracture Mechanics Toughness Tests. Method for Determination of K_{Ic}, Critical CTOD and Critical J Values of Welds in Metallic Materials BS 7448-3: 1997 Fracture Mechanics Toughness Tests. Method for Determination of fracture toughness of metallic materials at rates of increase in stress intensity factor greater than 3.0 MPa $m^{0.5}$ s^{-1} BS 7448-4: (1997) Fracture Mechanics Toughness Tests – Part 4: Method for Determination of Fracture Resistance Curves and Initiation Values for Stable Crack Extension in Metallic Materials

(*continued*)

Table 1.7 (*continued*)

Abrasion	
Attrition	ASTM D4058-96 (2006): Standard Test Method for Attrition and Abrasion of Catalysts and Catalyst Carriers ASTM D4060-01: Standard Test Method for Abrasion Resistance of Organic Coatings by the Taber Abraser
Tumble test	ASTM E279-97 (2005): Standard Test Method for Determination of Abrasion Resistance of Iron Ore Pellets and Sinter by the Tumbler Test ASTM C421-05: Standard Test Method for Tumbling Friability of Preformed Block-Type Thermal Insulation ASTM D441-86 (2002): Standard Test Method of Tumbler Test for Coal
Physico-mechanical	
General	BS 1377-7: 1990 (2003) Methods of Test for Soils for Civil Engineering Purposes. Shear Strength Tests (Total Stress) BS 1377-8: 1990 Methods of Test for Soils for Civil Engineering Purposes Shear. Strength Tests (Effective Stress) BS 1377-9: 1990 (2003) Methods of Test for Soils for Civil Engineering Purposes. In-Situ Test BS 1377-4: 1990 Methods of Test for Soils for Civil Engineering Purposes. Compaction related Tests BS 1377-6: 1990 (2003) Methods of Test for Soils for Civil Engineering Purposes. Consolidation and Permeability Tests in Hydraulic Cells and with Pore Pressure Measurement BS 1377-5: 1990 (2003) Methods of Test for Solids for Civil Engineering Purposes. Compressity, Permeability and Durability Tests
Bulk density	BS 6070-0: 1981 (2005) Methods of Sampling and Test for Sodium Carbonate for Industrial Use – General introduction and Determination of Pouring Density BS 5551 Part 3, Section 3.1 (1993). Replaced by BS EN 11236: 1995 Fertilisers. Physical Properties Method for the Determination of Bulk Density (Loose) BS 5551 Part 3, Section 3.2 (1983). Replaced by BS EN 1237: 1997 Fertilisers. Physical Properties Method for the Determination of Bulk Density (Tapped) BS 5551 Part 3, Section 3.3 (1989). Replaced by BS 5551-3.3: 1992 Fertilisers. Physical Properties. Method for the Determination of Bulk Density (Loose) of Fine-Grained Fertilisers BS 5551 Part 3, Section 3.6 (1989). Replaced by BS EN 12047: 1997 Fertiliser. Physical Properties Method for the Determination of Static Angle of Repose of Solid Fertilisers ISO 3923-1: 1979 Metallic Powders – Determination of Apparent Density – Part 1: Funnel method ISO 3923-2: 1981 Metallic Powders – Determination of Apparent Density – Part 2: Scott Volumeter Method

Source: ASTM eStandardsStore at www.webstore.ansi.org, British Standards at www.bsonline.bsi-global.com and ISO at www.iso.org.

(Ergun & Orning 1949). A modification of this fluidisation technique has been proposed by Geldart (1990, 2005).

In an attempt to classify the fluid behaviour of powders, Geldart (1972) used the simple method of grouping fluidisable powders into four groups. The three most important groups being:

Group A – Aeratable: material which shows a range of smooth particulate expansion between the minimum fluidisation velocity, Um_f and the velocity at which bubbles form, Um_b.

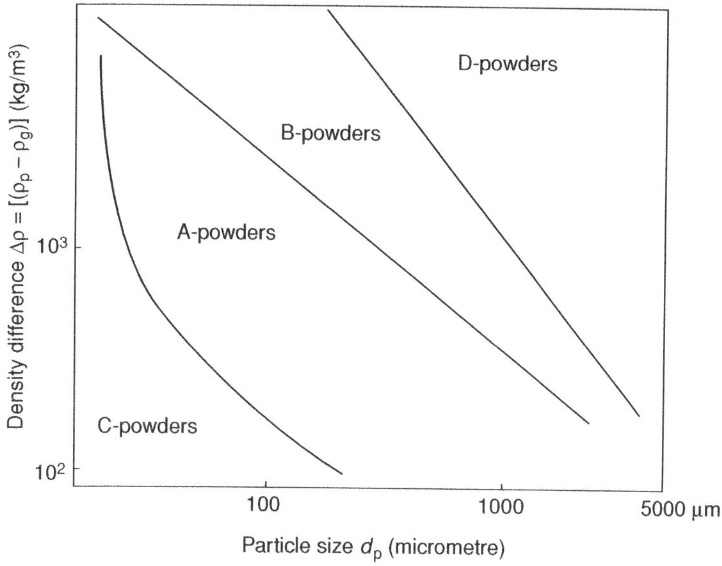

Figure 1.7 Geldart's empirical classification of aeratable (A), bubble (B), cohesive (C) and dense (D) fluidisable powders.

Group B – Sand like: material which shows bubbles for all gas velocities above Um_f.
Group C – Cohesive: material in which gas flow occurs through channels rather than distributed throughout the interstitial voids.

These empirical observations (Figure 1.7) give a rough indication of the categories of powders, which may possibly give storage and flow problems. This simple relationship between particle size and the difference between particle and fluid densities fails, however, to take into consideration other fundamental particle properties of cohesiveness, surface attraction or elastic/plastic deformability of materials. Rietema and Hoebink (1977) and Clift (1985) stated that the real differences between Groups A, B and C resulted from inter-particle forces. Seville and Clift (1984) showed that by increasing the loading on particles – using an involatile liquid to increase the capillary forces– the resultant increase in surface attraction caused a movement in powder behaviour from an initial Group B (freely fluidisable powder) through Group A and eventually to a Group C powder which showed various degrees of cohesiveness.

Inclusion of other bulk powder properties such as cohesion (C), gas viscosity (μ) and gravity (g), to Geldart empirical classification, can give a more informative dimensionless correlation. The correlation of the Archimedes number $[\rho_g d_p^3(\rho_p - \rho_g)g/\mu^2]$ with a modified Reynolds number $[C/(\rho_g g d_p)]$ can extend the empirical Geldart fluidisation classification model to other powder handling operations such as mixing, transportation and separation.

Both Molerus (1982) and Rietema (1984) extended the Geldart classification to a dimensionless correlation for bulk powder behaviour (Figure 1.8). A more sophisticated interpretation which takes into account particle interaction as well as particle forces has been proposed by Molerus (1980).

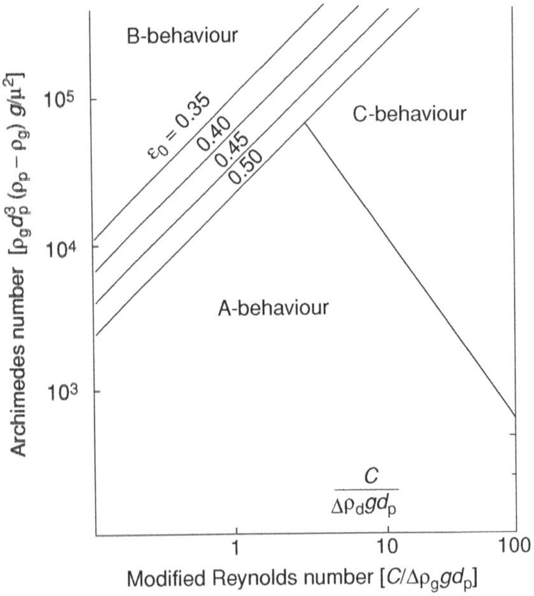

Figure 1.8 Archimedes number versus modified Reynolds number for aeratable (a), bubbling (b) and cohesive (c) powders at various voidages, ε (Molerus 1982; Rietema 1984).

1.4.5 *Application of the Hausner ratio to fluidised powder systems*

Geldart *et al.* (1984) in an extension of the work on the classification of fluidisable powders used the ratio of tapped density to a loosely packed density (Hausner ratio) to predict the type of particle movement that can occur with powders having various degrees of cohesiveness. Re-appraisal of this work shows that with a Hausner ratio below 1.27 (Table 1.8), all the powders investigated could be classified as aeratable. Powders with a Hausner ratio above 1.5 were, however, cohesive and difficult to fluidise. There was also an intermediate group between the Hausner ratios of 1.3 and 1.5 which showed intermediate fluidisable properties and behaviour.

1.4.6 *Floodability*

Floodability of a material is its tendency to have a liquid-like flow characteristic, which is caused by aeration within the powder. Powders can, when discharged from a hopper, behave in an unstable, liquid-like manner, which may be discontinuous, gushing and uncontrollable. The reasons for flooding are generally due to excessive aeration within the powder bed, which leads to easy powder fluidisation and ultimately to flooding.

Floodable powders tend to have:

- Large available external surface areas (i.e. small particles)
- Tends to adsorb gases

Table 1.8 Hausner ratio and ease of fluidisation.

Powder	Density (kg/m^3)	Mean particle size dp (mm)	Hausner ratio (–)	Flowability
Fly ash FO	364	79	1.001	Aeratable
Fly ash F3/7	369	68	1.002	Aeratable
Fly ash FSG	638	125	1.007	Aeratable
Fly ash F2/5	418	77	1.009	Aeratable
Glass B20	1597	26	1.140	Aeratable
Alumina 994	1810	55	1.156	Aeratable
Catalyst 11	1542	120	1.163	Aeratable
Metal oxide 1P	7360	30	1.165	Aeratable
Alumina FRFS	2430	0	1.195	Aeratable
Catalyst C1	1117	51	1.212	Aeratable
Alumina a320-0	3970	28	1.270	Aeratable
Alumina a320-N	3970	30	1.316	Intermediate
Alumina	2430	21	1.320	Intermediate
Alumina a360	3970	24	1.381	Intermediate
Alumina a500	3970	15	1.504	Cohesive
Metal oxide 10	5000	28	1.543	Cohesive
Alumina FRF20	2430	12	1.640	Cohesive
Plastic corvic	1637	30	1.762	Cohesive
Snowcal	2700	3	1.767	Cohesive
Aluminas FRF40	2430	10	1.797	Cohesive
Conic/U	1637	26	2.142	Cohesive
Alumina	2430	5	2.161	Cohesive
Lactose	1508	4	2.186	Cohesive
Perlite	2000	9	2.529	Cohesive

- Spherically shaped particles of a small uniform size
- Do not agglomerate and remain discrete particles
- Porous particles with large internal surface areas
- Low particle densities and are very dusty (10–75 μm)

Carr categorised the potential for powders to flood by using four parameters, each parameter being given a 'weighting' of 25 points. Thus 25 points were allotted to each of the following factors:

- Angle of fall (see Section 1.5.3 and Figure 1.9)
- Angle of difference (see Section 1.5.4)
- Dispersibility (see Section 1.4.8 and Figure 1.10)
- Flowability (see Section 1.4.7)

Summation of the above four factors contributes to a range of performance on a scale 0–100.

Very floodable:	80–100 points
Floodable:	60–79
Inclined to flood:	40–59
Could flood:	25–39
Non-floodable:	0–24

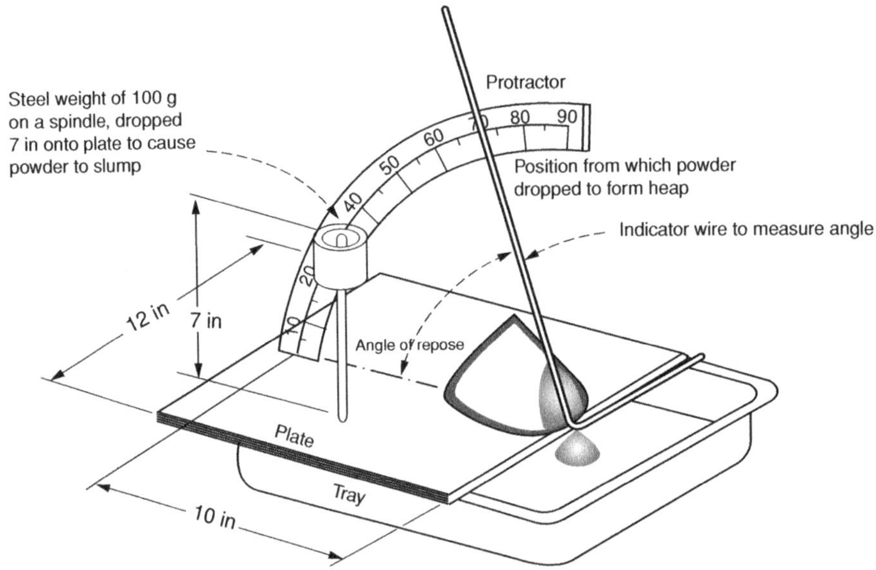

Figure 1.9 Apparatus for the measurement of angle of repose, angle of fall and angle of difference (Carr 1965a).

The more dispersible a powder is the dustier (see Section 1.4.8) it becomes and therefore the more floodable.

The degree of particle tensile strength was found by Yokoyama *et al.* (1982) to affect the floodability of a powder. The floodability index, using the various test methods for characterising bulk solids by the Carr's indices (ASTM D 6393-99: 1999) was used to evaluate an intermediate condition between an aerated powder and a packed dense powder (see Section 1.8.1).

1.4.7 *Flowability*

The flowability of a powder can be predicted from a series of laboratory tests using four different bulk powder flow parameters in a similar manner to Carr's floodability index. Flowability can thus be ranked from 0 to 100 using the four different bulk properties listed below:

Angle of repose
Compressibility
Angle of spatula
Either a uniformity coefficient or a cohesion factor

The choice as to whether a uniformity coefficient or a cohesion factor is selected is dependent upon the particle or granulate size. Cohesion can be defined, from the results of a sieve analysis, as the amount of powdered material expressed as a percentage of the initial charge, added to the top sieve, compared to the amount of powder which remains either on the top sieve or falls to the bottom pan of a set or nest of sieves. Thus, all particles remaining on

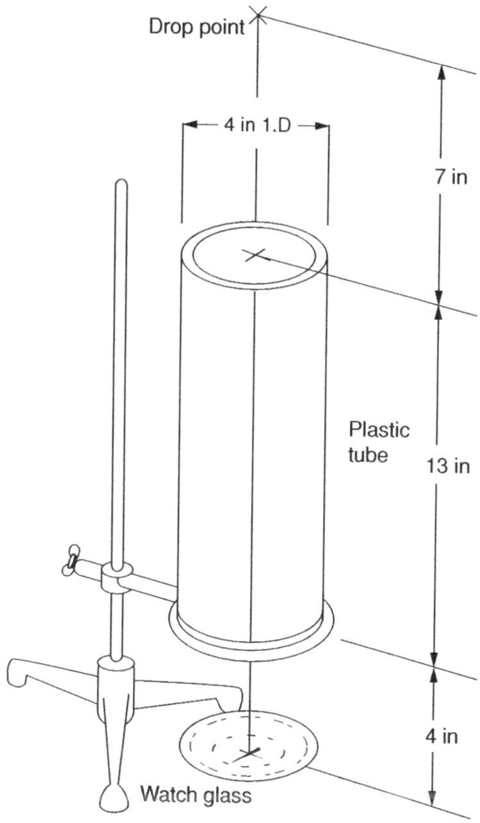

Figure 1.10 Dispersibility or dustiness apparatus.

a top sieve with a mesh size greater than the maximum particle size constitutes a being 100% cohesive, whilst in the situation where all particles pass through the nest of sieves, which has a sieve size larger than the minimum particle size, are regarded as free flowing (Figure 1.11).

Uniformity may also be defined from a sieve analysis. Uniformity is the ratio of the specific sieve size which allows 60% of the initial mass of particles to pass through the specific sieve size to the sieve size, which allows only 10% of the initial mass of particles to pass. The smaller the ratio, the greater the uniformity of the powder. Powders with particle sizes above 50–100 μm tend to have greater flowability due to having less cohesiveness and more uniformity.

1.4.8 *Dispersibility or dustiness*

Dustiness of a powder is its propensity to emit dust when undergoing stress in a process or handling operation. A wide variety of test methods exist, which range from the impact or a single drop test to fluidised beds and rotating drums. These and others have been reviewed by

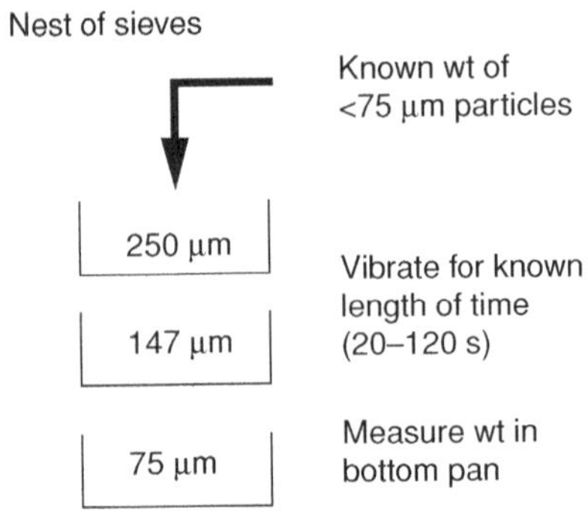

Figure 1.11 A method for the determination of the uniformity and/or cohesiveness of powders.

Hamelmann and Schmidt (2003) in terms of the applicability of any one test for the multitude of industrial powders produced in the powder industry. Different methods have generated a variety of standards from various and different industries but there is no definitive standard on the measurement of dustiness in the powder industries. Carr determined dispersibility or dustiness by simply dropping powdered material 'en masse' through a cylinder and weighing the remaining material collected on a watch glass (Figure 1.10). A selected starch powder was rated as having a 50% dispersibility because the starch sample lost 50% of its weight when falling down the tube and cascading off the catchment plate. Any material with 50% or more dispersibility was awarded 25 points.

Equipment available to measure dustiness varies from the drop method (Figure 1.10), Roach dust drop funnel (Lyons & Mark 1992), drop test (DIN 5599-2: 1999), rotating drum (Lyons & Mark 1994), inverse flow (Cowherd *et al.* 1989) and gas fluidisation (Schofield 1981).

1.4.9 *Permeability*

Since a powder assembly can generally be regarded as a biphasic system the permeability of a powder, or the ability of a gas to pass through a collection of particles or bed of powder, has now been recognised as a bulk powder property. The magnitude of permeability as a function of bulk density may be used in the determination of the powder flow rate and discharge of powders from hoppers.

The passage of gases (usually air) through fine powders may be a rate determining factor: too high a rate can lead to non-steady or erratic flow whilst too low a rate may be witnessed in 'hold-ups' or arching in hoppers, settling or deaeration.

Permeability is usually measured by the volume of airflow, q, at a specific pressure drop, Δp, through a bed of powder of known bulk density, ρ_b, and height, L (Carman 1937). From the knowledge of powder permeability the problems of non-steady or erratic flow may be circumvented and thus prevent poor weight control into a compaction die or a hazardous flooding discharge from a hopper. Permeability is generally determined using equipment

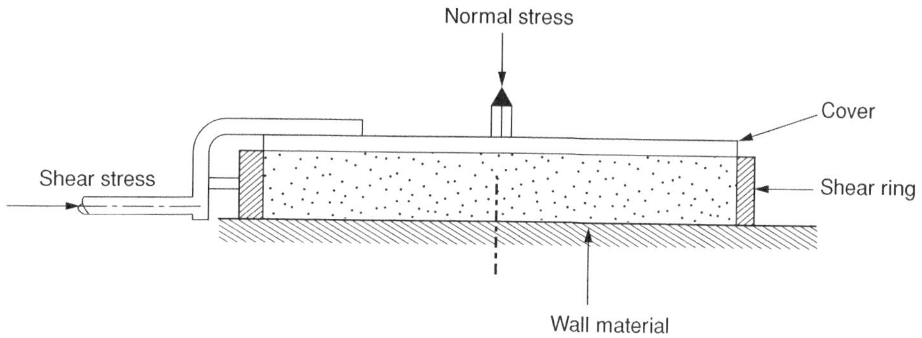

Figure 1.12 Apparatus to determine wall friction on plant wall material.

for the measurement of an external surface area by the use of the Kozeny–Carman equation (BS 4359 Pt 2: 1982) (see Section 1.4.4).

1.4.10 *Wall friction tester*

The main difficulty facing powder process engineers and powder plant designers is the need for information on two powder phenomena: firstly, the resistance to product flow generated by particles within the product and secondly, how easily a powdered product can flow over the metal, plastic or ceramic contact area or internal surfaces of plant equipment.

Fortunately, wall friction test data on materials used in powder plants can readily be obtained in a similar fashion to that of the Jenike shear test. A sample of wall material of approximately 120 × 120 mm may be used with a standard size shear cell and the wall sample taken should be in the same state of surface finish and 'final clean' condition as that used in the process. A shear ring is placed on top of the horizontal top surface of the test material and located against a horizontal shear transducer pin. A mould ring is placed on the shear ring and both are then filled with the particulate solid(s) under test. The twisting and consolidation lid is placed on the level powder and twisted to unite and compress the sample. The mould ring is removed and the sample levelled with the shear ring (Figure 1.12).

Weights are then stacked onto the hanger and shear lid corresponding to the initial consolidation stress. As shear stress is applied to the powder bed lying on the selected wall metal the shear stress will rise and approach steady state either directly or through a maximum. The maximum shear stress corresponds to a static wall friction whereas the steady state value is related to the kinematic wall friction. A series of shear stresses can be obtained from a series of normal stresses applied to the test material in contact with the wall material specimen (Figure 1.13).

When the values of shear stress are plotted against the normal applied stress the resultant line, the wall yield locus (WYL), may be a straight line or a convex curve upwards. If the WYL is linear and passes through the origin, the slope of the line is the angle of wall friction.

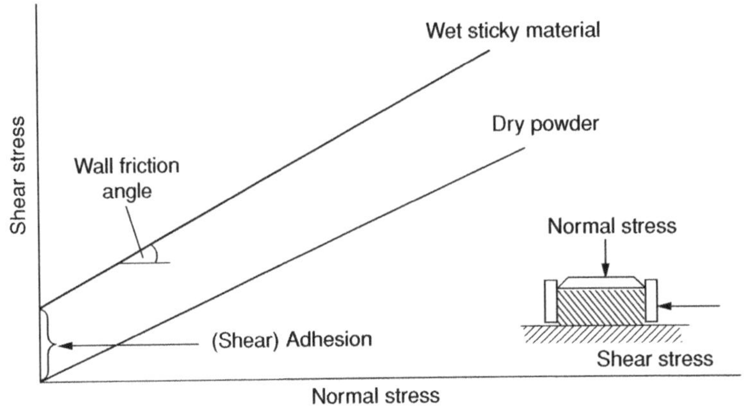

Figure 1.13 Graphical representation of the parameters of adhesion to, and wall friction on, hopper or plant material.

If the wall shear stress line is curved, a Mohr circle may be drawn in contact with the powder yield locus of a separate shear test obtained at a pre-shear normal stress identical or similar to that of the wall friction test (Akers 1992). The intersection of the curved WYL with the superimposed Mohr circle is then extrapolated to the origin to give the angle of wall friction.

In addition to wall friction, bulk density and shear strength are contributory factors to the resistance to flow of powder in plant. McGlinchey and McGee (2005) and McGee (2006) suggested that a visual diagram could be derived from the incorporation of the bulk characteristics of wall friction (ϕ_w), shear strength (τ) and bulk density (ρ_b), together with hopper wall angle (β_h), hopper outlet size (D_{crit}) and the Hausner ratio to produce a six axes 'spider web' diagram (Figure 1.14). The spider web diagram gives no numerical values for flowability but the position of a hatched area within three concentric circles, designated as easy (inner circle), moderate (middle circle) and poor (outer circle), gives an indication of bulk powder flowability.

Table 1.9 shows the suggested limits of the three 'webs', which were subsequently used to describe the flow of over 100 industrial powders.

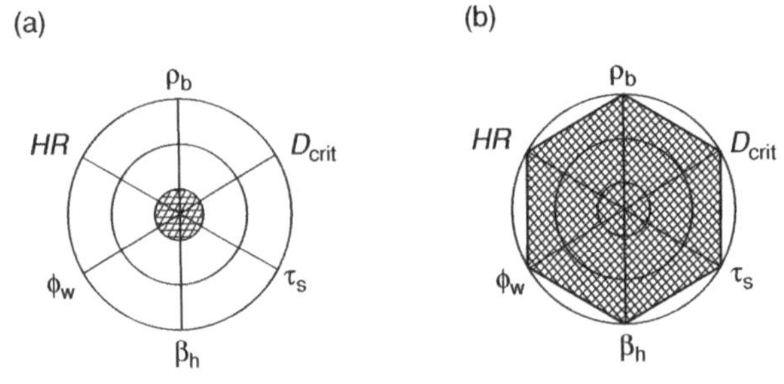

Figure 1.14 'Spider's web' diagram of easy (a) and poor powder flow (b) (McGlinchey & McGee 2005).

Table 1.9 Bulk powder parameters, which may characterise the degree of flowability of various industrial powders.

Flow	Wall friction (degree)	Bulk density (kg/m^3)	Shear strength (N/m^2)	Hausner ratio (−)	Outlet size (cm)	Mass flow wall angle (degree)
Easy	<20	1200	300	1.1	15	65
Moderate	25	800	1000	1.25	50	73
Poor	>30	400	2000	1.5	100	80

1.5 Powder angles

1.5.1 *Poured and drained angles*

There is much ambiguity when the term 'angle of repose' is used, because the angle of internal friction and angle of slide have, at times, been referred to as 'an angle of repose'. The angle of repose is commonly defined as the angle between the free surface of a pile of powder on a horizontal board to the horizontal plane. This angle is, however, dependent upon the method of formation of the pile of powder. The test is only relevant to slightly cohesive powders or non-cohesive powders, which can form a specific angle when either poured or drained. The *poured angle* of repose is the angle measured from a pile poured freely onto a flat surface.

The *drained angle* of repose is the angle measured on or from the conical surface of powder in a flat-bottomed container after the powder has been discharged or drained from the container via an orifice.

These two values are different, the latter being greater than the former, because the poured angle has particles sliding and rolling down the sloped powder surfaces causing separation, whilst the drained angle tends to achieve a convergence or mixing of particles within the remaining piled up material nesting in the container.

1.5.2 *Static angle of repose of a heap*

Angles of repose are measured in various ways. The angle of repose was defined by von Terzaghi (1925, 1943; Peck & Gholamriza 1996; von Terzaghi & Peck 1967) as the angle between the horizontal and the slope of a heap of soil dropped from a known elevation. The angle depends, however, on the diameter of the pile, higher slopes resulting from small heap diameters. With large diameter heaps segregation by sliding can also occur. Figure 1.15 shows a simple angle of repose measurement apparatus, while Figure 1.9 shows an apparatus designed by Carr which incorporates a protractor, an indicator wire and a jarring device to measure the angle of fall. The lower the angle of repose the more flowable a powder becomes because the angle of repose is a direct indication of the potential flowability of a powder. This technique is a relatively simple method and may be used, at times, to measure indirectly particle properties affecting flow such as:

- Shape and size
- Surface area
- Porosity
- Cohesion

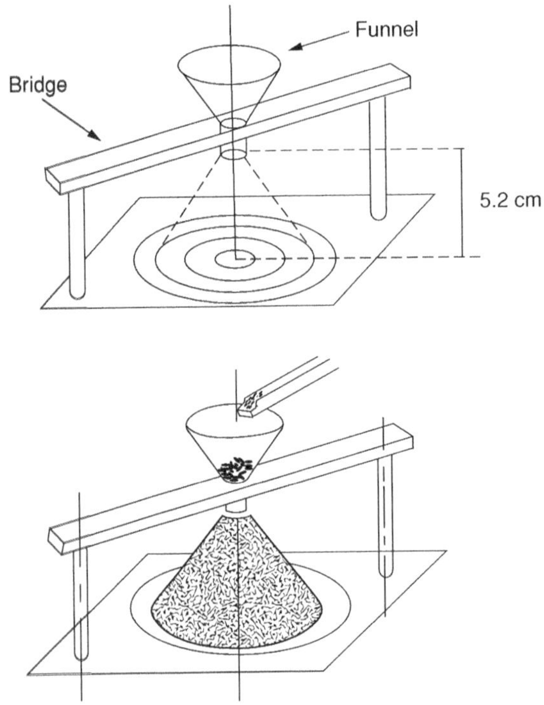

Figure 1.15 Simple apparatus for the measurement of angle of repose.

- Fluidity
- Bulk density

Although the angle of repose has been used to design feeders and storage bins, there is, nowadays, more sophisticated and accurate instrumentation available to measure other bulk powder factors such as:

- Jenike angle of friction
- Wall friction
- Cohesion
- Tensile strength

which are generally better criteria for use in mass flow hopper design.

Probably the only general powder standard to describe the procedure to measure an angle of repose is that described by British Standard (BS 4140-9: 1986) while there are three ISO standards for different powdered materials, ISO 902: 1976, for alumina; ISO 4324: 1977, for powders and granules and ISO 8398: 1989, for fertilisers. The apparatus consists of a glass or polythene funnel with a cut-off stem mounted on a tripod at a known height (5.2 cm) above a plate of metal or wood. This bottom plate has a series of concentric circles inscribed upon its surface to indicate the different angles of the poured conical surface. The diameters of the circles are calibrated in degrees with a step size of 2° (Figure 1.15). A commercial instrument for angle of repose measurement for use in the process control of semi-cohesive powders has been developed (Powder Research Ltd, Harrogate, UK) from

Table 1.10 Flowability indicators and categories of powder flow from USP 29-NF24 (2006).

Description of flow	Angle of repose (degree)	Carr's compressibility (%)	Hausner ratio (−)
Excellent/very free flow	25–30	<10	1.00–1.11
Good/free flow	31–35	11–15	1.12–1.18
Fair	36–40	16–20	1.19–1.25
Passable	41–45	21–25	1.26–1.34
Poor/cohesive	46–55	26–31	1.35–1.45
Very poor/very cohesive	56–65	32–37	1.46–1.59
Very very poor approx non-flow	>66	>38	>1.60

the work of Wouters and Geldart (1996). The magnitude of the angle of the heap can be used to quantify, in a similar fashion to the Carr's compressibility percentage, Jenike failure function and Hausner ratio, the ease of powder flow (Table 1.10).

1.5.3 *Angle of fall*

To measure the relative flow and stability of dry material, Carr dropped a small weight (approx. 100 g) at a distance of 17.8 cm (7 in) onto the horizontal surface upon which a heap of powder had been built in a manner akin to the determination of an angle of repose. Free-flowing material will slump or fall down to assume a different geometry from the original pile of powder – termed the angle of fall or slump. A fluid material, which can cause flooding, will, however, collapse under the impact of the weighted steel bush on to the horizontal surface, expelling the entrapped air. The lower the angle of fall the more floodable the material (Figure 1.9).

1.5.4 *Angle of difference*

The difference between the angle of repose and the angle of fall gives an indication of the potential floodability of materials. The greater the angle of difference the better the flowability, fluidisability and floodability.

1.5.5 *Dynamic angle of repose*

One of the adjuncts to angle of repose measurement is the 'slump test' (see Section 1.5.3) which simulates the movement of particles in a heap over each other and which can indicate a degree of dynamic cohesion or opposition to shear within a bulk powder. Following from the work of Kaye *et al.* (1995, 1997) on the sliding of particles over powder surfaces, the Aero-Flow Powder Flowability Analyzer (TSI Instruments Ltd) was developed. This dynamic angle of repose instrument uses a slowly rotating drum. Eventually, as the powder rotates within this transparent disk, an unstable condition is reached in which particles slide down the bulk surface of the powder under the influence of gravity to create an avalanche. A photoelectric detector monitors this avalanching behaviour to produce a series of 'saw-tooth' plots termed the 'phase space attractor map'. The periodicity, in terms of the time taken to achieve an avalanche, and shape of these plots, termed the 'strange attractor', provide information on powder flowability. The more scattered the data from the strange attractor the more cohesive the powder. Lee *et al.* (2000) used the mean time to avalanche as the criterion to quantify the flowability of six pharmaceutical excipients which were subsequently compared with other laboratory-derived powder flow indicators.

Table 1.11 Flow indicators for six pharmaceutical powders.

Powder	Static angle of repose	Hausner ratio	Carr's (%)	Rank order from dynamic angle of repose
Aspartame	52	1.92	48	1 cohesive
ML001	50	1.54	35	2 cohesive
Lactose	46	1.43	30	3 passable
Flour	43	1.35	33	4 fair
HPMC	33	1.32	24	5 free flow
Placebo	30	1.16	14	6 very free flow

Bodhmage (2006) following the work of Riley *et al.* (1978) correlated the physical properties and flowability indicators of six pharmaceutical powders. The six pharmaceutical powders were Aspartame (a non-carbohydrate sweetner), Respiose ML001 (an inhalation lactose), alpha-D-lactose monohydrate, Methocel HPMC (a water-soluble hydroxypropyl methycellulose), pastry flour and a granulated placebo mixture. The flowability indicators chosen were the static angle of repose, Hausner ratio, Carr's compressibility percentage or flow index and the dynamic angle of repose. The dynamic angle of repose was measured by a modified Aero-Flow Analyser. When the 'non-dynamic angle of repose' indicators were compared with the 'saw-tooth' frequency of avalanching data – mean time to avalanche – from Bodhmage's newly modified electrical capacitance tomographic instrument, it was found that all six pharmaceutical powders followed the same rank order as the more conventional indicators (Table 1.11).

Many of the empirical flow indicators currently used have been derived from the work of Carr (1965b,c). These diverse test methods have now been collected together in an American Society for Testing and Materials – ASTM D 6393-99: 1999 – standard which covers the apparatus and procedures for measuring the properties of free flowing and moderately cohesive bulk solids and which are collectively known as 'Carr indices'.

Test A – Measurement of Carr angle of repose
Test B – Measurement of Carr angle of fall
Test C – Calculation of Carr angle of difference
Test D – Measurement of Carr loose bulk density
Test E – Measurement of Carr packed bulk density
Test F – Calculation of Carr compressibility
Test G – Measurement of Carr cohesion
Test H – Measurement of Carr uniformity
Test I – Measurement of Carr angle of spatula
Test J – Measurement of Carr dispersibility

1.6 Bulk powder flow properties from internal angles and shear

1.6.1 *Failure properties*

Bulk powder flow properties can be used either as a qualitative comparison for product specifications or as quantitative numerical values in the design of plant.

The comparison of the qualitative behaviour of powders in terms of flowability, angle of repose, cohesiveness or tackiness gives a very general powder characterisation, or fingerprint for products. These terms may be sufficient to ensure the quality control and quality assurance of a process with repeated batch/continuous products but for quantitative design purposes, it has been established that the fundamental properties such as shear, Jenike internal angle of friction, wall friction and failure are essential for hopper design (Schwedes 1975). For powders to flow, powder strength must be less than the force applied to the assembly of particles. In such circumstances, the powders are then considered to have failed. The basic properties are thus termed failure properties. These failure parameters are derived from the mathematical manipulation of the yield loci of a consolidated powder sample obtained under specific conditions in a shear tester.

The parameters derived from shear tests generally are:

Jenike effective angle of internal friction	(δ)
The internal angle of friction (the internal angle of friction is a two-dimensional projection of the critical state line)	(δ_E)
Cohesion	(C)
Tensile strength	(T)
Unconfined failure stress	(f_c)
Major consolidation stress	(σ_1)
Failure function	(*ff*)

The above seven parameters are used to account for the amount of shear a powder has to be subjected to before that powder can move under stress. The degree of compaction or consolidation influences the flowability of power unless the powders are cohesionless (e.g. dry sand, gravel) and thus have no compaction strength. It would be incorrect to design powder handling and process plant on the quantitative fingerprinting characteristics of powders, such as angle of repose, because although the angle of repose is useful in the evaluation of hopper volume, the angle of repose does not measure the force or strength of attraction between particles, which is measurable by a shear test.

Failure properties are strongly influenced by humidity, temperature and time of consolidation. The above failure properties must, therefore, be measured under controlled conditions of temperature and humidity. For any prolonged storage, samples must be sheared on time consolidated samples.

1.6.2 *Flowability and failure*

One definition of flowability can be expressed as

> *a state of continuous deformation or relative displacement of a solid mass with time which involves movement of the centre of gravity of the solid mass and preservation of the continuity of the solid mass*

There is difficulty, however, in discussing powder flowability in terms of 'rheological phenomena' because although the powder mass consists of discrete particles there is an absence of continuity, as found in liquids. Flowability, although ill defined, can be used to describe the ease with which powders flow out of chutes or hoppers.

A direct way to assess the rate of flow is to time the passage of a standard amount of powder discharged from a specified funnel, opening or hopper (ISO 4490: 2000; ASTM B855-06).

1.6.2.1 *Interfacial sliding of particle assemblies* The frictional work required to cause particles to slide over each other and to initiate powder flow in a particle assembly depends upon both macroscopic and local microscopic particle involvement. The macroscopic level can be regarded as the level in which bulk flow of a powder occurs down container or hopper walls. This flow can then be regarded as analogous to the frictional flow of solid non-particulate bodies. Nedderman and Laohakal (1980) found that flow of particle assemblies down smooth walls could be described as virtual plug flow, with a constant velocity profile across the whole cross section of the body of powder undergoing movement.

At the microscopical particle level, the contribution to sliding and flowing is due to the interplay of particle surfaces and surface energies achieved by rolling particles. These modes of movement may, however, be co-incidental because a rolling particle will eventually be transposed, in an assembly of particles, by inter-particle sliding and therefore failure.

For rough walls, Stephen and Bridgwater (1978) found that the failure zone area, for the internal failure of powders, was twice the number of particle diameters reported by Nedderman and Laohakul. These failure zones can be related to the voidage fluctuations seen in randomly packed beds of spheres adjacent to container walls (Ridgway & Tarbuck 1968). It is generally found that the internal angle of friction is greater than the wall friction.

Contact surface roughness is therefore a significant factor in the frictional behaviour of solid surfaces and thus applicable to any moving particulate system. Particle movement can be conceptualised either as a model of smooth contact between particles and walls (Hertzian: Hertz 1882) or as a model which has rough contact area between the particulate system and wall (Archardian: Archard 1953, 1957). Both models have been used to predict the physical behaviour between particles and particles or between particles and a wall.

The more realistic powder technological model of Archard (1953, 1957) demonstrated that with rough contact surfaces the surface consisted of many micrometre-sized asperities or surface projections. The 'real' area of contact was therefore dependent upon two criteria: one criterion being the number of asperities that were brought, under certain loads, into contact with each other, whilst the other criterion was the degree of plastic deformation these asperities could undergo. The change in both the asperity size and height distribution on particle surfaces thus alters the contacting surface area and the sliding or friction between solid particles.

Kendall *et al.* (1971) and Kendall (1986) showed that it is not possible to obtain a value of Hertzian contact area based on particle geometry of fully roughened particles because the coefficient of friction increases with wet sands, lubricated by adsorbed moisture, with decreasing particle size.

Changes in the surface topography of rough surfaces need to be considered in all particle movements since contact forces are small. A small force of 100 nN has been shown to change the surface topography of the contact region and influence interfacial friction coefficients and interfacial shear strengths. Briscoe *et al.* (1984) investigated the influence of surface topography on the interfacial friction of powders from the flow behaviour of smooth (ballotini spheres) and rough (silica sand) particles rotating and sliding in a horizontal cylinder of borosilicate glass. They concluded that the interfacial frictional properties and physical behaviour of bulk powders could possibly be predicted from knowledge of microscopic particle properties. These predictions were valid if powder behaviour was approached from the viewpoint of an adhesion–friction model because with smooth surfaces, detailed analysis of rotating particulate systems are capable of accurately predicting frictional forces from measured values of torque. Measured torques could then be converted into forces acting over a known contact area, and the normal force evaluated from the weight of particles in the rotating tube at known packing densities which can lead ultimately to the failure property of the powder.

1.6.2.2 *Flowability based on hopper discharge* The prediction of the rate of discharge of granular materials from conical mass hoppers should take into account the friction of the materials and the effect of container wall friction.

A number of mechanical models have been proposed for powder discharge rates. Some of the earlier equations for discharge were based on models which were for coarse (>500 μm diameter) free-flowing particles. These cohesiveless powders either discharged in a manner regarded as non-mass (funnel) flow (Beverloo *et al.* 1961) or as mass flow (Brown & Richards 1970; Johanson 1965). Although both the Johanson and the Brown and Richards models are widely used with non-cohesive, free-flowing material, for cohesive powders the models of Davidson *et al.* (1966), Davidson and Nedderman (1973) and Williams (1977) have shown better success in terms of coincidence between theory and reality. Both the cohesive models – Davidson and Nedderman and Williams – and their predictive equations take into account the internal friction of powdered material. The Davidson and Nedderman model, built on the initial work and experimentation of Savage (1967), assumes a frictionless hopper wall whilst the Williams model takes into account both internal friction and hopper wall friction. The mathematical models of Johanson, Beverloo, Brown and Richards and Williams all show a relationship between discharge rate and hopper outlet diameter to the power of 2.5.

Johanson

$$Q = \rho \frac{\pi D^2}{4} \left[\frac{Dg}{4 \tan \theta} \right]^{0.5} \tag{1.4}$$

for conical hopper outlets

Beverloo

$$Q = 0.58 \rho g^{1/2} (D - k d_{\mathrm{p}})^{2.5} \tag{1.5}$$

Brown and Richards

$$Q = \rho g^{1/2} D^{5/2} \left(\frac{\pi}{6} \right) \left(1 - \frac{\cos^{3/2} \theta}{\sin^{5/2} \theta} \right) \tag{1.6}$$

Davidson and Nedderman

$$Q = \rho \frac{4\pi}{3} \left(1 - \frac{\cos^{3/2} \theta \times V_{\mathrm{w}}}{\cos^{5/2} \theta} \right) \tag{1.7}$$

Williams

$$Q = k_{\mathrm{c}} \rho g^{1/2} D^{5/2} \tag{1.8}$$

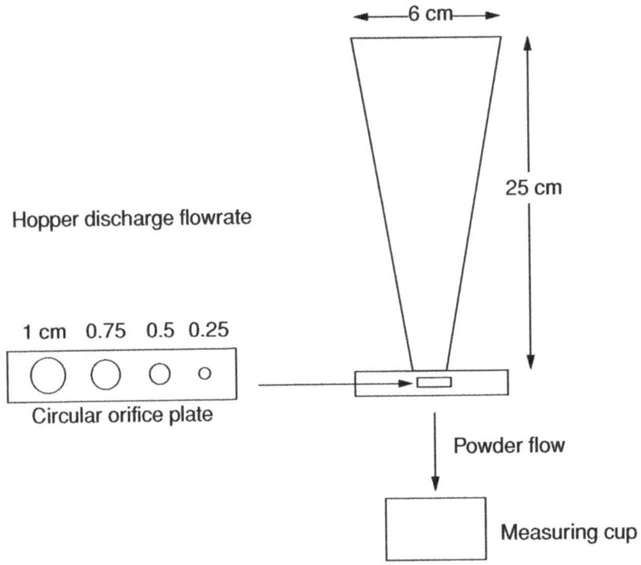

Figure 1.16 Direct measurement of flow using a simple discharge apparatus.

where Q is the mass discharge; θ is the semi-included angle of a conical hopper; ρ is the bulk density of granular or powdered material; g is the gravity; D is the diameter of hopper outlet; V_w is the function of the velocity along the hopper wall; k is the Beverloo constant which, with particle diameter, d_p, accounts for the wall effect when particles do not fully flow at the perimeter of the outlet. k has values between 1.3 and 2.9 with an average value of 1.4. k_c is the correction factor for hopper outlet diameter.

Direct measurement of flow through various-sized orifices may, however, be obtained by use of very simple laboratory equipment (Figure 1.16), and also instruments described in ASTM B855-06 and ISO 4490: 2000 or even with the patented Flodex commercial apparatus.

1.6.3 *Failure function from shear tests*

Jenike (1964, 1970) has published a fundamental and widely used definition and shear testing protocol on the flowability of powders. The failure properties of powders and thus flow are measured and calculated from a family of yield loci obtained from a number of shear tests.

The two parameters determined experimentally from the yield loci are as follows.

1.6.3.1 *The unconfined yield stress* (f_c) This is defined as the maximum principal stress which can act along a free surface to cause failure, instability to arching or the doming of powders in a hopper. Essentially the unconfined stress is used in the design of mass flow hoppers and is regarded as the stress which must be applied to any arch or dome to cause an arch to collapse under gravity from the weight of the powder above the arch or dome. A Mohr circle which describes the situation of normal stress on the powder surface of the

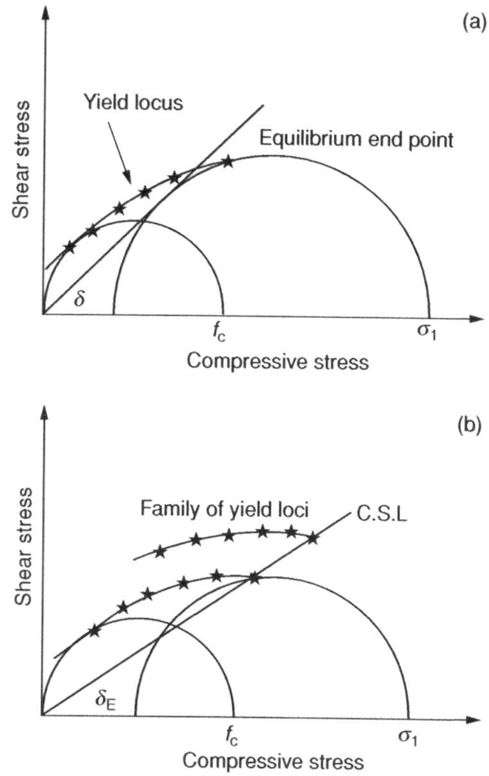

Figure 1.17 Yield loci for the determination of Jenike effective angle of friction (δ) (a) and internal angle of friction (δ_E) (b) from the critical state line (CSL) (f_c unconfined yield stress and σ_1 major consolidation stress).

arch accompanied with a tangential stress along the surface of the arch must pass through the origin of the shear–normal stress relationship graph and also touch the yield locus of the powder which has been compacted at a specific densified state. The unconfined yield stress is then the value of stress when the limit of the Mohr semi-circle cuts the maximum principal stress x-axis of the shear–normal stress graph (Figure 1.17).

1.6.3.2 *Major consolidation stress* (σ_1) This is the principal normal stress (σ_1) under which the sample has been consolidated in the principal stress plane. The major consolidation stress should not be confused with the initial compaction stress, σ_c, which is the stress that compacts the powder bed. Each different compaction stress, σ_c, leads to a different yield locus and becomes one of a family of yield loci at different densifications. The major consolidation stress is obtained by drawing a Mohr semi-circle through the equilibrium or end point of the yield locus and tangential to the yield locus.

This equilibrium stress is found when there is no volume change occurring in the powdered sample when stressed. The Mohr semi-circle, tangential to the locus, then cuts the normal stress axis to give σ_1.

1.6.3.3 *Jenike failure function (ff)* This is the relationship between the unconfined yield strength (f_c) and the major consolidation stress (σ_1). A plot of the values f_c versus σ_1 shows the possible relationship of the rate of flow of powdered material out of hopper orifices. Jenike could classify the flowability of powders from selected ratios of these values.

The Jenike failure function necessitates the measurement of a number of yield loci (minimum 3) to obtain both f_c and σ_1 values. The relationship between f_c and σ_1 may be linear or non-linear but must *not* pass through the origin of the graph.

If the failure function does pass through the origin then these powders are defined as simple powders (Williams & Birks 1965, 1967).

In order to achieve unique numbers for various and different degrees of flowability Jenike calculated the failure function at a specific value of the unconfined stress. The specific value of the ratio of the major consolidation stress to the unconfined stress was taken as 6.5 lbf/ft^3 (3.11 kPa) in the initial, Jenike designed, 4 inch diameter shear cell, because at this value the relationship between f_c and σ_1 showed only small deviations from linearity. Currently a value of 3 kPa is generally used with a standard shear cell tester.

Jenike failure function	Description of bulk powder flow
$ff < 2$	Very cohesive and non-flowing powders
$2 < ff < 4$	Cohesive
$4 < ff < 10$	Easy flowing
$10 < ff$	Free flowing

1.6.4 *Jenike effective angle of internal friction*

The Jenike effective angle of internal friction, δ, used in the Jenike method of hopper design is *not* identical to the internal angle of friction, δ_E. The internal angle of friction being a two-dimensional projection of the three-dimensional critical state line which indicates the demarcation of the phenomenon of compaction and failure of bulk solids (Roscoe 1970; Roscoe *et al.* 1958; Schofield & Wroth 1968).

The Jenike effective angle of friction is the angle of the straight line drawn through the origin of a normal stress–shear stress plot and tangential to the Mohr semi-circle, which inscribes the equilibrium, or end point of the yield locus when failure occurs at no sample volume change. The Mohr semi-circle represents the stresses in a powder consolidated under a major principal stress.

Although the Jenike effective angle of internal friction can be determined from one yield locus the internal angle of friction requires a family of yield loci, each locus corresponding to a different state of compaction (σ_c) (Figure 1.17).

A Working Party on the Mechanics of Particulate Solids of the European Federation of Chemical Engineers (WPMPS) was set up to investigate the reliability of measurement of the flow properties of powder using the Jenike shear cell tester. Although the initial shear tests on a calcite powder, distributed to 20 participating laboratories, indicated a wide scatter in the results, it was realised that although it was important to have a systematic and approved test protocol there was also a need to preserve the specific geometry of the Jenike shear cell.

To these ends, the WPMPS decided that the Jenike shear cell and the Jenike shear testing technique, as described in the well-known Bulletin 123, was of great practical use for bulk

Table 1.12 Flowability of pharmaceutical powders.

Material	Jenike flow function	Flowability
Micronised paracetamol (Acetamol) nominal size (2 μm)	1.6	Very cohesive
Griseofulvin nominal size (12 μm)	2.4	Cohesive
Finely powdered cephaloridine	4.7	Cohesive
Granulated cephaloridine batch 1 (<100 μm)	10	Free flowing
Granulated cephaloridine batch 2 (>100 μm)	12.3	Free flowing

powder characterisation and the standard shear testing technique (SSTT) (ASTM D6773-02) was thus developed. A commercial Standard Shear Tester is now manufactured by Jenike & Johnson Inc. whilst the protocol and technique for the standard shear test is published by the Institution of Chemical Engineers.

Novosad (1990) concluded that if the SSTT is strictly maintained in a geometrically specified SST or Jenike shear cell, then it should be possible to obtain reliable and reproducible measurement of the (Jenike) flow functions of cohesive powders. If the (Jenike) flow function obtained by SSTT is compared with flow functions obtained from different shear cells and different shear testing techniques there is no necessity that there should be agreement in the results. It is essential, however, that whenever shear tests are recorded the technique used must be quoted. Preference should be given to the SSTT together with the modifications used for filling the cell, the optimisation of the specimen (i.e. the creation of a critically consolidated sample), designation of the equilibrium endpoint on the critical state line (Schofield & Wroth 1968) and the twisting and inclusion of the shearing ring weight into the total vertical normal stress. These recommendations, outlined by Novosad, were used to certify a reference material of cohesive limestone powder – entitled CRM116 – by the Community Bureau of Reference (BCR) for use with the Jenike shear cell and SSTT (Akers 1992).

Enstad and Maltby (1992) concluded, from a series of shear tests carried out by five different laboratories on the certified material, CRM 116, that reproducible flow function results can be obtained only with skilled and experienced operators. Harwood (1971) applied the technique of Jenike to evaluate the flow of a range of pharmaceutical powders and found that as size of the powder increased the flowability became more free flowing (Table 1.12).

1.7 Instrumentation for the measurement of tensile strength and cohesion

1.7.1 *Cohesiveness and tensile strength*

1.7.1.1 *Cohesiveness and adhesion* Cohesiveness of a powder is the ability of individual particles to stick together. This attraction may be caused by the presence of moisture, electrostatic charges or the fineness of the particles within a powder. Fine powders generally have a high surface area, which is usually associated with an increase in the surface energies within the powder and thus enhancement of the binding forces between particles (Stanley Wood et al. 1990).

Powder cohesion is relevant to powder flow, mixing of powders and in the strength that a powder exhibits when subjected to a tensile stress. It can be argued that the only direct way to measure cohesion is via the determination of the minimum force needed to cause

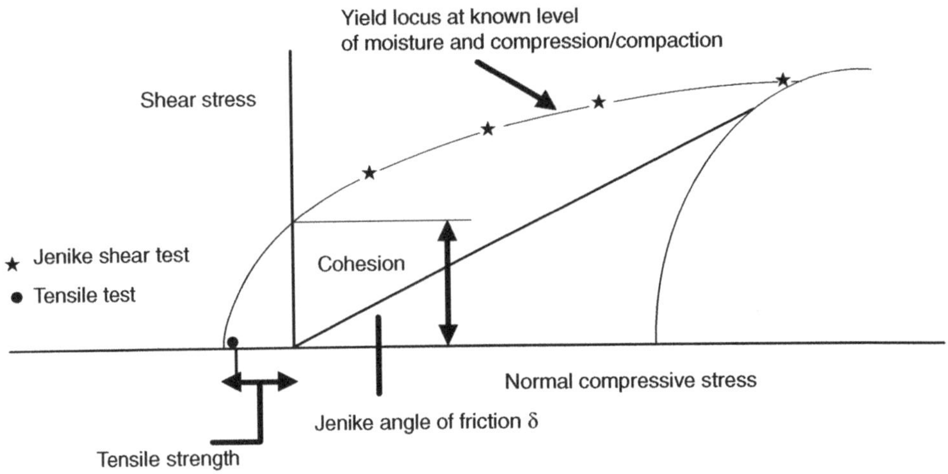

Figure 1.18 Schematic diagram to show the parameters of cohesion and tensile strength.

separation of a consolidated powder bed. All other ways are indirect since they measure properties related to, or as a consequence of, internal attraction and cohesive forces.

One approach to measure the cohesiveness of a powder is to determine the yield locus of the material, and then extrapolate the yield locus to zero normal stress. The intercept value on the shear stress axis is then the cohesiveness or adhesion of the material (Figure 1.18).

1.7.1.2 *Tensile strength* The tensile strength of a powder can be regarded as the completion of the yield locus in the shear stress–normal stress profile. The negative stress, at zero shear, to break apart the consolidated particulate powder is thus termed the tensile strength of a powder (T).

Measurement of tensile stress cannot be measured directly with a Jenike shear cell or an annular shear tester, although one approach in the measurement of tensile strength of a powder is to determine the yield locus of the material and then to extrapolate part of this locus to zero normal stress. The negative intercept on the normal stress axis is the tensile strength of the material under investigation (Figure 1.18).

Tensile strength is a fundamental failure property but tensile strength testing is dependent upon the direction of force necessary to cause separation of a bulk structure with respect to the direction of compaction or consolidation. Split cell testers pull the sample apart at 90° to the direction of compaction whilst the lifting lid or vertical shear testers pull in the same direction as the compaction/consolidation stress was applied. Results obtained from both methods differ greatly because tensile testing has a poor record of reproducibility, possibly due to the fact that consolidated powders in the tester cells may not be isotropic.

1.7.2 *Split cell and lifting lid (vertical shear) instruments for tensile strength and cohesiveness*

Although estimation of tensile strength, adhesion and cohesion from a Jenike shear test yield locus is the easiest and less demanding way of assessing powder stresses, there are other types of equipment which attempt to measure cohesion and tensile strength.

Basically there are two techniques, both of which are categorised by the direction of 'pull':

(i) Split cell testers which separate the consolidated powder bed at 90° to the direction of consolidation
(ii) Lifting lid testers which separate the consolidated bed in the same direction as the compaction force

1.7.2.1 *Split cell testers* Tensile strength measurements may also evaluate internal adhesion properties, independent of the mechanical interaction of shear plane, because T is measured at zero shear. Warren Springs Laboratories (WSL) designed two instruments: one to measure the cohesive strength of powders at varied states of compaction and the other to measure tensile strength. The tensile strength measurement equipment, developed by WSL, was the WSL tensile tester, a diametrically divided split shallow circular cell.

Before Ashton *et al.* (1964) developed the split cell apparatus for the measurement of the powder tensile strength the method used was a tilting plate. This method, however, made no provision for the consolidation of a powder at different bulk densities and the angle of inclination could only be measured, at best, with an accuracy of 5%.

(a) *Ashton, Farley and Valentin split cell tensile tester*: With the Ashton *et al.* (1964,1965) split cell, reproducibility in the results from tensile testing was found to be better than 0.5%. The apparatus consisted essentially of a shallow vertically split cylindrical cell with one half fixed and the other half movable upon ball bearings. Instead of using gravity force to fracture the sample, the tensile stress was applied horizontally through a pair of calibrated springs, which were extended using a motorised rack and pinion mechanism (Figure 1.19). The calibrated system recorded the force necessary to separate the powder compacted at various values of bulk density. Farley and Valentin (1965, 1967/68) used the WSL tensile tester, in conjunction with the Jenike failure function tester, to investigate the flow properties of a wide range of cohesive powders. The powders chosen were primarily those with a reasonably constant particle shape, wide range of sizes and densities, i.e. precipitated calcium carbonate, mined calcium carbonate, aluminium oxide and zinc dust (see Figure 1.30).

The correlation between cohesion and tensile strength, at identical packing densities, appropriated to the relationship:

$$C = 2T. \tag{1.9}$$

Hartley and Parfitt (1984) improved the split cell tensile strength apparatus by separation of the split cell of Ashton *et al.* (1964) to allow higher packing densities. The rack-and-pinion application was also replaced by a vibration-free variable speed mechanism. This was a vast improvement over the Ashton *et al.*'s original apparatus and that of the pulley and string mechanism of Yokoyama *et al.* (1982). This improved instrument thus allowed the tensile strength of carbon blacks to be measured to an accuracy of better than 0.1%. The apparatus once again consisted of three basic sections: a split cell and clamp, a compaction unit and a base unit with means to separate the split cell and measure the tensile stress as a function of time.

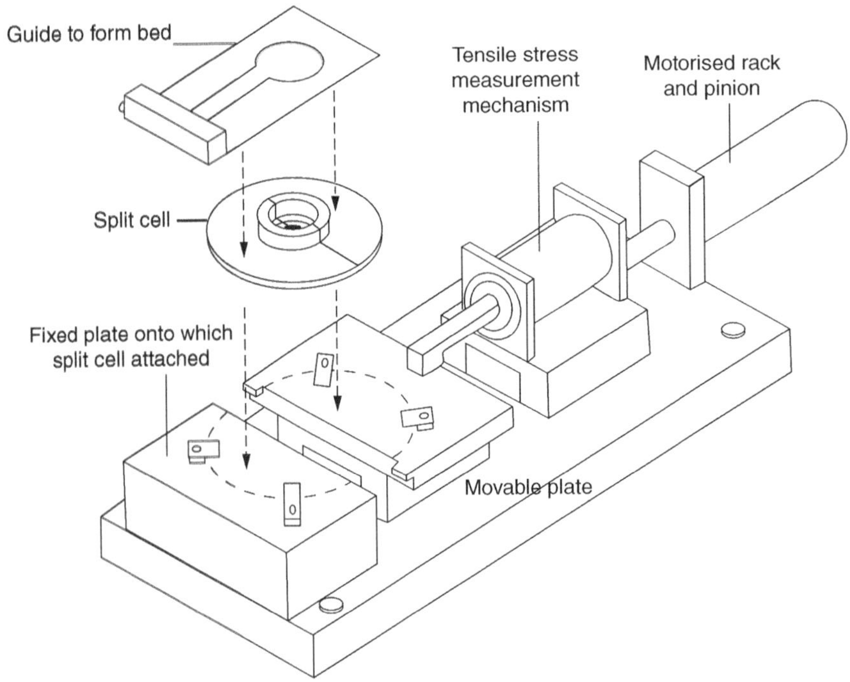

Figure 1.19 Ashton *et al.* (1964) split cell tensile tester.

The entire apparatus could be used in humid environments. The movable plate was suspended on low-friction ball bearings.

(b) *The Hosokawa Micron cohetester*: An alternative to the split cell instrument is the swing method developed by the company Hosokawa Micron Ltd. This measured bed expansion in addition to tensile strength by drawing and breaking a powder bed compressed in a split cell. The movable half-cell suspended by three plates, instead of moving over ball bearings, was pulled apart at a constant velocity of 2 mm/min. A linear variable differential transducer (LVDT) measured the linear displacement, the core of which is fixed to the movable half-cell and the coil to the stationary half-cell. The tensile stress and horizontal displacement was recorded on an *X–Y* plotter (Yokoyama *et al.* 1982) (Figure 1.20).

1.7.2.2 *Lifting lid testers for tensile strength* Boden (1981) developed a lifting-lid tester. This type of technique, recommended for the monitoring and comparison of cohesiveness, is easier to use and gives results with less scatter than with annular shear cells. A mould in the form of a ring, like a Jenike shear cell ring, is used with a metal lid just fitting inside. The base of the cell and the lower face of the lid are covered with adhesive tape, onto which glue is spread. The cell is filled with the powder to be tested and scraped level with top of the cell. The lid, with the lower face covered with glue, is placed on top of the sample and a compacting load applied to the lid and left in position until the glue has hardened. The lid is then slowly lifted by an electric motor via a tensile load cell. The stress required to break

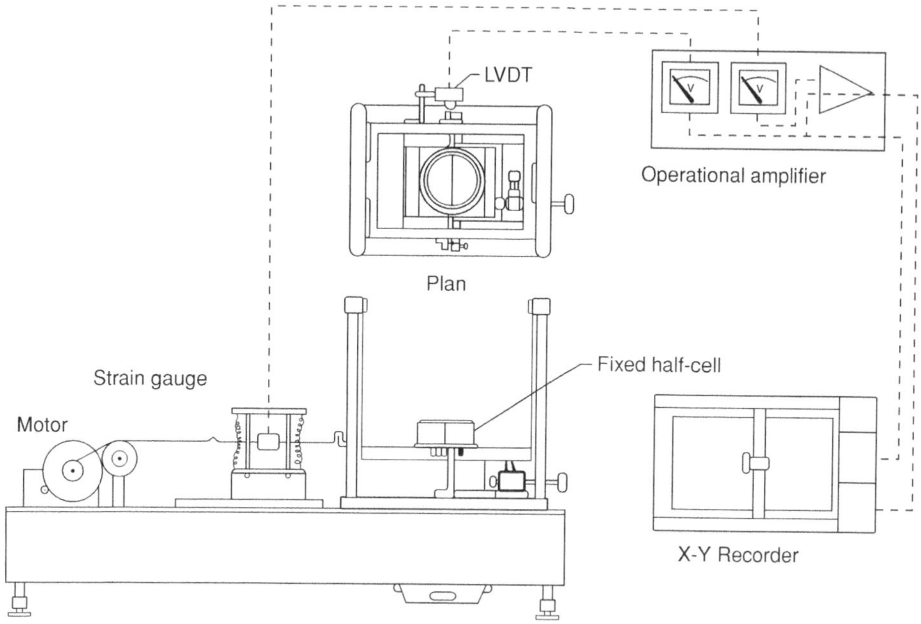

Figure 1.20 Hosokawa Micron cohetester.

the powdered specimen is recorded and both the lid and base of the cell are examined after failure to ensure that both are covered with powder. This indicates that failure has occurred within the powdered specimen and not at the powder-lid surfaces. If both surfaces are not covered with powder the test is rejected.

1.7.3 *Direct measurement of cohesion*

Investigation into the cohesiveness of bulk solids may proceed in two complementary ways.

One way which relies upon the fundamental knowledge of the stress–strain–volume behaviour of bulk solids is dependent upon the development of testers such as the biaxial and triaxial shear testers as well as the now universally accepted Jenike shear cell, or the standard shear test tester. Other instruments, such as the annular shear cells and the cross-sectional Peschl and Colijn (1977) tester, use the same stress–strain–volume principle. These annular shear cells may also be used to evaluate a bulk powder flow function. The powder flow function, having been discussed previously, still requires a family of yield loci before cohesion can be evaluated.

In view of the experimental errors normally affecting shear cell measurements and the amount of personal judgement required to draw Mohr stress circles tangential to a curved yield locus, there is always some uncertainty in the flow function derived from the Jenike-type shear yield loci method. A direct measurement therefore offers considerable advantage and, besides possibly giving better accuracy, may prove to be more rapid and reproducible.

The second way relies upon the development of specific powder testers, which will permit the straightforward measurement of cohesiveness and/or tensile strength of difficult flowing and poorly mixed cohesive powders.

The instruments available for direct measurements are mainly associated with two classes of devices. One class measures the torque or stress on an immersed vaned paddle in a bed of powder compressed at a known bulk density, whilst the other class is the mould or column failure method. The more practical mould devices are those based on the formation of a powder column in a cylindrical container (sometimes split into two halves) (see Section 1.7.3.2). The container walls are then removed to leave a freestanding cylindrical column of powder, which is then subjected to loads placed on the top of the column until the column fails.

1.7.3.1 *AJAX Equipment (Bolton) Ltd & Warren Springs Laboratory cohesion tester* It is appropriate to include in this chapter an instrument which was one of the first attempts to measure the cohesion of powders directly. It was initially developed at Warren Springs Laboratory, Stevenage, but is now commercially available from AJAX Equipment (Bolton) Ltd.

Cohesion of a powder is defined as the shear strength at zero normal load. The tester is thus designed to assess the cohesive strength of powder samples at varied states of compaction, from lightly settled conditions to firm uniaxial compressed compacts. The bench-mounted equipment consists of a radially finned vane mounted on a vertical spindle, which in turn is carried on low friction bearings (Figure 1.21).

The weight of the eight vaned 100 mm outer diameter head is supported mechanically by a small spring to give a 'floating head' arrangement. This is positioned above the powder, which has been compressed and prepared in a cell similar to the technique used with other bulk powder failure tests. The compaction pressure initially applied to the powder being of the order of 10 kPa. The vaned paddle is lowered until the radial fins penetrate and are just covered by the powder surface. The cell containing the powder being 'keyed' into the base

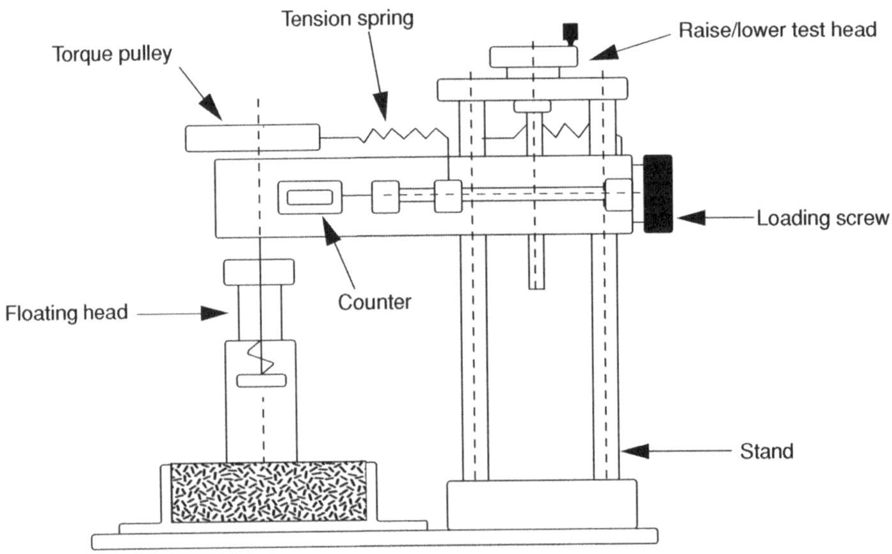

Figure 1.21 AJAX (Bolton) and Warren Springs Laboratory (WSL) cohesion tester.

of the tester by rotating the paddle into a position in which the forces acting normally to the powder surface are minimised and the vaned paddle weight is balanced by a spring to ensure that no external force is acting on the powder. Torque is then applied to the paddle via rotation and calibrated coil springs until a sudden deflection within the powder occurs. This signifies that failure at the yield point of the compressed powder has occurred between vane and the compacted powder in the cell. The area of the failure plane is taken as the horizontal area between the vanes of the paddle and cohesion may then be calculated from the recorded torque. Consolidation of the powder can be either instantaneous or over a period of time (time consolidated). The sample container may also be weighted with the powdered contents to ascertain the bulk density of the sample.

An alternative procedure, to ensure no external force is applied to the powder bed by the vaned paddle, is to place the compacted sample on a balance and when the paddle is immersed in the powder to raise the vaned head slowly until the balance reading is zero. This dynamic method of bulk powder characterisation is allied to the rheological method for measurement of the viscosity of non-Newtonian fluids and suspensions. Commercial instruments based on the WSL cohesion tester are now available in the form of the FT4 Powder Rheometer (Freeman Technology) and the Stable Micro Systems Powder Flow Analyser (Stable Micro Systems).

1.7.3.2 *Cylindrical mould testers* The ratio of the length of the cylindrical specimen (L) divided by its diameter (D) is of major importance since a compromise must be sought between two conflicting demands.

Firstly, friction between the mould wall and the powder causes a variation in vertical pressure throughout the specimen which can result in a variation in the bulk density throughout the column of powder. This bulk density variation can in turn cause variation in the strength measurement. If wall friction is too pronounced, this will affect the longitudinal strength of the powder column and the specimen will fail at its weakest point, usually near the lower end of the column. To obtain a uniform vertical stress distribution throughout the column, the L/D ratio should be as low as possible.

Secondly, the ratio L/D cannot, however, be made too low (Kesler 1959) since it has been found that the crushing strength of cylindrical specimens is a function of L/D. There is a critical value of L/D which must be exceeded for the measured crushing strength to be independent of the L/D ratio and to ensure the failure plane does not intersect the end plate of the powder column.

Monick (1966) assumed that failure planes are formed at an angle of 45° to the horizontal, which would give a critical L/D equal to unity. Specimens are known, however, to fail along slip planes at angles greater than 45°; therefore the critical value of L/D should be greater than unity.

The critical L/D dimension is one in which the failure slip plane coincides with or exceed either of the diagonals drawn within the vertical cross-sectional area of the freestanding column. The critical value of L/D is a function of the angle the diagonal within the freestanding column makes with the horizontal plane on which the column stands and the angle of internal friction (Endersby 1942). If the L/D ratio is lower than the critical L/D value, there is likelihood that for some powders, the failure plane will intersect one of the end planes. This mode of fracture (slippage) therefore requires extra work and produces a higher collapse or cohesive strength value. Bishop and Green (1965) and Williams *et al.*

(1970/71) found that end plate slippage became negligible when the L/D ratio was higher than 2, failure occurring within the powder column.

All direct mould testers consolidate material by uniaxial compaction and since the voidage within the compressed specimen may not be uniform, steady state deformation may not be achievable and the failure functions measured may have some degree of inaccuracy.

1.7.4 *Uniaxial compression*

1.7.4.1 *The Williams unconfined compression test equipment* The method developed by Williams *et al.* (1970/71), to eliminate the effect of wall friction on column formation, was one in which a compact or column of powder was formed in a segmented splitable mould. The mould is then removed to leave a freestanding cylindrical column. The compressive vertical uniaxial stress to cause breakage is measured and regarded as the unconfined yield stress for that consolidating stress. The failure function is assessed by the production of a series of compacts/columns under different consolidation stresses to give measurements of the corresponding unconfined yield stresses.

The mould is split into two halves and is made up of 13 sections. The powdered sample is poured loosely into the segmented split mould and consolidated by application of weights to stresses generally above 100 kPa for a period of time (Figure 1.22).

The mould used for forming a column of powder has a 7.5 cm square outside cross-section and 17.5 cm high with a 5 cm diameter axial hole drilled along its length. To achieve the elimination of wall friction the column is formed in a number of equal increments (N). Powder was poured into one section, compacted and then another section added. To assure good bonding between the several layers of compacted powder a compaction disk, with a profiled surface, is used instead of one with the usual flat surface. Once a complete column is formed and the mould removed, weights are added at timed intervals (circa 10–15 s) until

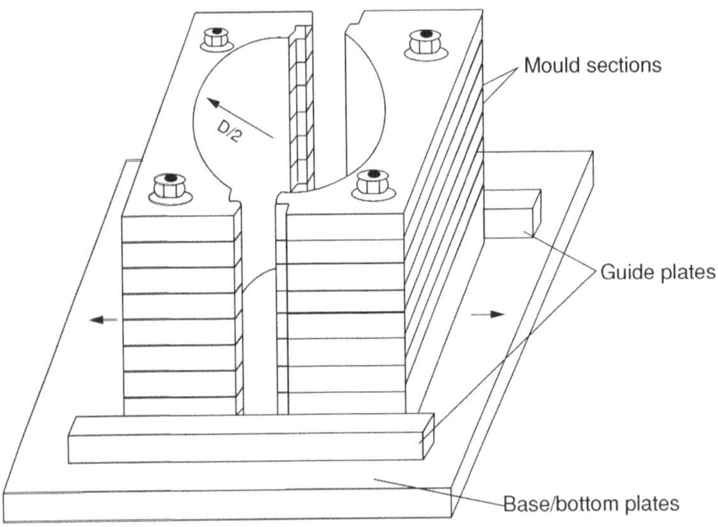

Figure 1.22 Williams mould for the determination of unconfined stress (Williams *et al.* 1970/71).

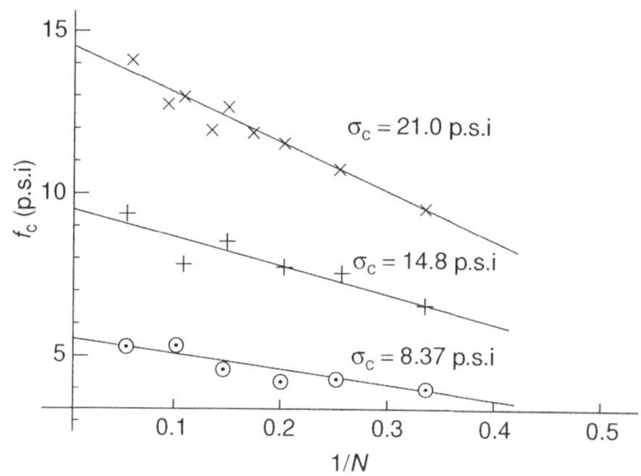

Figure 1.23 Williams mould data for the determination of a powder's unconfined yield stress (Williams *et al.* 1970/71).

the column collapses. The unconfined yield strength of the specimen was determined by the measurement of the vertical stress applied to the upper end by means of a strain rate device. From the measurement of the unconfined yield strength on columns produced with varying numbers of N increments, a plot of unconfined yield strength versus $1/N$ can be obtained (Figure 1.23). Extrapolation to $1/N = 0$, which corresponds to an infinitely large number of increments of zero thickness, gives a value of the unconfined yield strength. It is assumed, when the number of layers are greater than 10, that wall friction plays no role and probably gives a better estimate of the unconfined yield strength of the material.

The principal compaction stress is

$$\sigma_1 = \frac{L}{A} \tag{1.10}$$

where L is the compaction load and A is the cross section of the split mould.

The unconfined yield strength is

$$f_c = \frac{W - W_p}{A} \tag{1.11}$$

where W is the total applied weight to cause column collapse and W_p is the weight of powder initially in the mould.

Since the L/D ratio of the mould designed by Williams *et al.* was 3.5, the strength of the compact was deemed to be independent of the L/D ratio.

The failure function of titanium dioxide obtained from a family of yield loci measured in a Jenike shear cell was found to agree closely to those obtained with different values of N from the mould or unconfined compression test equipment. Although the consolidation stress needed to form a satisfactory column is higher than that normally found near the outlet of hoppers, it is within the range of stresses expected to occur in other powder flow systems such as screw conveyors, mixers and extrusion presses. Higher stresses of compaction and unconfined yield strength can be achieved with the use of an Instron Universal testing machine. A similar technique to Williams *et al.* has been used to measure the tackiness of detergent powders (Lappas & Dempski 1965; Labuza 2002).

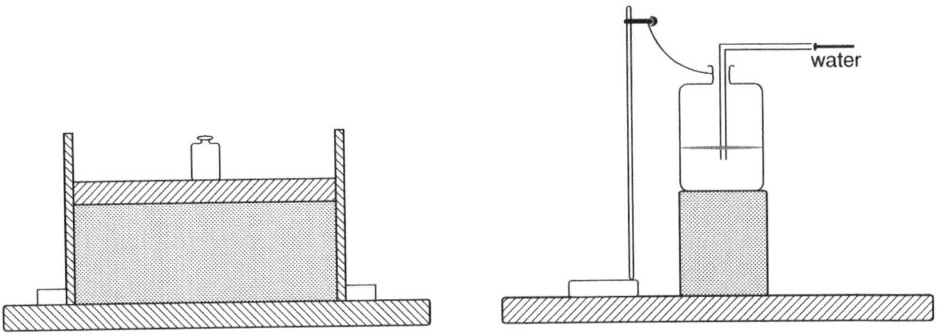

Figure 1.24 Gerristen's mould and method for the measurement of unconfined yield stress (Gerristen 1986, 1990).

1.7.4.2 *The Gerritsen method* Gerritsen (1986) pointed out that the densification of each of the increments in the Williams *et al.* method was not consolidated for the same period of time. Thus, the measured unconfined yield stress may not be applicable for materials which exhibit a time consolidation dependence.

Gerritsen thus designed an apparatus which allowed the determination of the unconfined yield strength at low consolidation stresses (Figure 1.24). The column formed, however, has to be able to withstand at least its own weight; thus there is a certain minimum major consolidation stress for each material and the height of the sample had to be as low as possible. In order to solve the contradictory requirements of an acceptable L/D ratio (circa 2) to avoid the failure plane intersecting the end plane and a low L/D ratio to avoid mould wall effects, it was proposed to form a compacted column using a mould with a low L/D ratio and cutting away the outer part of this column. This action left an inner cylinder with a sufficiently high L/D ratio for the sample to be acceptable for the subsequent test.

The mould has a diameter of 19 cm, which can be filled with a given weight of material to achieve a height after consolidation of approximately 7 cm. A thin aluminium disk, smaller in diameter than the mould, is placed at the centre of the sample surface. An airbag is placed between the aluminium disc, powder surface and consolidating lid. A consolidation stress is then exerted on the sample through the bag to achieve an even pressure/stress distribution. After consolidation the compaction force and mould are removed. Most of the bulk solid is then cut away to leave only the central part of the column under the aluminium disc. The unconfined yield strength of the test sample is obtained by the addition of weights to the top of the column until the column collapses. The value of f_c obtained by Gerristen, and subsequently used to determine a critical diameter to prevent arching in a hopper, was found to be less than the f_c values that were required to prevent arching in a practical experimental hopper and the f_0 values obtained with a triaxial tester.

1.7.4.3 *Peschl unconfined yield strength tester* A sample is vertically compacted in a rectangular mould, and after the top lid and normal stress are removed one of the sidewalls of the tester is driven horizontally into the sample up to the point of failure. The stress applied to fail the powder is an estimate of the unconfined yield strength. If a bulk solid shows anisotropic behaviour, it might be expected that the described tester would produce

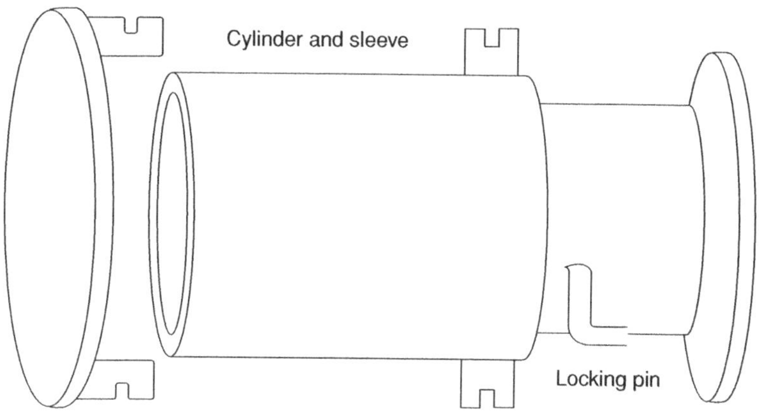

Figure 1.25 Procter & Gamble's apparatus for the determination of cake strength (Lappas & Dempski 1965; Labuza 2002).

small f_c values since its direction of application is different from that of the consolidating stress (Peschl 1988, 1989).

1.7.4.4 *Procter & Gamble device for cake strength measurements* The apparatus consists of a cylinder, sleeve, lid and locking pin (Figure 1.25). The locking pin is placed in the hole in the cylinder and the sleeve rests on the pin. The sleeve is filled with powder and the lid placed on top of the sleeve with both parts of the device being fastened together by elastic bands. The pin is then removed and a 12 lb weight remains on the lid for 2 minutes to compact the sample. After removal of the elastic bands and the weights, the sleeve is retracted down the cylinder to the base. A force gauge is applied to the lid and the force required to fail the exposed column of powder is noted. This is regarded as a direct measurement of the cohesiveness of the material (Labuza 2002).

The L/D ratio of this device is not fixed since it is dependent upon the powder compressibility. The sleeve is 15.25 cm long and has an inner diameter of 6.35 cm; the pinhole is located 6 cm below the top end of the solid cylinder, so that in case of an incompressible powder the L/D ratio will be roughly 1.5, and even less for compressible powders.

1.7.4.5 *Large-scale uniaxial tester* For coarse granular materials a split mould is used which has an L/D ratio of 2 (Tamura & Haze 1985). The height and diameter of the mould used are dependent upon the size of the granular material. The granules are compacted in three layers in three similar compression steps. The failure criterion is regarded as the value of the compaction stress at which the column could not remain freestanding.

1.7.4.6 *Compression tackiness tester* The compression tackiness tester, developed by Colgate-Palmolive (Piscataway, NJ), is a simple version of the uniaxial compression test (Monick 1966). The mould is placed on a stand with the two halves clamped together and the material compressed at 22 lbf (98 N). A plunger is placed on top of the freestanding column after mould removal. At 30-second intervals, weights in half-pound increments (227 g) are added to the plunger until column collapse. The compression force of 22 lbf mimics the force generated at the bottom of a pile of bags filled with soap intermediates or detergents and approximates to a stress of 16 kPa (Figure 1.26).

Although the amount of weight to cause column collapse prepared at 98 N is defined as 'tackiness', other loads may be used to form columns which subsequently collapse at known added weight. The shape of the curves generated from these experiments gives an indication of the 'tackiness' or non-tackiness of materials.

Sticky material may also be classified by the determination the measurement of the wall friction on plant equipment (Figure 1.13). Moisture increases tackiness as does dust content and the melting point of detergents and soaps. Techniques and the methodologies used to measure the tackiness of detergent powders and pharmaceuticals were reviewed by Lappas and Dempski (1965), while recently Adhikari *et al.* (2001) have reviewed the mechanisms of stickness in foods.

1.7.4.7 *Compressibility* A powder sample can be compressed in a shear cell ring, mould or punch and die and the powder displacement recorded. The slope of the line generated by the volume of powder, void ratio (volume of voids/volume of solid) or displacement versus the logarithmic applied stress is a measure of the compressibility of bulk solids. The simplest method to test compressibility (wet or dry) is the uniaxial compression of a powder in a consolidometer or oedometer (BS 1377-7: 1990; ISO/TS 17892–5:2004).

1.7.4.8 *The Johanson indicizer system* Johanson (1987, 1992) developed a novel system called the Hang-up indicizer to determine a powder flow function and to replace the tedious evaluation method of Jenike. The bulk sample is contained within a cylindrical test cell (Figure 1.27). The cell bottom is supported during consolidation to obtain a flat sample surface. The sample is consolidated by a two-part upper piston. The consolidation pressure is applied for a pre-set time. The outer part has a sliding fit to the cylinder. The inner piston, which is attached to a load cell, measures the force on the inner piston and is independent of the friction between the cylinder and the outer part of the piston. The two-part upper piston is then retracted and the bottom support removed, leaving the sample supported only by a ledge around the cylinder's circumference at the bottom end of the cylinder. The inner piston is then moved downward to fail the sample and to measure its strength. This empirical failure strength is multiplied by a factor of about 2.2 chosen from comparable tests with other methods, to obtain the unconfined yield strength of power.

Johanson claims that this new approach produces a similar powder flow function in one tenth of the time required by the Jenike test. However, Enstad and Maltby (1992) and Schwedes and Schulze (1990, 1992) criticised the Johanson method and doubted whether the measured failure stress was representative of a material's unconfined yield strength, even after the application of a correction factor. This observation was confirmed by Marjanovic *et al.* (1995, 1998) with five powders (Table 1.13).

Bell *et al.* (1994) compared the Johanson indicizer – a unixial direct shear tester – with a Jenike shear cell and a Peschl rotational shear cell on nine various powders, having an overall size range of 600–0.61 μm, from a certified reference powder (BCR limestone, CRM-116) to white wheat, sub-micrometre mineral oxides and a polymer. The values of the unconfined yield strength (f_c) and the arching and rathole diameters determined by the indicizer were often appreciably smaller than those determined by the Jenike and Peschl shear cells (Figure 1.28). Table 1.14 shows the variation of an unconfined yield stress, f_c,

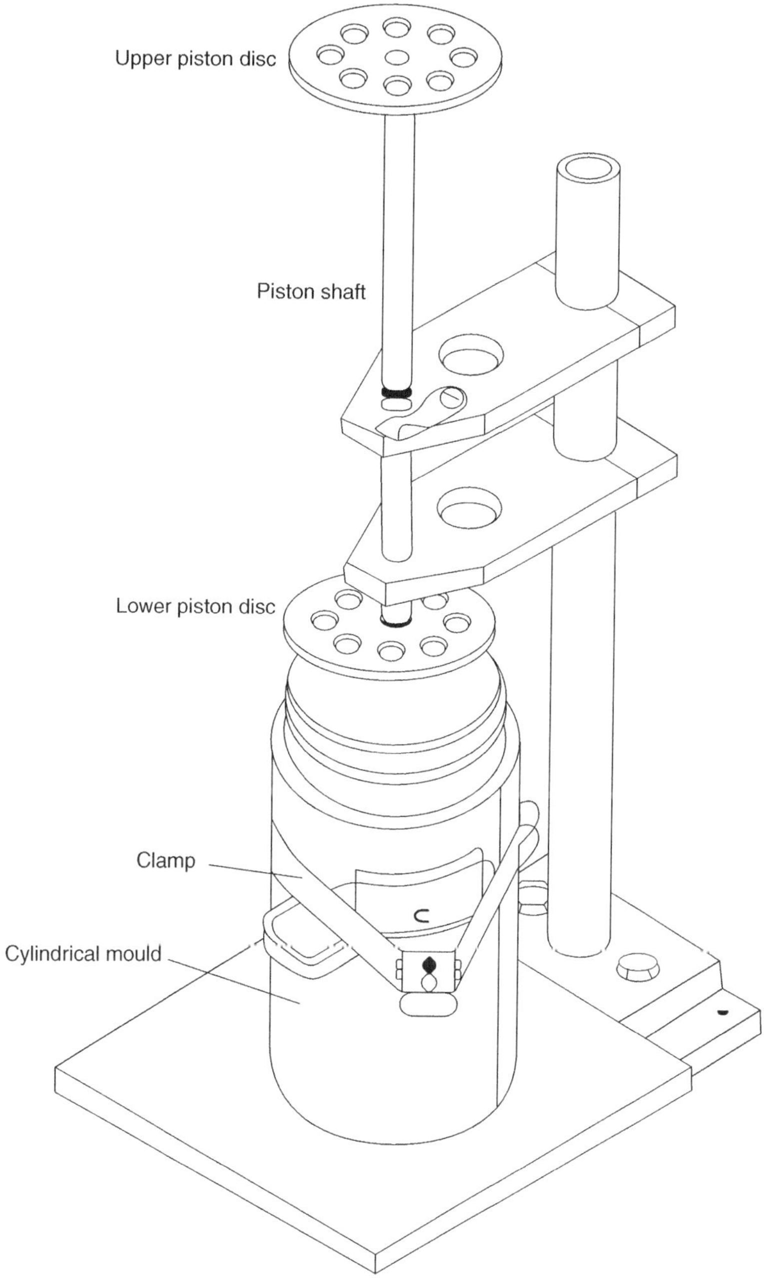

Figure 1.26 Compression tackiness tester (Monick 1966).

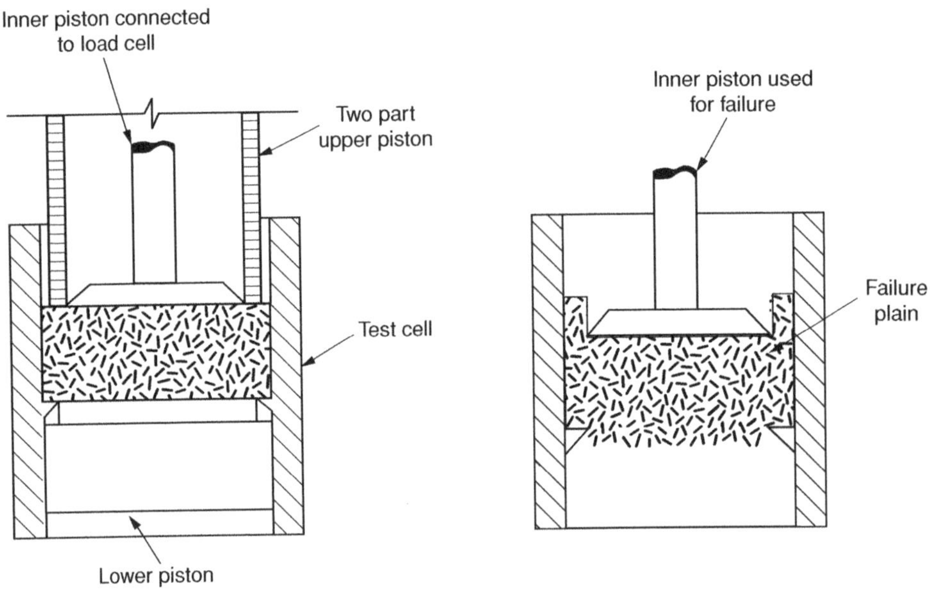

Figure 1.27 Johanson indicizer (Johanson 1987, 1992).

at various major consolidation stresses (σ_1 equal to 2.87, 9.58 and 19.15 kPa) for a selected number of powdered materials.

1.7.4.9 *POSTEC – unaxial research tester* In the POSTEC – uniaxial tester, discussed by Maltby and Enstad (1993), the sample is confined in a cylindrical die and wrapped in a flexible membrane which is stretched between the outer periphery of the piston and the inner perimeter of the lower part of the die. Since the membrane is stretched and in contact with the wall and powder, the sample is compacted homogeneously; thus the wall friction between the specimen and the die is reduced. Comparison of the POSTEC uniaxial tester with a biaxial and Jenike-type shear cell testers, with the standardised CRM-116 limestone powder, indicated that the f_c values obtained with the POSTEC are slightly less than those obtained with Jenike-type shear cells and a biaxial tester. Since the total time for

Table 1.13 Comparison of shear and powder bulk data obtained from a Jenike shear cell and a Johanson indicizer.

			Jenike		Johanson	
Material	Particle (μm)	Hausner ratio (–)	f_c (kN/m^2)	Internal angle (degree)	f_c (kN/m^2)	Internal angle (degree)
Soda ash NR2	68	1.48	0.54	37	0.39	43
GS9	41	1.92	3.17	30	2.32	40
GS14	30	2.16	4.30	27	2.73	36
Lactose L2	38	1.71	2.21	30	0.96	44
L1b	27	1.69	2.02	28	1.58	41

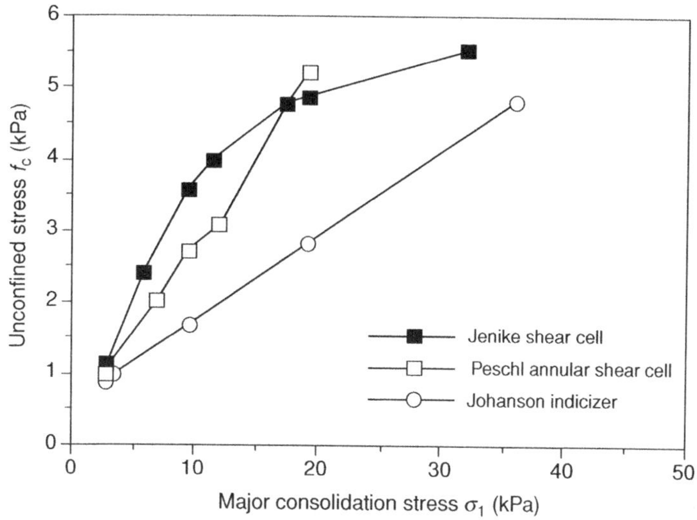

Figure 1.28 Comparison of the degree of flowability from two shear cells and a cylindrical mould tester (Bell *et al.* 1994).

one experiment is of the order of 20 minutes, this uniaxial tester should be considered to have great potential for the measurement of the behaviour of powders in the stress region of <100 kPa (Maltby *et al.* 1993).

1.7.4.10 *Jenike and Johanson quality control tester* The test apparatus consists of a sample container, which has a small cylindrical section above a converging conical section, mounted on a frame (Figure 1.29). Three different containers, all of the same height but of

Table 1.14 Comparison of the unconfined stress, f_c, from the shear cells of Jenike and Peschl with the Johanson indicizer on BCR 116, a mineral oxide, corn starch and flour.

		Unconfined yield stress f_c (kPa)		
Material	σ_1 (kPa)	Johanson	Jenike	Peschl
BCR 116	2.87	0.86	1.12	0.7
	9.58	1.66	3.60	2.72
	19.15	2.81	4.87	5.21
Mineral oxide B	2.87	1.46	1.66	1.70
	9.58	4.38	6.60	7.27
	19.15	8.55	13.66	20.71
Corn starch	2.87	0.85	2.47	0.98
	9.58	1.54	3.24	3.25
	19.15	2.59	5.01	6.42
Flour	2.87	0.91	2.17	1.12
	9.58	1.61	3.34	3.30
	19.15	2.59	5.01	6.42

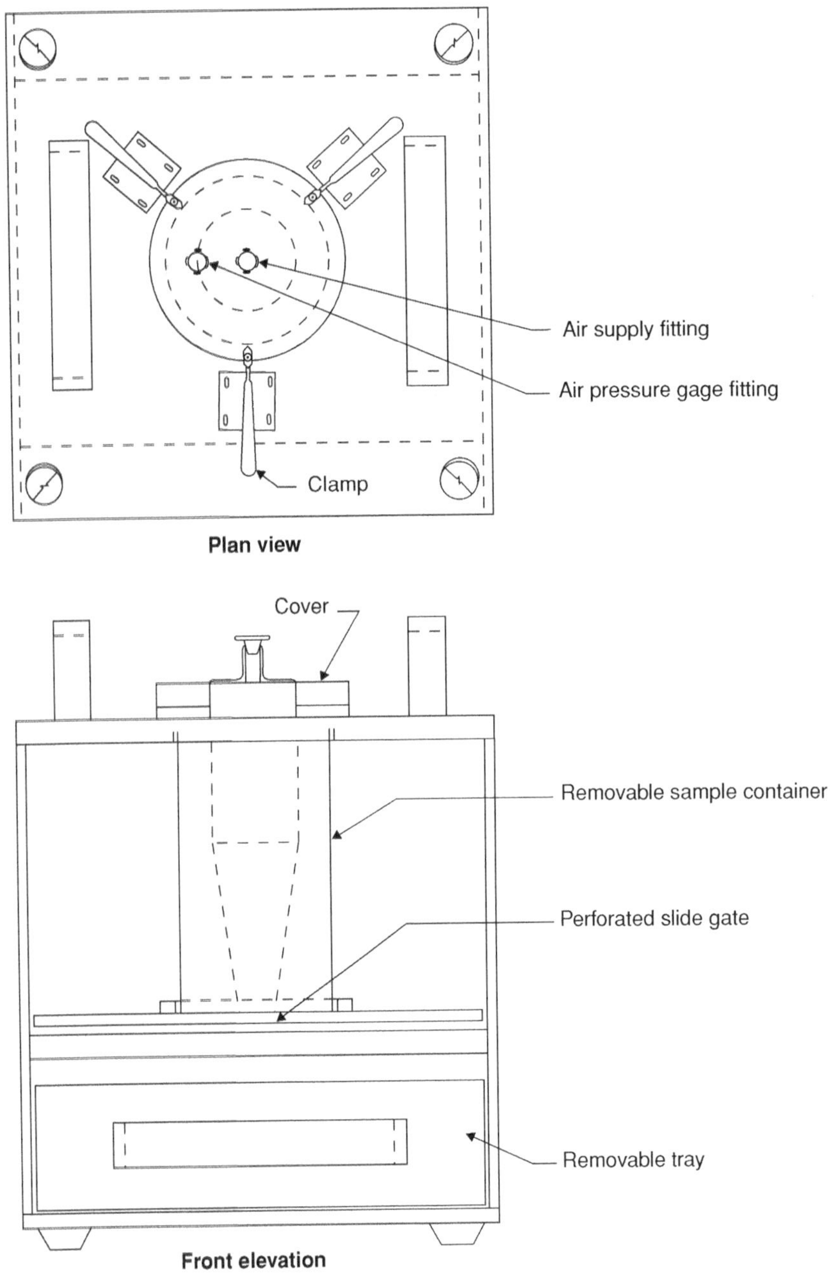

Figure 1.29 Jenike and Johanson Quality Control Tester (JJQC) (Jenike and Johanson Inc).

different outlet sizes and conical section with different wall angles, are currently available. These containers can be fitted through a hole in the top of the frame and are supported by a flange at the top of the container. Below the outlet of the conical section, there is a perforated slide gate, which allows air to pass through the powder within the container, but prevents powdered material from leaving. A gasketed cover clamped to the top of the container forms an airtight seal and has two fittings. An air supply is connected to one of the fittings and a pressure gauge to the other fitting. These are subsequently connected to a control panel to measure the air supply pressure and test pressure. The minimum recommended air supply is circa 500 kN/m^2 (60 psig).

About 4000 cm^3 of aerated powdered material is gently filled into the container with the smallest outlet situated on top of the closed perforated slide. The lid, connected to the air supply and test gauge, is then clamped on top of the container and a specific pressure applied for 30 seconds, to compact the powder. Pressure is then reduced to atmospheric pressure and the perforated slide gate removed from the bottom of the conical section. Air pressure is then increased until the arch, formed in the outlet, fails and the powdered cohesive material flows from the container at a recorded pressure.

The consolidation and failure process can then be repeated for a minimum of four times. The entire stepwise process can be repeated at different values of consolidation pressure.

This tester is designed for cohesive materials so a limit is placed on the maximum particle size of powder used in the tester. Tests can be conducted at both ambient temperature and humidity or under conditions created in an environmental chamber. Tests conducted with the Jenike and Johanson Quality Control Tester (JJQC) have been compared with results obtained from a conventional Jenike shear cell tester using cohesive titanium dioxide, laundry detergent, baking soda and table salt. A plot of the 'failure' pressure obtained from the JJQC tester and the Jenike shear cell tester strength showed that the JJQC test values are significantly lower at high consolidation pressures but were comparable to the Jenike values at low consolidation pressures (Ploof & Carson 1994).

Tests with 'dry' limestone dust (0.07% moisture) and with 0.35% and 1.36% moisture added showed that as the percentage of moisture in the powder increased the average failure and thus the cohesiveness of the material increased.

1.7.4.11 *Penetrometer method* Knight and Johnson (1988) introduced a penetrometer technique for measuring the cohesive strength of particulate material. They found that the unconfined yield strength f_c could be expressed in terms of the penetration force, F, and the penetration depth, P, and was dependent on the angle of wall friction and the internal angle of powder friction. A powder sample was compacted into a container 120 mm in diameter and 25 mm deep. An Instron tester was used to drive the penetrating cone into the compressed powder bed at a rate of 10 mm/min, and the Instron tester also continuously recorded the force. Since the container was sufficiently large and the stress caused by the weight of powder above the failure zone was low, the failure force was effectively unconfined.

1.8 Bulk power correlations

Plinke *et al.* (1994) investigated the dependence of cohesion on moisture content, particle size distribution and the melting temperature of materials. Four industrial materials were chosen – titanium dioxide, limestone reference powder (CRM116), glass beads and lactose – with a range of sizes $-d < 5$ μm, $5 < d < 25$ μm and $d > 25$ μm – and tested for cohesiveness over a range of moisture values of 0–6% w/w.

The base or zero value of the moisture content of the material was taken as a 'as received material' condition. The actual moisture content of each material was determined at the time of the test by averaging the moisture measurement taken before and after the test to account for any moisture lost to, or taken up from, the environment during the test.

Comparison of the results showed that with the Jenike shear cell test, higher cohesion values were obtained when the powder beds were compacted at 6 kPa than when the powder beds were compacted at 3 kPa. The Jenike shear cell test gave a higher cohesion value, when compacted at 6 kPa, than that with the Peschl rotational split level shear test compacted at 5 kPa. The higher cohesive values being attributed to an increase in contact points and binding forces. The moisture content increased the cohesion for limestone. The materials titanium and glass beads behaved in a similar manner. Since titanium and glass beads do not adsorb water the increase in cohesion must be due to the increase in the thickness of the liquid film surrounding the non-wetted particles resulting in an increase in capillary binding forces between particles; a phenomenon seen in the process of granulation and agglomeration. The lactose powder was relatively unaffected, in terms of cohesion, by moisture. These observations showed that cohesion, C, may be dependent upon the material factors of

- Moisture content – M (%)
- Mass median diameter – d_{50}
- Melting temperature – Tm
- Voidage – e
- Specific surface area – BET (m^2/g)
- MOHS material hardness – MOHS

to postulate the algebraic relationship

$$C = e^{1.2} M^{0.2} d_{50}^{-0.2} Tm^{0.3} \tag{1.12}$$

The statistical confidence of this correlation can be quantified knowing that this relationship showed a coefficient of correlation of $R^2 = 0.67$.

Inclusion of the specific surface areas of the powders and the Mohs hardness of the particles into the above correlation increased the coefficient of correlation, R^2, from 0.67 to 0.69 which illustrated an increase in reliability.

The relationship of cohesion with particle size was found to be

$$C = 1.25\, d_{50}^{-0.6} \tag{1.13}$$

which can be allied to the shear flowability index of the Warren Spring correlation between cohesion and tensile strength derived by Farley and Valentin (1967/68).

1.8.1 *Relationships between cohesion and tensile strength*

Farley and Valentin (1967/68) were the first to correlate the bulk powder properties of cohesion C, and tensile strength T, for five different powders, containing a range of particle sizes. It has been shown for a range of powders (calcium carbonate, precipitated and mined, aluminium oxide and zinc dust) that cohesion appears to be approximately equivalent to $2T$

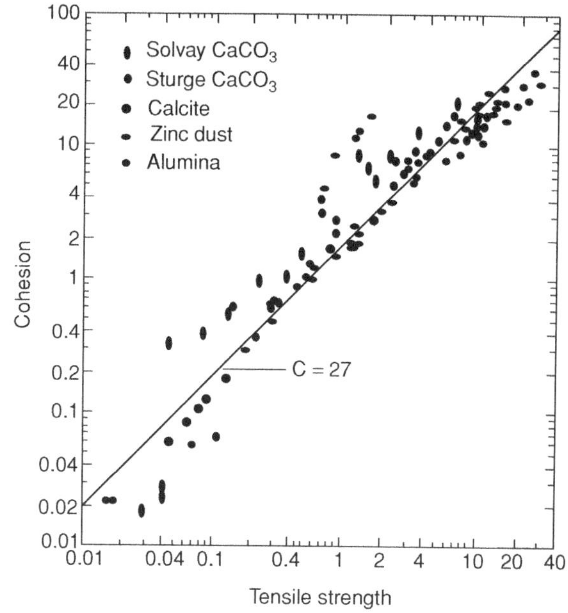

Figure 1.30 Relationship between cohesion and tensile strength for a range of industrial powders (Farley & Valentin 1967/68).

(Figure 1.30). This relationship has also been confirmed for pharmaceutical intermediates and products by Eaves and Jones (1971) and Pilpel and Walton (1974) (Figure 1.31).

It is now believed, however, that more powder flow properties can be incorporated into this simple relationship of $C = 2T$. Incorporation of the parameters of particle size, bulk and powder densities leads to the relationship expressed as the Warren Spring equation

$$\left(\frac{\tau}{C}\right)^n = \left(\frac{\sigma}{T}\right) + 1 \tag{1.14}$$

where $n = 1 + (0.53/d^{2/3})$ and $T = A(\rho_b/\rho_p)^m$; d is the volume/surface mean diameter, ρ_b the bulk density of powder, ρ_p the particle density, m is an index relating T, the tensile strength, to packing density and A is a constant.

The shear flowability index, n, was found, from past observations (Farley & Valentin 1965, 67/68), to be independent of the bulk density of sheared compacted powder. Because of this independence of particle size from bulk density it is now realised that the shear flowability index, n, from the Warren Spring equation and the Jenike internal angle of friction may be the preferred parameters to characterise and quantify the flowability of powders. Jenike and others (Williams *et al.* 1970/71; Williams & Birks 1965; Hill & Wu 1996; Cox & Hill 2004) selected the Jenike failure function to be one of the best indicators to predict the ease of powder movement and powder flowability.

Rajendran Nair *et al.* (1990, 1993) examined the cohesiveness of mixtures of a coal and 10% clay mix, which had a residual moisture content of 0.15% moisture, with mixes having various amounts of added water. The parameter of cohesiveness (C_m) was derived from the yield loci measured in a Jenike shear cell tester and the tensile strength measured was from

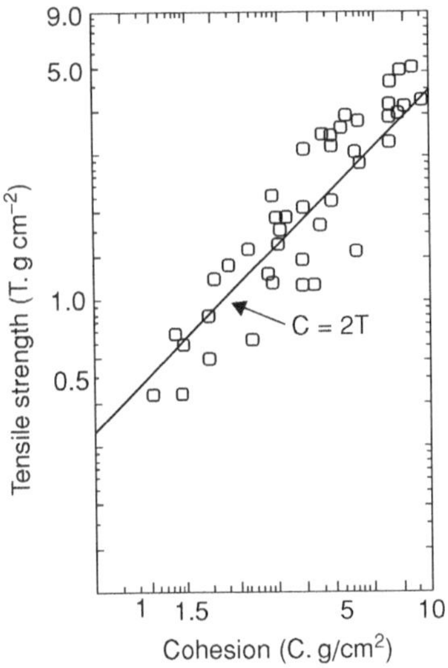

Figure 1.31 Relationship between cohesion and tensile strength for pharmaceutical powders (Pilpel & Walton 1974).

a split cell tester (Hosokawa Micron cohetester). As the percentage of moisture increased from 0 to 15% w/w, both the cohesiveness and the shear stresses increased (Figure 1.32).

From the initial linear part of these series of yield loci, Rajendran Nair and co-workers calculated the internal friction, ϕ_{RN}, from the slope of the yield loci and also computed the tensile strength, T_{RN}, from the extrapolated yield loci of the coal–clay–water mixtures, to arrive at the relationship:

$$T_{\mathrm{RN}} = \frac{C_m}{\phi_{\mathrm{RM}}} \tag{1.15}$$

Direct measurements of tensile strengths of the coal–clay–water mixtures obtained from the Hosokawa Micron cohetester, T_{H}, were then compared with those obtained from the extrapolated Jenike yield loci to propose the linear relationship:

$$T_{\mathrm{RN}} = 1.5\, T_{\mathrm{H}}^{-0.5} \tag{1.16}$$

Yamatmoto (1990) also used a Hosokawa cohetester to investigate the tensile strength of sub-micrometre limestone and silica when exposed to various relative humidities and found an increase in tensile strength at different degrees of packing or bed porosity.

Yokoyama *et al.* (1982) measured the tensile strength with the swing method of the Hosokawa cohetester, for a range of particle sizes (1–90 μm) of different densities (0.928–6.01 g/cm^3) over a range of moisture percentages (0.06–15.2% w/w), to evaluate the potential floodability of powders. Yokoyama *et al.* defined their dimensionless floodability, as opposed to Carr's definition of flowability, as the ratio of the cohesive force (C) – measured by cohetester – to the gravity force on particles ($m_{\mathrm{p}}g$).

$$\frac{C}{m_{\mathrm{p}}g} = \frac{3T}{(1-\varepsilon)d_{\mathrm{p}}\rho g} \tag{1.17}$$

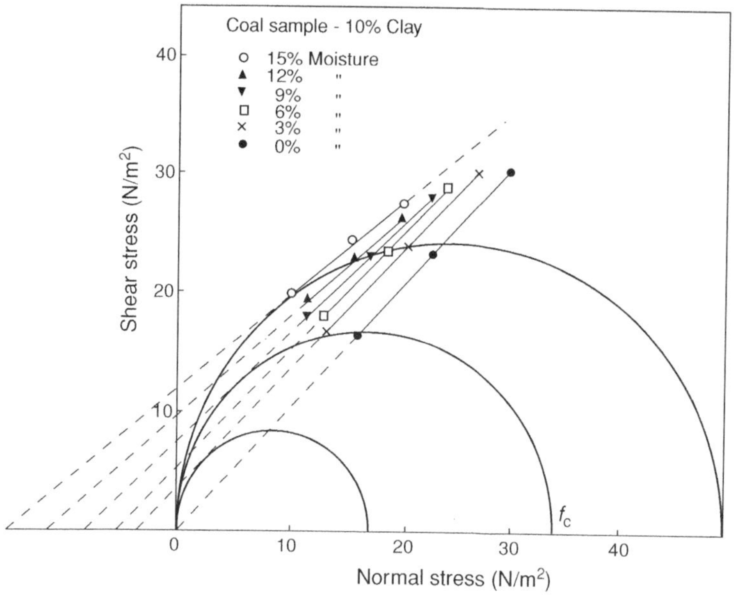

Figure 1.32 Rajendran Nair data of shear stress versus normal stress for coal–water mixtures (Rajendran Nair *et al.* 1990).

where T is the tensile strength of a powder bed; d_p is the particle diameter; ε is voidage of the bed and ρ is the particle density.

Figure 1.33 shows the calculated dimensionless term $C/(m_p g)$ versus the ratio ρ_b/ρ_{av} where ρ_b is the bulk density of the bed and ρ_{av} is the arithmetic average of aerated and packed densities of the powder.

1.8.2 *The influence of particle characteristics on bulk powder properties*

The packing variation in packed products and goods, together with the rate of packing, has to be constantly monitored to maintain production assurance.

Bulk and pack densities are known to be dependent upon powder feed rates, height of fall and particle morphology when directed into specific volume containment. Thus powder flowability, cohesiveness, impact pressure and powder yield strength together with other physical particle properties such as particle size and particle shape contribute influentially to powder bulk behaviour.

From Carr's (Carr 1965a,b) suggestion that fine cohesive powders might behave as a liquid and have a tendency to fluidise and possibly flood, Yokoyama *et al.* (1982) correlated Carr's floodability index with a dimensionless ratio of the cohesive–gravity forces $(C/m_p g)$. As the particle size of the powdered materials decreased there was an increase, as expected, in the cohesive–gravity force ratio. The Carr's floodability index with materials having small particle sizes [$BaTiO_3$ (1 μm), Kanto loam (2 μm) and $CaCO_3$ (5.6 μm)] produced below average floodability whilst materials with particle sizes above 80 μm had a high degree of floodability (60–75%). Figure 1.34 shows, however, an excellent similarity between the dimensionless term of Yokoyama *et al.* $(C/m_p g)$ with that of Carr's floodability index.

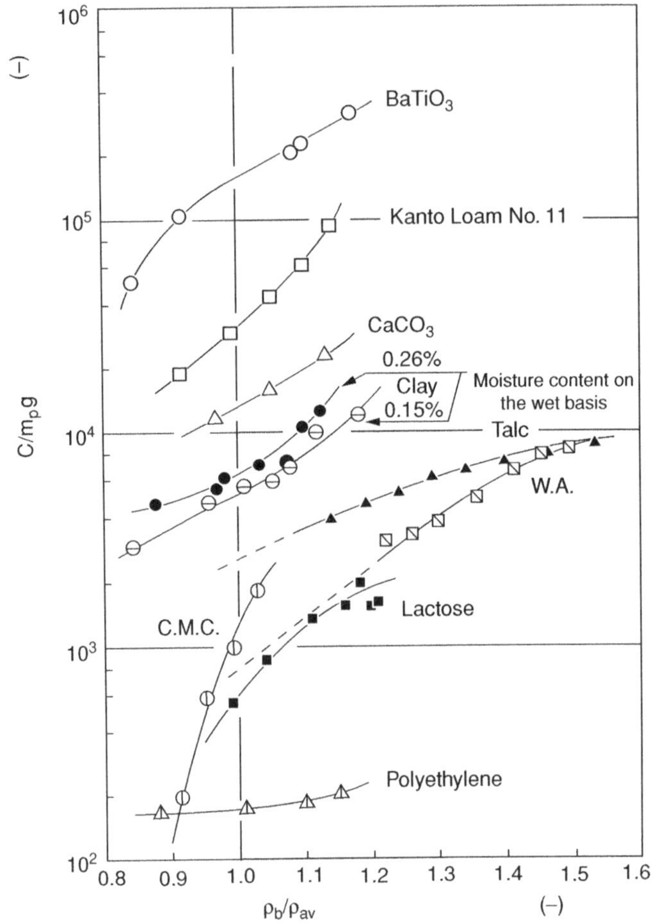

Figure 1.33 Yokoyama *et al.* (1982) data for the potential floodability of powders.

Kurz and Minz (1975) investigated the flowability of powders in terms of the relationship between the unconfined yield strength, f_c, and the major consolidation, σ_1 – the failure function of Jenike — for different-sized limestone powders (3.1–55.0 μm) having either a narrow or a wide particle size distribution. The width of the distribution was defined by a variation coefficient, Cv, where $\mathrm{Cv} = \sigma_{\mathrm{stat}}/x_{1.3}$, with σ_{stat} as the standard deviation of the particle size profile and $x_{1.3}$ the average particle size of the number–volume diameter distribution of the limestone particles. A narrow distribution was considered to have a Cv < 0.5 while a wide distribution had a Cv > 0.5.

It was concluded that average particle size was not the only criterion, which could be used to characterise the cohesiveness of powders, but also the spread of particle sizes in a size distribution was also a significant contributory factor. Limestone powders, with average sizes above 8 μm and a narrow size distribution freely flowed but with the same-sized limestone particles which had a wide distribution, cohesive flow behaviour existed.

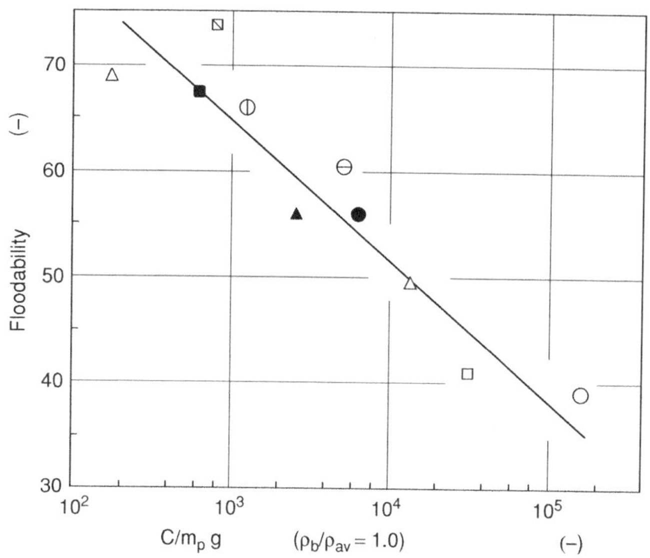

Figure 1.34 Similarity between Carr's floodability factor (Carr 1965b,c) and Yokoyama *et al.* parameter, C/m_pg (Yokoyama *et al.* 1982).

The shape of particles is now regarded as having as much influence upon the mechanical properties, the degree of flowability and the cohesiveness of powders as particle size.

Pilpel and Walton (1974) and Walton and Pilpel (1972a,b) measured the shear and tensile properties of differently shaped materials and moisture (Pilpel 1970) to assess their failure and flow behaviour into various filling and storage units.

1.8.3 *Overview of the instrumentation and techniques available for the determination of bulk powder properties*

In a survey of the various bulk powder tests developed and in use for the determination of the failure properties and measurement of the degrees of flowability of powders, Schwedes (2003) indicated that the equipment available varied from highly theoretical and technical instrumentation to very simple empirical techniques.

The choice given to the production engineer or research powder technologist is thus dependent upon the level of information required. In the design of mass flow hoppers there tends to be a greater need for knowledge on yield loci, stresses and failure criteria than that possibly needed for routine powder quality control.

The instrument and technique which may be chosen is also governed by the type of bulk solid (which may range from free flowing to very cohesive) undergoing investigation and also the frequency of information needed, cost of the device, time and skill allocated for the test together with the availability of equipment for the measurement of both particle and bulk solids characteristics.

Table 1.15, taken from the work of Schulze (1995, 1996a,b), indicates a wide range of apparati and protocols to measure a powder yield strength or a (Jenike) failure function for use in the determination of bulk powder flowability.

Table 1.15 Overview of the various methods for the characterisation of the flowability of bulk solids.

Flowability testers	Figure	Criterion 1	Criterion 2	Criterion 3	Criterion 4	Criterion 5	Criterion 6	Criterion 7
Simple flow equipment								
Aero-Flow		no						no
Angle of repose	1.9 and 1.15	no						no
Flow discharge (Flodex)	1.16	no						no
Indirect & direct shear								
Bi- and tri-axial testers	1.35	□	□	□	□/□	□/□	□	□
Jenike's shear tester and the Standard Shear Tester Technique	1.35	□	□	□	□/□	□/□	□	□
Ring shear tester	1.35	□	□	□	□/□	□/□	□	□
Torsion shear tester		□	□	□	□/□	□/no	□	□
Uniaxial columns								
Procter & Gamble caking tester	1.25	□			□/□	□/no	□	□
J & J Quality control tester	1.29	□			□/□	no/no	□	□
Johanson Hang-up indiciser	1.27	□			□/□	no/no	□	□
Monoaxial tester		□			□/□	□/no	□	□
Penetration test					□/□	□/no	□	□
Powder bed tester	1.22 and 1.24	□			□/□	□/no	□	□
Tackiness	1.26	□			□/□	□/no	no	□
Tensile and cohesion								
WSL tensile tester	1.19	□			□/□	□/no	□	□
Hosokawa Micron cohetester	1.20	no	no			no/no	□	no
Lifting lid vertical shear		no	no		□/□	□/□	□	no
Tensile strength tester	1.19	□			□/□	□/no	□	□
AJAX and WSL cohesion tester	1.21	□			□/□	□/no	□	□

□, acceptable; no, not acceptable.
From Schulze (1995, 1996a,b) [see Schwedes (2003)].

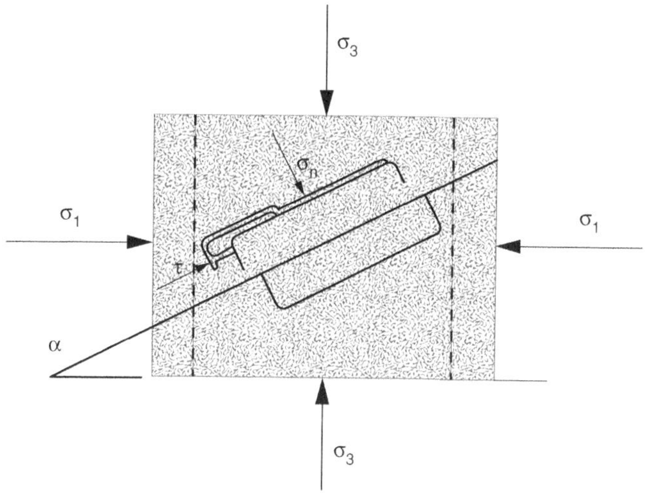

Figure 1.35 Difference between direct and indirect shear testers. (Jenike direct shear tester inside square and biaxial tester solid rectangle.)

Direct and indirect shear stress equipment is usually limited to research projects because of the expense, the skill and time evolved to determine the yield loci and failure functions of particulate materials from bi- and tri-axial shear testers. The designation of which type of shear tester has been used may be defined by the location of the shear failure zone. With direct shear testers – Jenike type – the major principal stress rotates during shear (Figure 1.35), whilst with indirect shear testers – tri- and bi-axial – the directions of the principal stresses are fixed and orientated in either three or two dimensions, respectively, and remain constant during the test.

The Jenike (or Standard Shear Tester) tester together with commercially available ring and torsional shear testers are, however, relatively easier to use than the tri- and bi-axial shear testers. In some cases ring, torsional and the Standard Shear Tester (Jenike) are now automated and computer controlled for ease of use and speed of determination. These instruments measure parameters for use in mass flow hopper design and calculation of critical hopper outlet diameters needed to ensure mass flow as opposed to plug or 'rat-holing' flow (Hill & Cox 2002). A simple approach for the determination of flow from an orifice can, however, be achieved with either a Flodex (Hanson Research Corporation), a Jenike and Johanson Quality Control Tester or a 'home made' model (Figures 1.16, 1.29, 1.9 & 1.15). The uniaxial and monoaxial tester approach relies upon the formation of a column of powder, either constrained or freestanding, to measure an unconfined yield shear stress together with the normal consolidation stress to give a failure function. The uniaxial tester, with a vertically applied failure stress, should have an L/D ratio greater than 2 in order to achieve accurate unconfined stresses. The monoaxial tester, which is sheared in a horizontal direction, does not have this limitation. Column formation and subsequent failure are relevant to the simple, rapid results and user-friendly equipment seen with the Johanson hang-up indicizer, caking or tackiness testers. Cohesion and tensile strength are also of great importance to the production powder technologist or plant engineer. Tensile strength cannot, however, be directly measured by Jenike and ring-type shear cells. This lack has been overcome by the development of the WSL tensile tester (Ashton *et al.* 1964, 1965; Farley & Valetin 1967/68) and the Hosokawa cohetester split cell tensile tester. Although cohesion and adhesion may be evaluated from a shear stress versus compaction stress relationship, the Warren Spring-Bradford tester (Orband & Geldart 1997) has now been marketed as the Ajax (Bolton) cohesion tester, a dedicated instrument for cohesion measurement.

The simplest method to indicate the flowability of bulk powders is to use any technique that measures the static angle of a heap of powder – angle of repose – using either the Carr's angle of repose equipment (Figure 1.9) or a simple apparatus (Figure 1.15). A dynamic angle of repose can now be measured with an Aero-Flow Analyzer.

Schulze judged the acceptance of testers for various research and industrial problems by highlighting the different shear approaches and the various types of bulk powder testers available, based on seven criteria. These criteria are numerated in Table 1.15:

1 The ability to compact or consolidate the powder, in a movable cell which, with subsequent horizontal shear measures the powder failure or yield strength.
2 Maintenance of an applied normal consolidation stress whilst undergoing shear to steady state flow.
3 To ensure mass flow in any hopper design the direction of the major principal stress during steady state flow should coincide with the direction of failure of the powder along the hopper wall.
4 Reproducible stress conditions on the bulk solid sample both at compaction (consolidation) and failure illustrated as 'item/item'.
5 Knowledge that average stresses and uniform stress distribution, to ensure no anisotropy at the compaction plane and at the failure plane, is achievable.
6 Possibility to vary the magnitude of the consolidation stress applied to the powder sample.
7 Possibility for the measurement of shear on 'time consolidated' samples.

Possibly the most important criteria for quality control, flowability and bulk powder handling are the items highlighted under numbers 1, 4, 6 and 7.

Scrutiny of the type of instrumentation, which has been used in the diverse field of bulk powder handling and outlined in this chapter, should allow personnel in the many industries which produce, handle and package powders, an opportunity to select one or more instruments and techniques which may aid the prevention of some of the production problems which could arise from day to day.

References

Abdelkarim, A.M. (1982) *Shear and normal stresses in uniaxial compaction*. Ph.D. thesis, University of Bradford, Bradford, UK.

Adhikari, B., Howes, T., Bhandari, B.R. & Truong, V. (2001) Stickiness in foods: a review of mechanisms and tests methods. *Int. J. Food Properties*, **4**(1), 1–33.

Adler, A. (1969) Flow Properties of Metal Powders. *Int. J. Powder Metallurgy*, **5**(1), 7–20.

AJAX Equipment (Bolton) Ltd. *Milton Works*. Mule St, Bolton, Lancashire. www.ajax.co.uk.

Akers, R.J. (1992) *Certification of a Limestone Powder for Jenike Shear Testing. CRM116*. Commission of European Committees, BCR, Office for Official Publications of the European Communities, Brussels, Luxembourg.

Archard, J.F. (1953) Contact and rubbing of flat surfaces. *J. Appl. Phys.*, **24**(8), 981–988.

Archard, J.F. (1957) Elastic deformation and laws of friction. *Proc. R. Soc.*, **A243**, 190–205.

Ashton, M.D., Farley, R. & Valentin, F.H.H. (1964) An improved apparatus for measuring the tensile strength of powders. *J. Sci. Instum.*, **41**, 763–765.

Ashton, M.D., Cheng, D.C-H., Farley, R. & Valentin, F.H.H. (1965) Some investigation into the strength and flow properties of powders. *Rheol. Acta*, **4**(3) 206–218.

ASTM B329-98 (2003) *Apparent Density of Metal Powders*. American Society for Testing Materials, International, West Conshohocken, PA. www.astm.org.

ASTM B855-06. *Arnold Meter & Hall Funnel*. American Society for Testing Materials, International, West Conshohocken, PA. www.astm.org.

ASTM D 4180-03. *Hall Test Cup Method*. American Society for Testing Materials, International, West Conshohocken, PA. www.astm.org.

ASTM D 6393-99 (1999) *Carr Flowability Indices*. American Society for Testing Materials, International, West Conshohocken, PA. www.astm.org.

ASTM D 6773-02. *Standard Shear Test Technique*. American Society for Testing Materials, International, West Conshohocken, PA. www.astm.org.

ASTM eStandardsStore at www.webstore.ansi.org.

Bell, T.A., Ennis, B.J., Grygo, R.J. *et al.* (1994) Practical evaluation of the Johanson hang-up indicizer. *Bulk Solids Handling*, **14**(1), 117–125.

Beverloo, W.A., Leniger, H.A. & van de Velde, J. (1961) The flow of granular solids through orifices . *J. Chem. Eng. Sci.*, **115**, 260–269.

Bishop, A.W. & Green, G.E. (1965) The influence of end restraint on the compression strength of cohesiveless soils. *Geotechnique*, **15**, 243–266.

Bockstiegel, G. (1966) Relations between pore structure and densification mechanism in the compacting of iron powder. *Int. J. Powder Met.*, **2**, 13.

Boden, J. (1981) *The tensile strength and anisotropy of powder compacts*. Ph.D. thesis, University of Bradford, Bradford, UK.

Bodhmage, A. (2006) *Correlation between physical properties and flowability indicators for fine powders*. M.Sc. thesis, University of Saskatchewan, Saskatchewan, Canada.

Briscoe, B.J, Pope, L, Adams, M.J. (1984) Interfacial friction of powders on concave counterfaces. *Powder Technol.* **37**, 169–181.

British Standards at www.bsonline.bsi-global.com

Brown, L. & Richards, J.C. (1970) *Principles of Powder Mechanics*. Pergamon Press, London.

BS 1377-7 (1990) *Methods of Tests for Civil Engineering Puposes. Shear Strength Tests (Total Stress)*. British Standards Institution (BSI), Gunnersbury, London, UK.

BS 2955 (1993) *Glossary of Terms Relating to Particle Technology*. BSI, Gunnersbury, London, UK.

BS 4140-9 (1986) *Methods of Test for Aluminium Oxide. Measurement of Angle of Repose*. BSI, Gunnersbury, London, UK.

BS 4140-23 (1987) *Methods of Test for Aluminium Oxide. Determination of Absolute Density Using a Pyknometer*. BSI, Gunnersbury, London, UK.

BS 4359: Pt 2 (1982) *Determination of Specific Surface Area of Powder. Permeametry & the Kozeny–Carman Equation*. BSI, Gunnersbury, London, UK.

BS EN 543 (2003) *Apparent Density of Powder and Granulues*. BSI, Gunnersbury, London, UK.

BS EN 725-8 (1997) *Bulk Tapped Density of Ceramic Powders*. BSI, Gunnersbury, London, UK.

BS EN 993-17 (1999) *Density of Dense, Shaped Refractory Products by Mercury Methods*. BSI, Gunnersbury, London, UK.

BS EN 1097-3 (1998) *Bulk Density of Loose Aggergates*. BSI, Gunnersbury, London, UK.

BS EN 23923-2 (1993) *Apparent Density of Metallic Powders*. BSI, Gunnersbury, London, UK.

BS EN ISO 7837 (2001) *Bulk Density of Loose Fertilizers*. BSI, Gunnersbury, London, UK.

Carman, P.C. (1937) Fluid flow though granular beds. *Trans. Inst. Chem. Eng.*, **50**, 150–166.

Carr, R.L. (1965a) Angle of repose. *Chem. Eng.*, **18**(1), 163–169.

Carr, R.L. (1965b) Classifying flow properties of solids. *Chem. Eng.*, **72**, 69–72.

Carr, R.L. (1965c) Evaluating flow properties of solids. *Chem. Eng.*, **72**(2), 163–169.

Carson, J.W. & Marinelli, J. (April 1994) Characterize bulk solids to ensure smooth flow. *Chem. Eng.*, **101**(4), 78–90.

Clift, R. (1985) Particle-particle interactions in gas-particle systems. *Powtech. 85. No. 91*. Institution of Chemical Engineers, Rugby, UK.

Cowherd, C., Grelinger, M & Englehart, P.J. (1989) An Apparatus and methodology for predicting the dustiness of materials. *Am. Ind. Hyg. Assoc. J.*, **50**, 123–130.

Cox, G.M. & Hill, J.M. (2004) The limiting ideal theory for shear-index cohesionless granular materials. *ANZIAM J.* **45**, 373–392.

Davidson, J.F. & Nedderman, R.M. (1973) The hour-glass theory of hopper flow. *Trans. Inst. Chem. Eng.*, **51**, 29–35.

Davidson, J.F., Nedderman, R.M. & Walker, D.M. (1966) An approximate theory for pressures and arching in hoppers. *Chem. Eng. Sci.*, **21**, 975.

DIN 5599-2 (1999) *Standardised Single-Drop Dustiness Apparatus to Generate a Dustiness Index of Fillers and Pigments.* Deutsches Insitut fur Normung, Berlin, Germany. www2.din.de.

Eaves, T. & Jones, T.M. (1971) Cohesion and tensile strength of bulk solids. *Rheol. Acta*, **10**(1), 127–134.

EN 23923 (1992) *Scott Volumeter.* BSI, Gunnersbury, Middlesex, UK.

Endersby, V.A. (1942) Fundamental research in bituminous stabilization. *Proc. Highway Res. Board*, **22**, 442–459.

Engelsmann, J. Ludwigshafen, Str 137-141, D-67059. Ludwigshafen am Rhein, Germany. www.engelsmann.de/.

Enstad, G.G. & Maltby, L.P. (1992) Flow property testing of particulate solids. *Bulk Solids Handling*, **12**, 451.

Ergun, S. (1952) Fluid flow through packed columns. *Chem. Eng. Prog.*, **48**, 89–94.

Ergun, S. & Orning, A.A. (1949) Fluid flow through randomly packed columns and fluidized beds. *Ind. Eng. Chem. Fundamentals*, **41**, 1179.

Farley, R. & Valentin, F.H.H. (1965) Problems associated with storage hoppers. *Trans. Inst. Chem. Eng.*, **43**, T193–198.

Farley, R. & Valentin, F.H.H. (1967/68) Effect of particle size upon the strength of powders. *Powder Technol.*, **1**, 344–354.

Flodex. Hanson Research Corporation, Chatsworth, CA. USA www.hansonresearch.com and Acil Sarl, 1, place de Gaulle 78400 Chatou, France.

Freeman Techology, Worcestershire, UK. www.freemantech.co.uk.

Geldart, D. (1972) The effect of particle size and size distribution on the behaviour of gas-fluidised beds. *Powder Technol.*, **6**, 201–215.

Geldart, D. (1990) Estimation of basic particle properties for use in fluid–particle process calculations. *Powder Technol.*, **60**, 1–13.

Geldart, D. (2005) The characterisation of bulk powders. In: *Characterisation of Bulk Solids* (ed D. McGlinchey). Blackwell Publishing, Oxford, UK.

Geldart, D., Harnby, N. & Wong, A.C. (1984) Fluidization of cohesive powders. *Powder Technol.*, **37**, 25–37.

Gerritsen, A.H. (1986) A Simple Method for Measuring Powder Flow Functions with a view to hopper design. *Partec '86*. Nuremberg, Germany, Part 3, pp. 257–279.

Gerritsen, A.H. (1990) *Proc 2nd World Congress Particle Technology*, Kyoto, Japan, Part 1, pp. 61–70. [See Schulze, J.]

Hamelmann, F. & Schmidt, E. (2003) Methods of dustiness estimation of industrial powders – a review. *KONA Powder Part.*, **21**, 7–18.

Hamelmann, F. & Schmidt, E. (2004) Methods for characterizing the dustiness estimation of powders. *Chem. Eng. Technol.*, **27**, 844–847.

Hartley, P.A. & Parfitt, G.D. (1984) An improved split-cell apparatus measurement for the measurement of tensile strength of powders. *J. Phys. E Sci. Instrum.*, **17**, 347–349.

Harwood, J. (1971) Compaction effect on flow property indexes for powders. *J. Pharm. Sci.*, **60**, 161–3.

Hausner, H.H. (1966) *Proceeding Conference Particle Size Analysis.* Society for Analytical Chemistry, London.

Hausner, H.H. (1967) Friction conditions in a mass of metal powder. *Int. J. Powder Met.*, **3**(4), 7–13.

Hertz, H. (1882) *Uber die Beruhrung fester elastischer kurper – On the contact of elastic solids*, pp. 146–162 and *Uber die Beruhrung fester elastischer kurper und uber die harte – On the contact of elastic solids and on hardness*, pp. 163–183 from *Miscellaneous Papers by H. Hertz* (English translation, eds D.E. Jones & G.A. Schott) 1986. Macmillian, London.

Hill, J.M. & Cox, G.M. (2002) Rat-hole stress profiles for shear-index granular materials. *Acta mech.*, **155**, (3-4). 157–172.

Hill, J.M. & Wu, Y-H. (1996) The punch problem for shear-index granular material. *J. Mechanics & Applied Mathematics*, **49**, (1) 81–105.

Hosokawa Micron Ltd. Runcorn, Cheshire, UK. www.hosokawa.co.uk.

Ileleji, K.E. (2005) Report on *Investigation of methods to improve the flowability of distillers dried grains.* Agri & Biological Dept., Purdue University, USA.

ISO 902 (1976) *Angle of Repose for Alumina.* International Organisation for Standardisation, Geneva.

ISO 3923 Part I *Flowability.* International Organisation for Standardisation, Geneva.

ISO 3923-1 (1979) *Apparent Density Funnel Method.* International Organisation for Standardisation, Geneva.

ISO 3923-2 (1981) *Scott Volumeter.* International Organisation for Standardisation, Geneva.

ISO 4324 (1977) *Angle of Repose for Powders and Granules.* International Organisation for Standardisation, Geneva.

ISO 4490 (2000) *Hall Flowmeter.* International Organisation for Standardisation, Geneva.

ISO 6770 (1982) *Flowability & Bulk Density of Tea.* International Organisation for Standardisation, Geneva.

ISO 8398 (1989) *Angle of Repose for Fertilers*. International Organisation for Standardisation, Geneva.

ISO www.iso.org.

ISO/TC 17892-5 (2004) Geotechnical investigation and testing. Laboratory testing of soil – Part 5: Incremental loading oedometer test.

Jenike, A.W. (1964) *Storage and Flow of Solids Bulletin 123*, 1st edn. University of Utah, Salt Lake City, USA. [See also (1970), 6th edn.]

Jenike & Johanson Inc. Tyngsborough, MA, USA. www.jenike.com.

Johanson, J. (1965) The use of flow-corrective inserts in bins. *Trans. ASME.*, **232**, 69–80.

Johanson, J.R. (1992) The Johanson Indicizer System vs. the Jenike Shear Tester. *Bulk Solid Handling*, **12**(April–June), 273–240.

Johanson, J.R. (29 December 1987) *Bulk Solids Property Tester.* U.S. Patent 4715212.

Kaye, B.H., Faddis, N. & Gratton-Liimataienen, J. (May 1997) Characterising the flowability of a powder using avalanching studies. *Proc. Powder & Bulk Solids Exhibition*, Chicago, IL, pp. 123–133.

Kaye, B.H., Gratton-Liimatainen, J. & Faddis, N. (1995) The effect of flow. Agents on the rheology of plastic powders. *Part. Part. Syst. Charact.*, **12**, 232–236.

Kendall, K. (1986) Flocculation clustering and weakness of ceramics. *Nature*, **319**, 203–205.

Kendall, K., Alford, McN. & Birchall, J.D. (1971) Elasticity of particle assembles as a measure of the surface energy of solids. *Proc. Roy. Soc.*, **A324**, 301.

Kesler, C.E. (1959) Effect of length to diameter (L/D) ratio on compressive strength. *Proc. ASTM*, **59**, 1216–1229.

Knight, P.C. & Johnson, S.H. (1988) Measurement of powder cohesive strength with a penetration test. *Powder Technol.*, **54**(4), 279–283.

Kostelnik, M.C. & Beddow, J.K. (1970) *Modern Developments in Powder Metallurgy*, Vol. 4 (ed H. Hauser). Plenum Press, New York, pp. 29–48.

Kostelnik, M.C., Kludt, F.H. & Beddow, J.K. (1968) The initial stage of compaction of metal powders in a die. *Int. J. Powder Metallurgy*, **4**(4), 19–28.

Kurz, H.P. & Minz, G. (1975) The influence of particle size distribution on the flow properties of limestone powders. *Powder Technol.*, **11**, 37–40.

Labuza, T.P. (2002) *Rapid Testing Symposium. Paper on Prediction of Caking*. Procter and Gamble, Cincinnati, OH, USA.

Lappas, L. & Dempski, R.E. (1965) Pharmaceutical sciences – 1964. A literature review. *J. Pharm. Sci.*, **54**(7) 931–958.

Lee, Y.S.L., Poynter, R., Podezeck. F. & Newton, J.M. (2000) Development of a dual approach to assess powder flow from avalanching behaviour. *AAPS Pharm. Sci. Tech.*, **1**(3), article 21.

Lyons, C.P. & Mark, D. (1992) *HSE Contract Research Report 40*. Warren Spring Laboratory, Stevenage, Hertfordshire, UK.

Lyons, C.P. & Mark, D. (1994) *HSE Contract Research Report 62*. Warren Spring Laboratory, Stevenage, Hertfordshire, UK.

Maltby, L.P. & Enstad, G.G. (1993) Uniaxial tester for quality control and flow property characterisation of powders. *Bulk Solids Handling*, **13**, 135.

Maltby, L.P., Enstad, G.G. & de Silva, S.R. (1993) Measurements from the flexible-boundary biaxial tester. *Proc. Powder & Bulk Solid Conference*. Reed Exhibition Company, London, UK, pp. 15–27.

Marjanovic, J., Geldart, D. & Orband, J.L.R. (1998) Review on testers for measuring flow properties of bulk solids. *World Congress Particle Technology 3*, Brighton, UK. CD-Rom Institution of Chemical Engineers, George E Davis Building, Railway terrace, Rugby, UK, pp. 167–171.

Marjanovic, J., Geldart, D., Orband, J.L.R. & Mooney, T. (1995) *Partec 95 (Janssen Centennial)*. Nurenberg, Germany, pp. 69–78.

McGee, E. (August 2006) Powder behaviour. *Solid Bulk Handling*, 12–14.

McGlinchey, D. & McGee, E. (July 2005) A multifaceted approach to characterising powders for flow. *7th World Congress of Chemical Engineering*, 10–14 July 2005, Glasgow, Scotland.

Molerus, O. (1980) A coherent representation of pressure drop in fixed beds and of bed expansion for particulate fluidized beds. *Chem. Eng. Sci.*, **35**(6), 1331–1340.

Molerus, O. (1982) Interpretation of Geldart's Type A, B, C and D powders taking into account interparticle cohesion forces. *Powder Technol.*, **33**, 81–88.

Monick, J.A. (May 1966) Alcohols: Their chemistry, properties and manufacture. *Proc. Chem. Manufact.*, 108–112.

Nedderman, R.M. & Laohakal, C. (1980) The thickness of the shear zone of flowing granular materials. *Powder Technol.*, **25**, 91–100.

Novosad, J. (1990) *Proc. 2nd World Congress Particle Technology.* Kyoto, Japan, Part 1, p. 54; Part 111, p. 167.

Orband, J.L.R. & Geldart, D. (1997) Direct measurement of powder cohesion using a torsional device. *Powder Technol.*, **92**(1), 25–33.

Peck, R.B. & Gholamriza, M. (1996) *Theoretical Soil Mechanic in Engineering Practice*. J. Wiley & Sons, Inc. Corporate Headquarters, NJ.

Peschl, I.A.S.Z. (1988) Mechanical properties of powders. *Bulk Solids Handling*, **8**(5), 615–624.

Peschl, I.A.S.Z. (1989) Equipment for the measurement of mechanical properties of bulk powders. *Powder Handling Process.*; **1**(1), 73–81 and Quality control of powders for industrial application. *Powder Handling Process.*, **1**(4), 357–368.

Peschl, I.A.S.Z. & Colijn, H. (1977) New rotational shear testing technique. *J. Powder Bulk Solid Technol.*, **1**, 55–60.

Pilpel, N. (April 1970) Some effects of moisture on the flow and cohesiveness of powders. *Manuf. Chem. Aerosol News.*, **41**, 19–22.

Pilpel, N. & Walton, C.A. (1974) The effect of particle size and shape on the flow and failure properties of procaine penicillin powders. *J. Pharm. Pharacol.*, **26**, lP–l0P.

Plinke, M.A.E., Leith, D., Hathaway, R. & Loffer, F. (1994) Cohesion in granular materials. *Bulk Solids Handling*, **14**(1), 101–106.

Ploof, D.A. & Carson, J.W. (1994) Quality control tester to measure relative flowability of powders. *Bulk Solids Handling*, **14**(1), 127–132.

Powder Research Ltd, Harrogate, UK.

Prescott, J.K. & Barnum, R.A. (2000) On powder flowability. *Pharm. Technol.*, **24**(10), 60–84.

Rajendran Nair, P.B., Ramanujan, T.K. & Ramanujam, M. (1990) *Proc. 2nd World Congress Particle Technology*. Kyoto, Japan, Part 1, pp. 124–133.

Rajendran Nair, P.B., Ramanujan, T.K. & Ramanujam, M. (1993) Flow characteristics of raw coal fines. *Adv. Powder Technol.*, **4**(4), 297–310.

Ridgway, K. & Tarbuck, K.J. (1968) Voidage fluctuations in randomly-packed beds of spheres adjacent to a containing wall. *Chem. Eng. Sci.*, **23**(9), 1147–1155.

Rietema, K. (January–February 1984) Powders, what are they? *Powder Technol.*, **37**(1), 5–23.

Rietema, K. & Hoebink, J. (November–December 1977) Transient phenomena in fluidized beds when switching the fluidizing agent from one gas to another. *Powder Technol.*, **18**(2), 257-265.

Riley, G.S., Mann, S. & Jesse, R.O. (1978) Angle of repose of cohesive powders. *J. Powder Bulk Solids Technol.*, **2**(4), 15–18.

Roberts, T.A. & Beddow, J.K. (December 1968) Some effects of particle shape and size upon blinding during sieving. *Powder Technol.*, **2**(2), 121–124.

Roscoe, K.H. (1970) 10th Rankine lecture. *Geotechnique*, **20**, 122.

Roscoe, K.H., Schofield, A.N. & Wroth, C.P. (1958) On the yielding of solids. *Geotechnique*, **8**, 22.

Savage, S.B. (September 1967) Gravity flow of a cohesionless bulk solid in a converging conical channel. *Int. J. Mech. Sci.*, **9**(9), 651–659.

Schofield, A.N. & Wroth C.P. (1968) *Critical State Soil Mechanics*. McGraw Hill, New York.

Schofield, C. (1981) Dust generation and control in materials handling. *Bulk Solids Handling*, **1**(3), 419–442.

Schulze, D. (1995) Zur Fließfähigkeit von Schüttgütern – Definition und Meßverfahren. *Chem. Eng. Technol.*, **67**(1), 60–68 (in German). www.dietmar-schulze.de.

Schulze, D. (1996a) Flowability of bulk solids – Definition and measuring techniques, Part I. *Powder Bulk Eng.*, **10**(4, Part 1), 45–61. [See Schwedes, J. (2003)]

Schulze, D. (1996b) Flowability of bulk solids – Definition and measuring techniques, Part II. *Powder Bulk Eng.*, **10**(6, Part 11), 17–28. [See Schwedes, J. (2003)]

Schwedes, J. & Schulze, D. (1990) *Proc 2nd World Congress Particle Technology*. Kyoto, Japan, Part 1, pp. 61–70.

Schwedes, J. (1975) Shearing behaviour of slightly compressed cohesive granular materials. *Powder Technol.*, **11**, 59–67.

Schwedes, J. (2003) Review on testers for measuring flow properties of bulk solids. *Granular Matter*, **5**(1), 1–43. www.texturetechnologies.com.

Schwedes, J. & Schulze, D. (1992) Letter to the editor: Measurement of flow properties of solids. *Bulk Solids Handling*, **12**(3), 454–455.

Seville, J.P.K. & Clift, R. (January–February 1984) The effect of thin liquid layers on fluidisation characteristics. *Powder Technol.*, **37**(1), 117–129.

Stable Micro Systems. Surrey, UK. www.stablemicrosystems.com.

Standard Shear Testing Technique (SSTT) ASTM D6773-02. American Society for Testing Materials, International, West Conshohocken, PA.

Standard Shear Testing Technique for particulate solids using the Jenike shear cell (1989) Institution of Chemical Engineers, George E Davis Building, Railway Terrace, Rugby, UK, pp. 167–171.

Stanley-Wood, N.G. (1983) *Enlargement and Compaction of Particulate Solids*. Butterworth, London.

Stanley-Wood, N.G. (1987) Uniaxial powder compaction. In: *Tribology in Particulate Technology* (eds B.J. Briscoe & M.J. Adams). Adam Hilger, Bristol, UK.

Stanley-Wood, N.G. (2000) Particle size analysis: introduction. In: *Encyclopaedia of Analytical Chemistry: Applicatons, Theory and Instrumentation*, Vol. 6 (ed R.A. Myers). Wiley, Chichester, UK, pp. 5301–5337.

Stanley-Wood, N.G. (2005) Particle characterisation in bulk powders. In: *Characterisation of Bulk Powders* (ed D. McGlinchey). Blackwell Publishing, UK.

Stanley-Wood, N.G. & AbdelKarim, A.M. (1982) *J. Powder Metallurgy Int.*, **14**, 135.

Stanley-Wood, N.G. & Abdelkarim, A.M. (1983a) Application of critical state to uniaxial compaction. *Powder Technol.*, **35**, 185.

Stanley-Wood, N.G. & Abdelkarim, A.M. (1983b) *Proc of RILEM/CNR Int. Symp on Principles & Applications of Pore Structural Characterisation*, Milan (eds J.M. Haynes & P. Rossi-Doria). Arrowsmith, Bristol.

Stanley-Wood, N.G., AbdelKarim, A., Johansson, M.E., Sadeghnejad, G. & Osborne, N. (1990) The variation in, and correlation of, the energetic potential and surface areas of powders with degree of uniaxial compaction. *Powder Technol.* **60**, 15–26.

Stanley-Wood, N.G. & Johansson, M.E. (1980). Variation of intra- and inter-particle porosity with degree of compaction. *Analyst*, **105**, 1104–1112.

Stanley-Wood, N. & Sarrafi, M. (1988) Variations in, and relationships of surface area, internal angle of friction and compact diametral fracture strength with degree of compaction. *Particle & Particle Systems Characterisation*, **5**(4), 186–192.

Stanley-Wood, N.G., Sarrafi, M., Mavere, Z. & Schaefer, M. (1993) The relationships between powder flowability, particle-arrangements, bulk densities and Jenike failure properties. *Adv. Powder Technol.*, **4**(1), 33–40.

Stephens, D.J. & Bridgwater, J. (September–October 1978) The mixing and segregation of cohesionless particulate materials. Part I: Failure zone formation. *Powder Technol.*, **21**(1), 17–28.

Tamura, T. & Haze, H. (1985) Determination of the flow of granular materials in silos. *Bulk Solids Handling*, **5**(3), 663–640.

TSI Instruments Ltd. High Wycombe, Buckinghamshire, UK. www.tsiinc.co.uk.

USP29-NF24 (2006) *General Information Chapter '<1174> Powder Flow'*. US Pharmacopeial Convention, Rockville, MD, USA, p. 3017.

von Terzaghi, K. (1925) *Erdhaumechanik auf Bodenphysikalischer Grundlage*, Deuticke, Vienna.

von Terzaghi, K. (1943) *Theoretical Soil Mechanic in Engineering*. J. Wiley & Sons, Inc. Corporate Headquarters, Hoboken, NJ.

von Terzaghi, K. & Peck, R.B. (1967) *Theoretical Soil Mechanic in Engineering Practice*, 2nd edn. J. Wiley & Sons, Hoboken, NJ.

Walton, C.A. & Pilpel, N. (1972a) *British Pharm. Conference*, Keele, Paper 3.

Walton, C.A. & Pilpel, N. (1972b) The effects of particle size, shape and moisture content on the tensile properties of procaine penicillin powders. *J. Pharm. Pharmacol.*, **24**(Suppl), 10P–16P.

Williams, J.C. & Birks, A.H. (1965) The preparation of powder specimens for shear cell testing. *Rheol. Acta*, **4**, 170–182. Paper presented at a meeting of the British Society of Rheology, University of Nottingham, April 6–8, 1965.

Williams, J.C. & Birks, A.H. (December 1967) The comparison of the failure measurements of powders with theory. *Powder Technol.*, **1**(4), 199–206.

Williams, J.C. (1977) The rate of discharge of coarse granular materials from conical mass flow hoppers. *Chem. Eng. Sci.*, **32**(3), 247–255.

Williams, J.C., Birks, A.H. & Bharracharya, D. (September 1970/71) The direct measurement of the failure function of a cohesive powder. *Powder Techol.*, **4**(6), 328–337.

Wong, A.C. (2000) Characterisation of flow by bulk density ratios. *Chem. Eng. Sci.*, **55**, 3855–3859.

Woodcock, C.R. & Mason, J.S. (1987) *Bulk Solid Handling: An Introduction to the Practice and Technology*. Kluwer Academic Publishers, Dordrecht, The Netherllands.

Wouters, I.M.F. & Geldart, D. (1996) Characterising semi-cohesive powders using angle of repose. *Part. Part. Syst. Charact.*, **13**, 254–259.

WSL Tensile Tester (March 1977) *Chem. Eng.*, 161–164.

Yamatmoto, H. (1990) Relationship between adhesive force of fine particles and their dispersibility in gas. *Proc. 2nd World Congress Particle Technology*. Kyoto, Japan, Part III, pp. 167–173.

Yokoyama, T., Fujii, K. & Yokoyama, T. (1982) Measurement of tensile strength of a powder by a swing method measuring instrument. *Powder Technol.*, **32**, 55–62.

2 Hopper/bin design

JOHN W. CARSON

2.1 Fundamentals

Bins and silos are used to store bulk solids by virtually every industry around the world. There are many confusing terms in bulk solids technology, beginning with naming the storage vessel. To some people the term *silo* connotes a larger size vessel than a *bin*, but what is larger – and therefore a silo – to one person could be small to another. Some refer to the storage vessel as a *tank*, which at least avoids the distinction between a bin and a silo, but to others this term is reserved for liquid storage. Further adding confusion are specialised terms used in various industries. For example, grain storage vessels are often referred to as an *elevator*, whereas in a coal-fired power plant the vessel is often called a *bunker*. Smaller, usually portable, vessels often go by the name Tote® or IBC.

To avoid confusion in this chapter, I will use the term *bin* as a general descriptor of a bulk solids storage vessel of any size.

A bin can be divided into two major sections: cylinder and hopper (see Figure 2.1). The *cylinder* section is that portion of the vessel that has constant cross-sectional area over its height, i.e. vertical sidewalls. Note that by this definition a cylinder is not necessarily circular, although this is the most common shape used for bins.

In contrast to the cylinder section, a *hopper* has changing cross-sectional area over its height. In most cases the hopper is converging, so its cross-sectional area is smaller at the bottom than at the top. However, some hoppers used to store poorly flowing materials such as wood chips have a diverging geometry.

In most industrial situations a *feeder* is used to control discharge from a hopper outlet. Control involves not only stopping and starting flow but also metering the rate of discharge. It is important to note that a feeder is quite different from a conveyor in terms of operation. A conveyor simply transports material from one location to another, so essentially whatever mass per unit time enters it comes out at the discharge end. Thus, a conveyor provides no ability to meter or control the throughput rate, but a feeder does since it is flood-loaded. This distinction is important in terms of design of a feeder.

Bins vary tremendously in size, from ones that store only a few kilograms of material to those with storage capacities in excess of 10 000 metric tonnes. The range of discharge rates can similarly vary from a few grams per minute to several thousand tonnes per hour.

Bins are typically used to store incoming raw materials, intermediate products as part of the manufacturing process and/or outgoing products to be shipped to customers. The most common function of a bin is to provide surge capacity to compensate for effects of changes in production rate or frequency, and variations in rate or frequency of incoming or outgoing shipments. Another somewhat less common but still important function of a bin is to process, condition, age or blend the bulk solid stored therein.

Bins can be very problematic, resulting in increased cost of plant operation, low product yield, extended time to market with new products, quality/product performance problems and safety and health issues. Typical problems include process stoppages, feed rate and process control issues, and product quality concerns due to particle segregation, attrition, degradation or buildup. Other common problems are: difficulty in properly cleaning the vessel, difficulty of dust containment and shortened equipment life due to wear, corrosion or inadequate structural integrity.

Many injuries and several deaths associated with bins occur each year. Sometimes these are the result of structural failure caused by the stored bulk solid applying loads to the vessel walls that they were not designed to withstand. Collapsing arches or ratholes, or falling into a bin and becoming engulfed in the bulk solid, can also cause serious injury or death. In addition, injuries may occur while poking on moving parts inside a feeder or swinging a mallet to encourage flow.

2.2 Flow patterns

Two primary flow patterns can develop when a bulk solid discharges from a bin: funnel flow and mass flow. *Funnel flow*, also called *core flow*, is defined (Jenike 1964) as a flow pattern in which some material is stationary while the rest is moving (see Figure 2.2). According to

Figure 2.1 Subdivisions of a bin: cylinder (vertical section at top), hopper (converging section below).

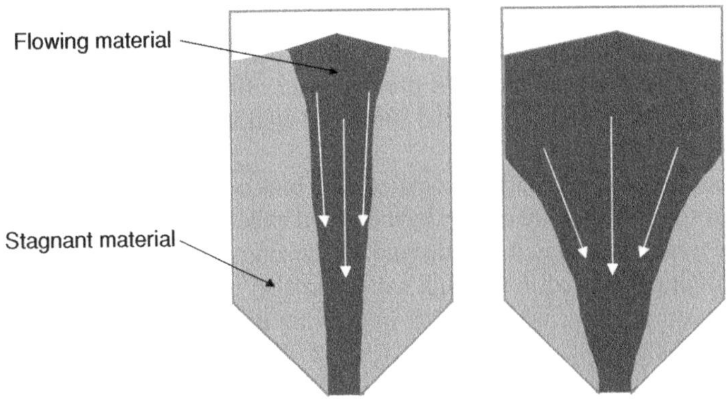

Figure 2.2 Examples of funnel flow.

this definition, a bin that contains any stagnant material during discharge qualifies the flow pattern as funnel flow.

Rotter (2005) further refined the definition of funnel flow into *pipe flow* and *mixed flow*. The distinction between the two is whether or not the flow channel intersects any portion of the walls of the bin (usually in the cylinder section). If there is no intersection (pipe flow), the flow channel walls may be vertical (*parallel pipe flow*), converging from top to bottom (*taper pipe flow*), vertical along the sidewall of the bin (*eccentric parallel pipe flow*) or converging along the sidewalls of the bin (*eccentric taper pipe flow*) (see Figure 2.3). If the flow channel intersects the cylinder wall (mixed flow), the converging flow channel may be symmetric about the centreline of the bin (*concentric mixed flow*), fully eccentric if the hopper opening is to one side of the vessel (*fully eccentric mixed flow*), or intersecting the cylinder walls at varying elevations because of a partially eccentric outlet (*partially eccentric mixed flow*) (see Figure 2.4).

In contrast to funnel flow, Jenike defined *mass flow* as a flow pattern in which all of the material is in motion whenever anything is withdrawn (see Figure 2.5). Note that this definition *does not* require that all material at a given evaluation across the cross section be moving at the same velocity. Indeed this is nearly impossible elevation to achieve, since particles near the centreline of a bin have virtually no shearing resistance, whereas particles sliding along converging hopper walls flow move slowly. This velocity distribution can be minimised by appropriate choice of hopper angle and hopper wall material of construction, or through the use of an insert.

One should beware of vendors claiming they can supply ‘mass flow’ bins with no consideration of the hopper wall material or stored bulk solid. The term ‘mass flow’ is often misused.

A third flow pattern, *expanded flow*, is a combination of funnel flow and mass flow (see Figure 2.6). Usually this is achieved by placing a small mass flow hopper below a funnel flow hopper. The mass flow hopper section expands the flow channel from the outlet up to the top cross section of the mass flow hopper. It is important to ensure that this cross-sectional area is sufficiently large so as to avoid ratholing in the funnel flow hopper section. Expanded flow designs are generally considered only when the cylinder diameter exceeds 6 m or so.

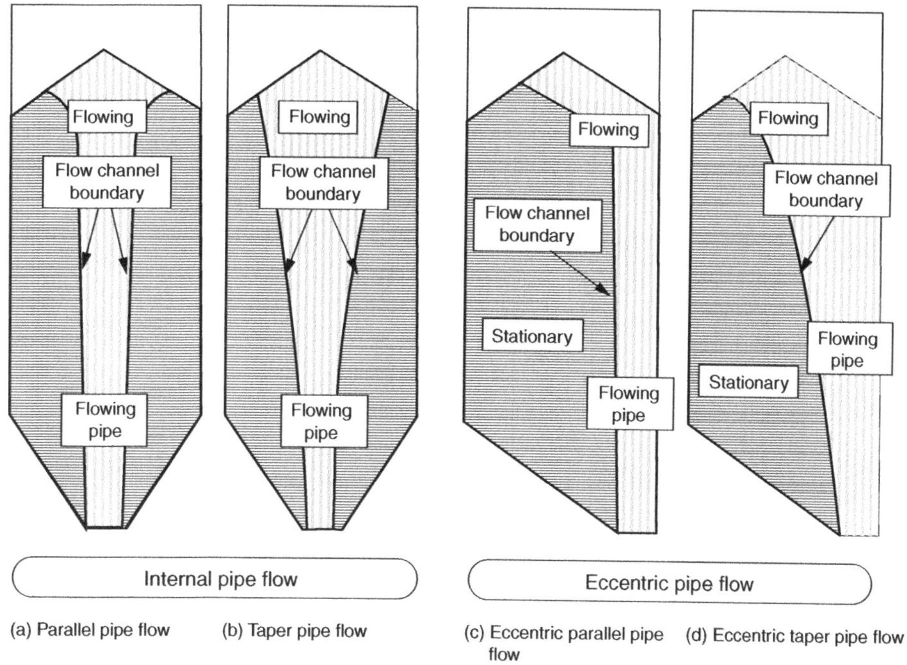

Figure 2.3 Forms of pipe flow. (Courtesy Rotter 2005.)

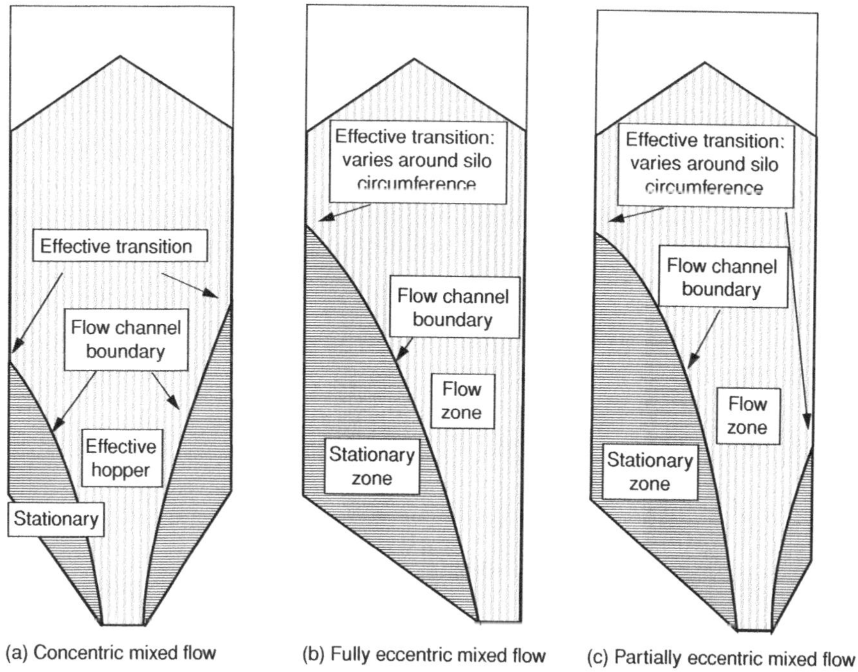

Figure 2.4 Forms of mixed flow. (Courtesy Rotter 2005.)

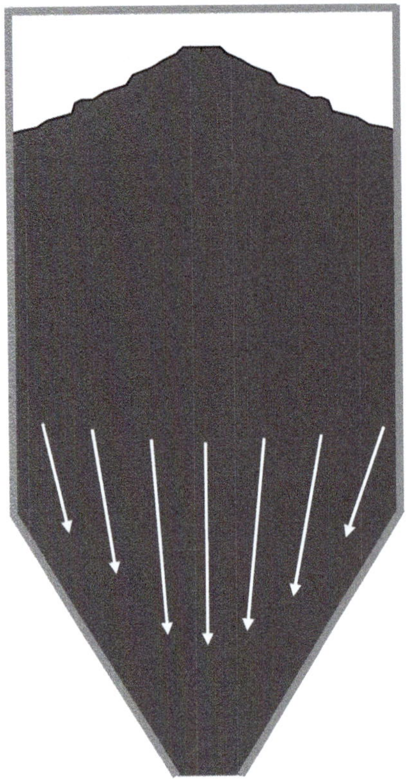

Figure 2.5 Mass flow.

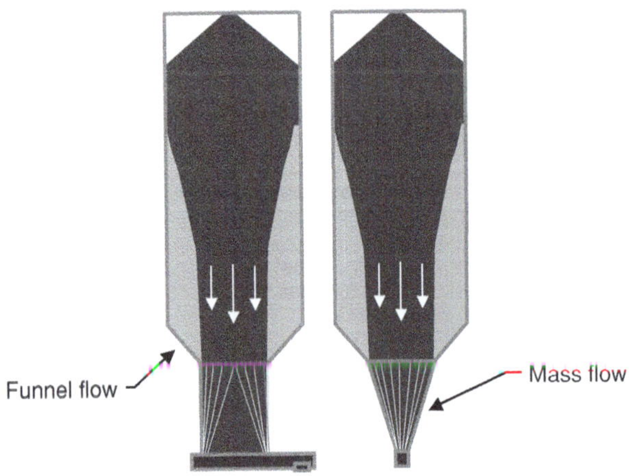

Figure 2.6 Example of expanded flow.

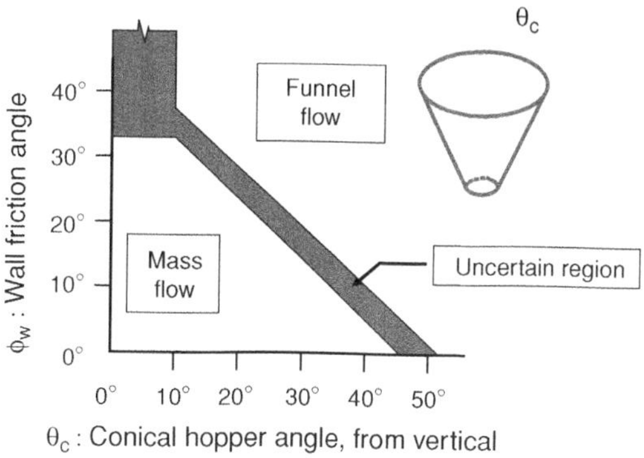

Figure 2.7 Example of conical hopper design chart (based on data from Jenike 1964).

Jenike (1964) presented a series of graphs defining the limits of hopper angle within which mass flow can develop. The major variables that determine these limits are:

- Wall friction angle, ϕ_W. This is the arc tangent of the coefficient of sliding friction between the bulk solid and hopper wall material. This value often changes with pressure, usually decreasing as pressure increases. In order to achieve mass flow throughout a hopper, it is important (and often sufficient) to determine the wall friction angle at the pressure level expected near the hopper outlet, since this is the region of minimum pressure during flow.
- Effective angle of internal friction, δ.
- Hopper geometry, e.g. cone, wedge.

Jenike's graphs were developed for axi-symmetric cones (hopper angle measured from vertical denoted as θ_c) and for infinitely long wedge-shaped hoppers (θ_p). Examples of design charts are shown in Figures 2.7 and 2.8. The two primary bulk solid parameters that affect limiting hopper angles for mass flow ϕ_W and δ are in turn often affected by temperature (of the bulk solid and the hopper wall material), moisture, content of the bulk solid, hopper

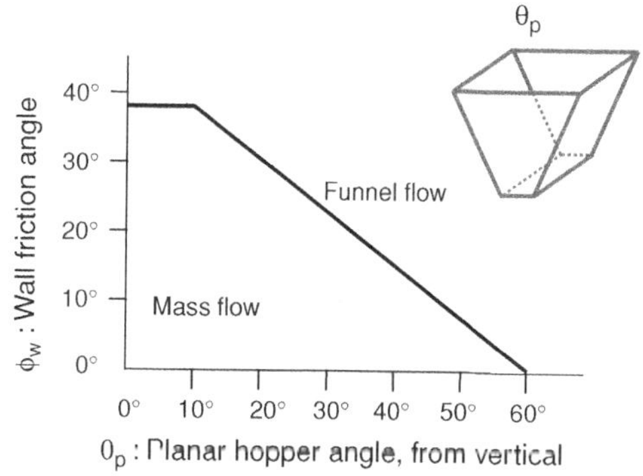

Figure 2.8 Example of planar hopper design chart (based on data from Jenike 1964).

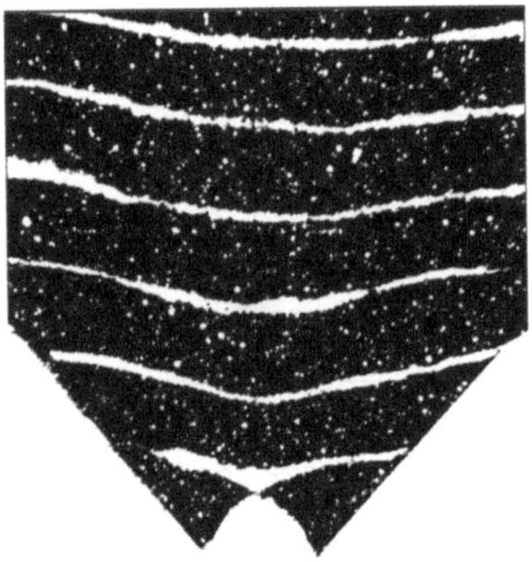

Figure 2.9 Cohesive arch.

wall material including its surface finish, and time of storage at rest. A secondary effect is the particle size distribution of the bulk solid.

It is important to emphasise that ϕ_w values can be determined only by shear testing (as described in Chapter 1). No other so-called flow tests (e.g. angle of repose) can provide this information.

2.3 Arching

Arching (sometimes called *bridging* or *doming*) can result from either mechanical interlocking or cohesive strength (see Figure 2.9). Mechanical interlocking occurs when particles are large relative to the outlet opening, whereas cohesive arching occurs because of bonding between particles.

To avoid interlocking arching, the outlet dimension must exceed some critical multiple of the characteristic particle dimension. While this characteristic dimension is not well defined, a conservative approach is to take this as the maximum length of a chord that spans the particle in any direction. The minimum values for outlet size are as follows:

- For a circular or square outlet, the outlet size should be at least 6–8 times the characteristic particle dimension.
- For an elongated outlet, its width must be at least 3–4 times the characteristic particle dimension and its length must be at least 3 times the width.

Mechanical interlocking usually governs minimum outlet dimensions only when the bulk solid's mean particle size is greater than about 6 mm, there are few 'fines', and the material has little to no surface moisture or other condition that would cause particles to adhere to each other.

Sizing an outlet to avoid cohesive arch formation is not as simple as considering only particle size. Flow properties tests need to be run to determine the material's *flow function*, which is the relationship between cohesive strength and consolidating pressure. Once this has been obtained, the hopper's *flow factor* must be determined. The original source for

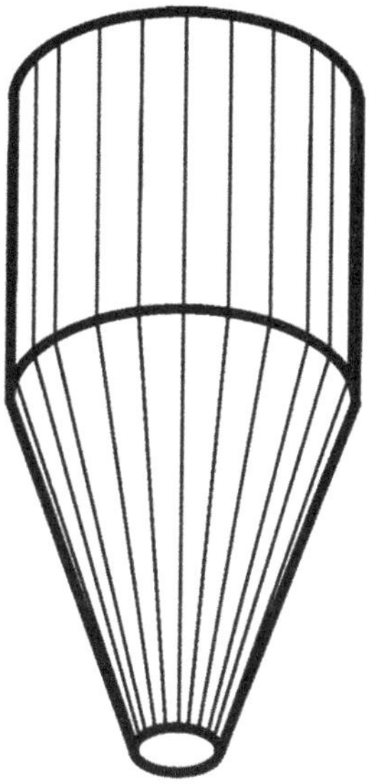

Figure 2.10 Mass flow conical hopper.

flow factors, and still a convenient way of obtaining them, is through graphs presented by Jenike (1964).

To avoid arching in a mass flow hopper, the outlet diameter BC (see Figure 2.10), or the width of wedge-shaped outlet BP (see Figure 2.11), must be at least the following:

$$B = \frac{\bar{\sigma_1} H(\theta)}{g\rho_b} \tag{2.1}$$

where B is BC or BP, m; $\bar{\sigma_1}$ is the critical stress required to cause arch to fail, N/m^2; $H(\theta)$ is the dimensionless function dependent on hopper angle (Jenike 1964). Value typically ranges from 2.0 to 2.3 for conical mass flow hoppers, half this for wedge-shaped hoppers; g is the acceleration due to gravity, 9.81 m/s^2; ρ_b is the bulk density at consolidating pressure calculated at hopper outlet, kg/m^3.

For a wedge-shaped hopper, the outlet length must be at least two times its width if the end walls are vertical and at least three times its width if the end walls are converging.

To avoid arching in a wedge-shaped funnel flow hopper, the outlet width, BF, must be at least that calculated using Equation (2.1) with the appropriate values of $\bar{\sigma_1}$ and ρ_b and $H(\theta) = 1$.

A material's flow function, which can be determined only by shear tests, is often strongly influenced by its temperature, time of storage at rest, moisture content and particle size distribution. A hopper's flow factor is a function of the effective angle of internal friction (δ), wall friction angle (ϕ_w) and hopper geometry.

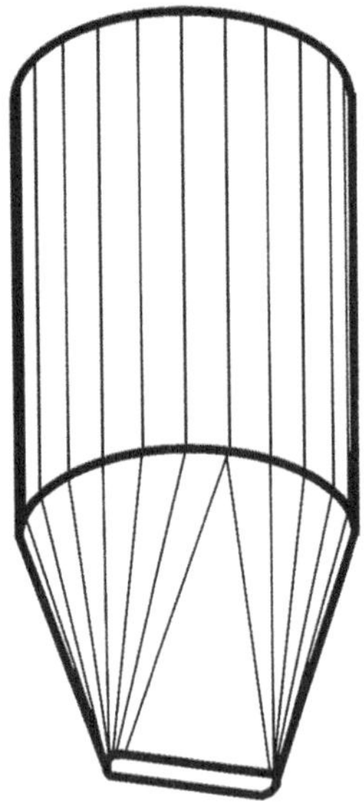

Figure 2.11 Mass flow transition hopper.

2.4 Ratholing

Also called *piping*, this is a phenomenon in which a more or less vertical flow channel develops above the hopper opening and, once emptied, remains stable (see Figure 2.12).

In order to avoid the formation of a stable rathole, it is necessary that the size of the flow channel exceed the critical rathole diameter, DF, which is calculated as follows:

$$\mathrm{DF} = \frac{\bar{\sigma_1} G(\phi)}{g\rho_b} \quad (2.2)$$

where $\bar{\sigma_1}$ is the cohesive strength of bulk solid at major consolidating pressure calculated at bin outlet, N/m^2; $G(\phi)$ is the dimensionless function dependent on bulk solid's angle of internal friction (Jenike 1964). Typical values range from 2.5 to over 7.

The two key parameters in this equation, flow function and angle of internal friction, can be strongly affected by the bulk solid's temperature, time of storage at rest, moisture content and particle size distribution.

2.5 Flow rate

The maximum discharge rate of a coarse, freely flowing material through an orifice (i.e. hopper outlet) can be calculated using the following equation:

$$Q = \rho_b A \left\{ \frac{Bg}{[2(1+m)\tan\theta]} \right\}^{1/2} \quad (2.3)$$

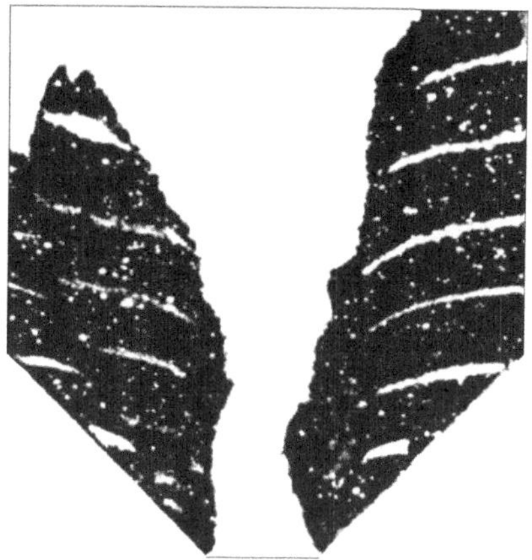

Figure 2.12 Rathole.

where Q is the maximum steady discharge rate, kg/s; ρ_b is the bulk density, kg/m^3; A is the cross-sectional area of outlet, m^2; B is the outlet diameter or width, m; $m = 1$ for conical hopper, 0 for wedge-shaped hopper; θ is the hopper angle (measured from vertical) in degrees.

This equation can be modified to take particle size into account, but this alteration is important only if the particle size is a significant fraction of the outlet size (Beverloo *et al.* 1961).

The limits of this equation are as follows:

- The bulk solid must be *coarse*. This is usually interpreted as a material having a mean particle size of at least 6 mm with a minimal amount of fines.
- The bulk solid must be *free-flowing*. Essentially this means that there are no arching or ratholing concerns.

Fine powders behave dramatically different than coarse, free-flowing bulk solids do. The discharge rates of fine powders through a hopper opening may be excessively high (a phenomenon called *flooding*) or excessively low. Which of these extremes occurs depends to a large extent on the flow pattern in the bin. A funnel flow bin will often exhibit flooding as a result of a stable rathole collapsing and the material becoming fluidised. Another way that flooding can occur in a funnel flow bin is when fresh material added to the bin channels through a stable rathole, resulting in a short residence time for the solid to deaerate.

Mass flow bins, on the other hand, often have a limited rate of discharge when handling fine powders. This is due to interstitial gas flow through the voids as a result of (often small) gas pressure gradients that develop naturally as material flows through a converging hopper. Such gradients are not present with coarse, free-flowing materials because of their much larger void space. Often the gas pressure gradient at the outlet of a mass flow hopper acts upward, drawing in air through the outlet. Acting counter to gravity, this retards flow, and the result can be limiting rates of discharge that are several orders of magnitude smaller than a coarse, free-flowing material having the same bulk density.

The discharge rate of fine powders from mass flow bins can be significantly increased by using low levels of aeration. As discussed by Royal and Carson (1991), one technique is to use an air permeation system, which consists of a sloping shelf or insert that allows a small amount of air to be added to the bin. This air permeates into the bulk solid and lessens the need for air to be drawn in through the outlet. One must be careful to distribute the air as uniformly as possible to maximise its effectiveness, limit the air flow rate and pressure so not to cause fluidisation, and avoid impediments to solids flow through proper design of the sloping shelf or insert.

Another technique is to aerate the bulk solid as described below.

Two material parameters that affect flow rate and aeration characteristics are a powder's permeability and compressibility. Tests are available by which each can be measured. Analysis of the results of these tests, including calculations of critical, steady state rates of discharge, is quite complex.

2.6 Segregation

Segregation is a common problem in many industries, resulting often in unacceptable quality variations and sometimes in safety concerns. Segregation typically occurs by one of three common mechanisms (Bates 1997; Carson *et al.* 1986).

2.6.1 *Sifting*

This involves the movement of smaller particles through a matrix of larger ones (see Figure 2.13). In order for sifting to occur, four conditions must exist:

- Range of particle sizes, typically a minimum ratio of 2:1 or more
- Mean particle size larger than about 500 μm
- Free-flowing material
- Interparticle motion

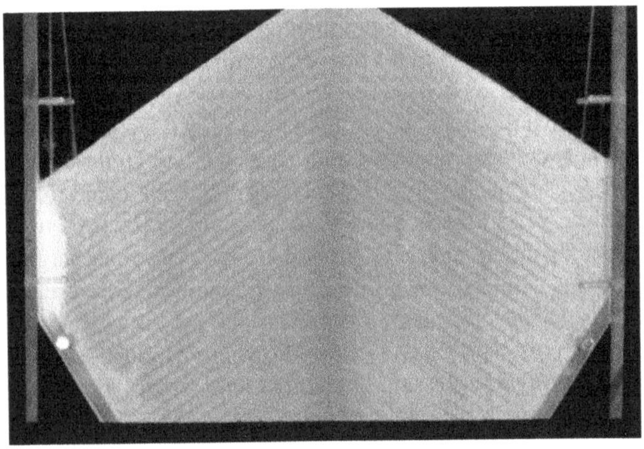

Figure 2.13 Sifting segregation.

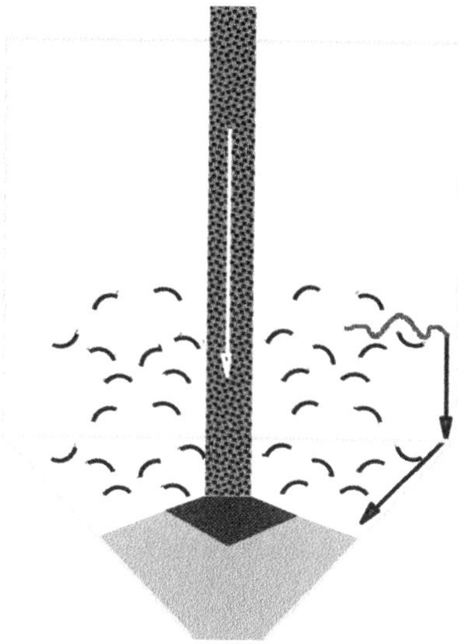

Figure 2.14 Particle entrainment during filling of a bin.

If any one of these conditions does not exist, segregation by the sifting mechanism is unlikely. Testers are available (ASTM 2003a) by which the sifting segregation tendency of a material can be determined.

2.6.2 *Dusting (particle entrainment)*

Dusting occurs when small and/or light particles are carried by air currents within a vessel, resulting in a segregation profile that is difficult to predict and to avoid (see Figure 2.14).

2.6.3 *Fluidisation (air entrainment)*

Fluidisation occurs when larger and/or denser particles settle through a fluidised matrix of finer and/or lighter particles (see Figure 2.15). The result is a top-to-bottom segregation pattern that is difficult to overcome. Test apparatus is available (ASTM 2003b) to measure the fluidisation segregation tendency of a bulk solid.

In addition to these three main segregation mechanisms, there are others that occur less frequently. These include mechanisms of trajectory and dynamic effects.

2.7 Importance of outlet and outlet region

The outlet region of a hopper is more important than any other region of a bin in determining both the flow pattern that develops as well as the type and extent of flow problems.

While it is obvious that, when gravity alone is acting, particles flow downward from the top of a bin to the outlet, flow is actually initiated at the outlet when a gate is opened or a feeder is started. Flow then propagates from the outlet upward, so what happens in the outlet

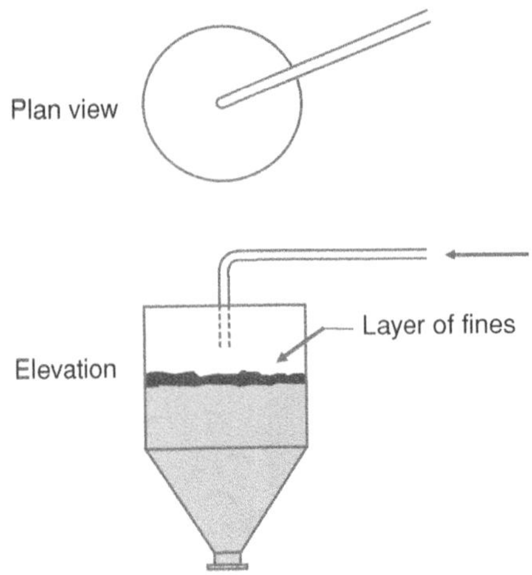

Figure 2.15 Example of fluidisation segregation.

region is extremely important. Furthermore, if a mass flow pattern develops in the hopper, the solid's pressure acting on the particles as well as between the particles and the hopper wall is at its minimum value at the outlet. Since the wall friction angle ϕ_W is often not a constant and usually varies such that its highest value occurs at the lowest pressure, this results in the outlet region requiring the steepest hopper angle for mass flow to develop.

Arching or ratholing, when they occur, almost always do so at a hopper outlet. The reason is obvious, since the outlet is (almost always) the smallest flow channel through which the material must move.

Another reason why the outlet region is so important is that it directly affects the maximum discharge rate achievable from the hopper. No matter whether critical discharge is calculated using Equation (2.3) or is limited by fine powder flow, critical discharge rates vary more or less linearly with outlet area.

Avoiding flow impediments throughout a bin is important if full mass flow is to be achieved. Such impediments are particularly troublesome if they are present in the outlet region and include mismatched bolted flange connections (see Figure 2.16), sight glasses, poke holes and partially opened slide gates. The importance of good weld quality and finishing in this region cannot be over-emphasised, even while recognising that polishing near small openings is difficult.

When flow problems occur, many operators resort to a mallet to encourage flow, and the blows are often directed to the outlet region since that is where most flow problems originate. Unfortunately, mallet blows can distort a hopper, thereby creating even more of an impediment to flow.

Two other causes of flow problems that can occur due to problems at a hopper outlet are improper feeder design and upward gas pressure gradients. If a feeder is not designed to fully activate the outlet area,[1] this will prevent mass flow from developing in the vessel, thereby potentially creating significant flow problems. Gas pressure gradients within a bin

[1] See Figure 2.17, Chapter 5 and Carson (2000a).

Figure 2.16 Example of flow obstruction (lip) caused by misalignment.

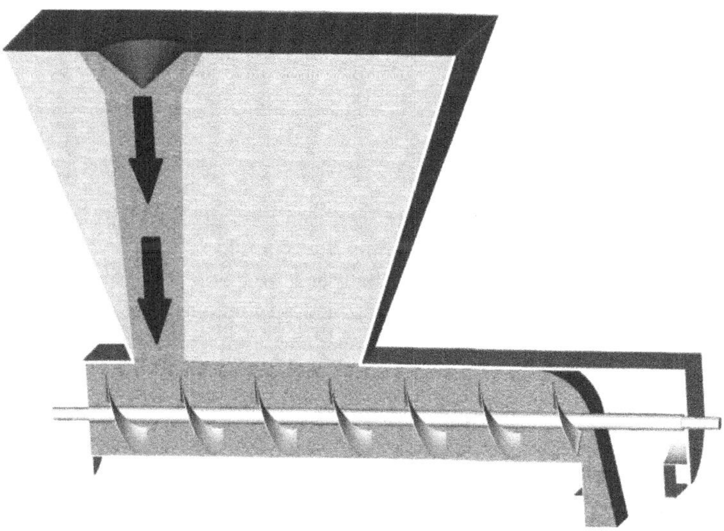

Figure 2.17 Flow pattern caused by a constant pitch screw feeder.

are often concentrated and most detrimental when they occur at the hopper outlet. An upward pressure gradient can significantly retard the bulk solid's discharge rate, particularly if a fine powder is being handled, and can also create arching problems with cohesive bulk solids.

2.8 Aerated versus non-aerated

When handling a fine powder, a mass flow pattern to avoid ratholing and accompanying flooding is usually advisable. However, as noted above, this often results in limited discharge rates. If higher discharge rates are required and two common approaches (adding an air permeation system or increasing the outlet size – and accompanying gate and feeder size) cannot satisfy the demand or are impractical, there are two alternatives: fluidised discharge and air-assisted discharge (Troxel *et al.* 2005).

These techniques are similar in that they involve use of a permeable membrane that covers most or all of the hopper section. Air (or some other gas) is added into the bin through a plenum located below the permeable membrane. These techniques differ in the quantity of material fluidised and whether or not a feeder is used.

A fluidised discharger is used when the bin size is such that its entire contents can be fluidised, the bin walls are capable of withstanding hydrostatic pressures, a low bulk density as the solid exits the bin is acceptable, and gas collection and cleaning are not a concern. Discharge is controlled through use of a feeder suitable for metering fluids, such as a tightly sealing rotary valve.

An air-assisted discharger is designed to aerate only a thin layer of bulk solid above the permeable membrane. For such a system to work properly the air must be able to easily exit the bin through its outlet when solids are discharged. This usually means that discharge cannot be restrained, such as is the case with a flood-loaded feeder. Instead, a full port ball or dome valve is used to discharge slugs of bulk solid from the bin.

Examples of these two techniques are given in Figures 2.18–2.19.

2.9 Selection criteria

2.9.1 *How to set bin size*

There are several considerations in determining the required size of a new bin. One of the first considerations is the mass of material that needs to be stored. This value may be set based on consumption rate of the bulk solid, frequency of incoming or outgoing shipments, production rate or frequency, and/or required residence time if the vessel is to be used for aging or some related purpose. All of these are process considerations that may set the desired mass. Another consideration involving mass is limitations of the foundation upon which the bin is to be placed.

Once the desired mass has been chosen, the next step is to convert it into a volume. This involves estimating the bulk density of the solid to be stored. Unfortunately, bulk density is not a constant for most bulk solids and often varies over a wide range depending on consolidating pressure.

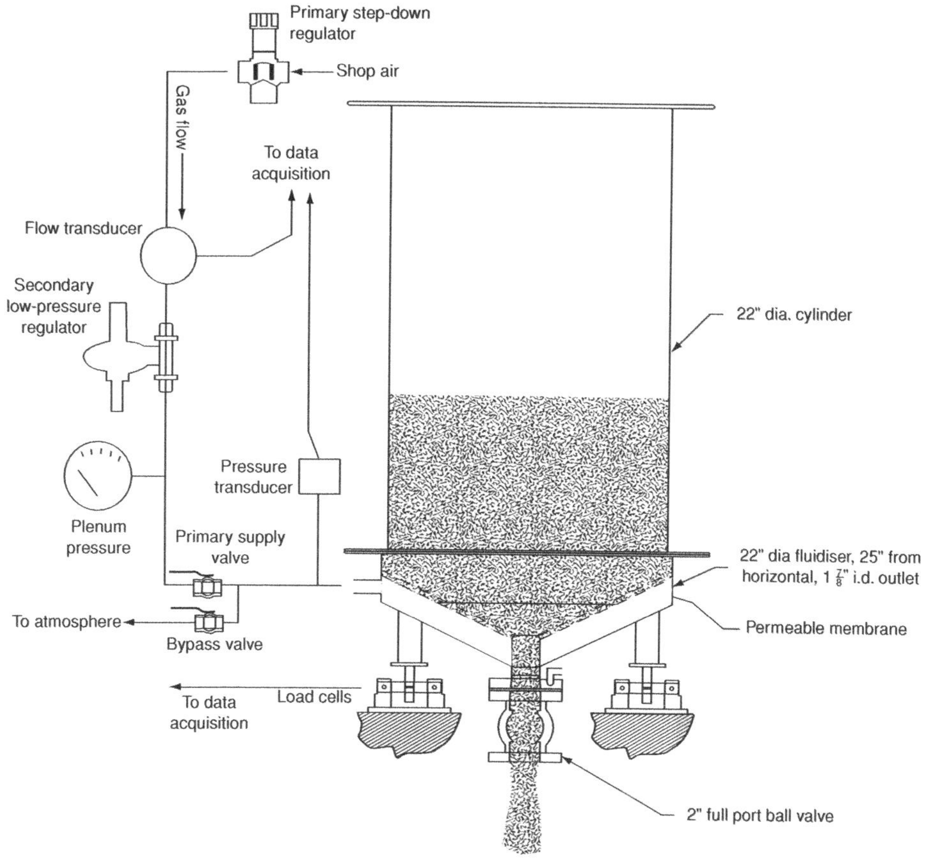

Figure 2.18 Example of air-assisted discharger.

Bulk density values as a function of pressure can be determined from a compressibility test. It is important that such a test be run on the actual particle size distribution of material to be stored in the bin, not just the fine fraction (which is usually the approach taken to measure wall friction and cohesive strength). A good approach is to estimate the consolidating pressure at the bottom of the cylindrical section of the bin. Then, assuming that the top surface has no surcharge acting upon it, divide the maximum consolidating pressure by 2 to get an average value within the cylinder and use the bulk density at this average consolidating pressure to size the vessel. Use a conservative (i.e. low) bulk density value to ensure that the bin volume is sufficient for the desired use. (*Note*: This applies only to volume calculations, not bin loads.)

Once the volume has been determined, the next step is to determine individual overall dimensions. In the case of a circular cylindrical vessel, this means setting the diameter and height of the cylinder section, since in most cases the majority of storage volume is contained therein. Whenever possible, it is desirable to fabricate the vessel in a shop because this is more economical than doing so in the field and also provides better quality control. The limit then becomes the size of vessel that can be transported from the shop to the plant

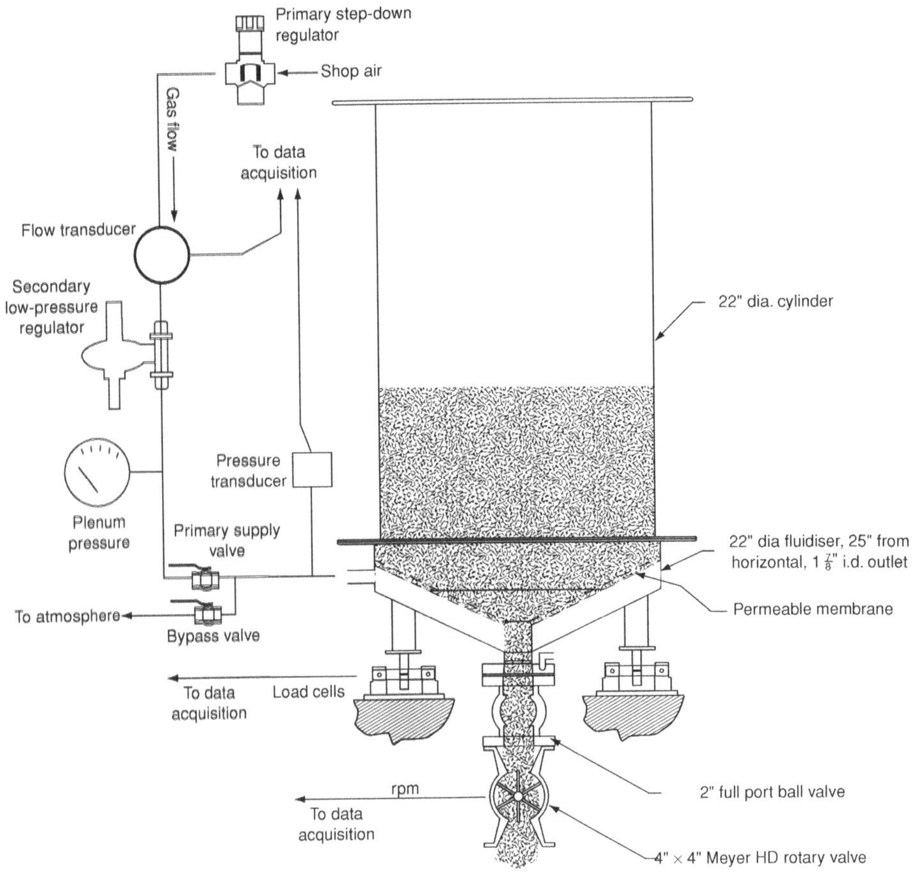

Figure 2.19 Example of fluidised discharger.

where the bin is to be used. In many parts of the world this limit on cylinder diameter is approximately 4.3 m.

Whether or not the bin is to be shop fabricated, plant space considerations usually need to be taken into account. For example, plant layout may limit the maximum bin diameter. Access within the plant may dictate the maximum size that can be transported to the location where the bin is to be placed.

Another consideration regarding bin size involves obtaining a reasonable height-to-diameter ratio, H/D. Usually bins are most economical when this ratio is between roughly 1.5 and 4.0. While these are not firm limits, they do provide a good guideline for economical construction. Vessels with smaller H/D values have much of their volume in the hopper section, which is more expensive to fabricate than a cylinder. At the other end of the spectrum, vessels with large H/D values may be taller than necessary or be more subject to wind or seismic considerations than a shorter vessel.

Vessel height limits may need to be considered in sizing the vessel. If the bin is to be located inside an existing building, ceiling height limits the bin height unless a portion of the cylinder can protrude through the roof. If the bin is to be located outside a building, the height of surrounding structures or zoning requirements may limit the maximum height. Beyond these considerations, it is generally wise to limit vessel height in order to minimise the height to which material must be elevated in order to fill the vessel. In considering vessel height, it is important to include the added height of other equipment that must go on the roof, such as bin vents, dust collectors, etc.

2.9.2 *Flow pattern selection*

Once the overall size of a vessel has been determined, the next step is to select the appropriate flow pattern. Mass flow should be chosen if the bulk solid is likely to experience ratholing if stored in a funnel flow vessel, or if there is concern about particle degradation (caking, spoilage, oxidation) with use of a first-in-last-out funnel flow pattern. If the bulk solid segregates side-to-side as the bin is being filled (due, for example, to the sifting or dusting mechanisms), a mass flow vessel will minimise the degree of segregation upon discharge, whereas a funnel flow vessel will exaggerate this segregation pattern. Mass flow vessels generally avoid problems of flooding of aerated powders that are typical with funnel flow vessels, and their uniformity of discharge is far superior to that which is possible with a funnel flow vessel. This can be important, for example, when designing a surge vessel to refill a loss-in-weight feeder.

If the bulk solid's characteristics are such that a mass flow pattern is not required, a funnel flow bin should be designed. Not only will this save on headroom and maximise storage volume for a given headroom, it will also minimise the effects of sliding abrasion on the bin walls.

An expanded flow bin should be designed if partial mass flow is acceptable in terms of the bulk solid's properties and, as a general rule of thumb, the cylinder diameter is greater than about 6 m.

2.9.3 *Inlets and outlets*

The location of the bin's inlet point (or points) is not as critical as the location and number of outlet points. However, inlet location may have some effect on particle segregation and/or bin loads.

The next consideration is whether to use one or more outlets. In general, it is always preferable to use a single outlet that is centred on the vessel's centreline. This is particularly true if mass flow is to be achieved, particle segregation is a concern and/or there are bin loads considerations in designing the vessel.

If a multiple outlet bin cannot be avoided, one option is to use a single, centred outlet and split the stream directly below, as shown in Figure 2.20. For larger vessels with multiple outlets in a row, it is desirable to use an odd number of outlets and put limits on their operation (see Figure 2.21). For example, if the vessel has three outlets, limit discharge to only the centre outlet if discharge from only one outlet is required, and to the two end outlets if dual discharge is required. In this way, the effects of segregation and eccentric bin loads will be minimised. If multiple outlets are to be located below a circular cylinder, it is helpful to locate each outlet equidistantly from the bin centreline and operate all outlets

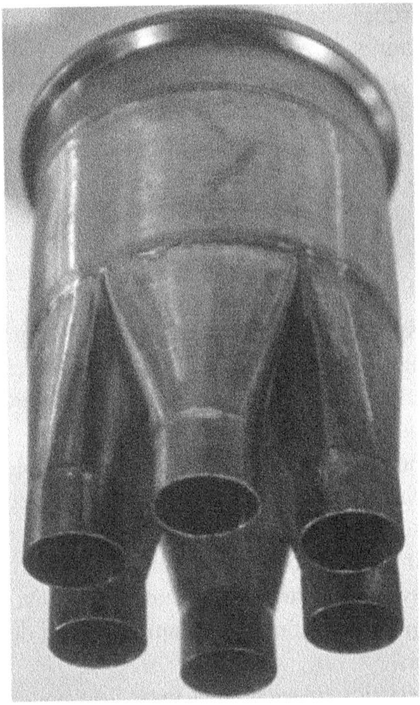

Figure 2.20 Example of splitting single, centred outlet into multiple outlets at same radius.

simultaneously, or as close to this as possible, to minimise segregation and eccentric loads (see Figure 2.22).

There are a number of considerations when sizing the bin outlet. If arching is a concern (either in mass flow or funnel flow), the discharge frequency must be considered since this sets the maximum time that the bulk solid will remain at rest within the vessel. If a funnel flow vessel is being designed, the critical rathole diameter will also be needed to set the minimum outlet dimensions.

If neither arching nor ratholing is a concern, perhaps the required flow rate will set the outlet size. The instantaneous rate of discharge is the most important parameter, but consideration must also be given to minimum and maximum design rates. Often flow rate is specified on a mass per unit of time basis, and this usually needs to be converted to a volume per unit of time to size the downstream feeder. A good rule of thumb in selecting the appropriate bulk density value to be used for such a calculation is to assume that the *effective head*[2] at the outlet is equal to the outlet width if a slotted outlet configuration is used, or equal to half the outlet diameter for a circular outlet. These rules of thumb apply only if a mass flow design is being used.

In addition to arching, ratholing and flow rate considerations, the size of the feeder must be considered. Most commercial feeders come in standard dimensions, so it may be necessary

---

[2] *Effective head* is the height of a column of the bulk solid contained in a cylinder with frictionless walls. It can be calculated by dividing the consolidating pressure by the solid's bulk density.

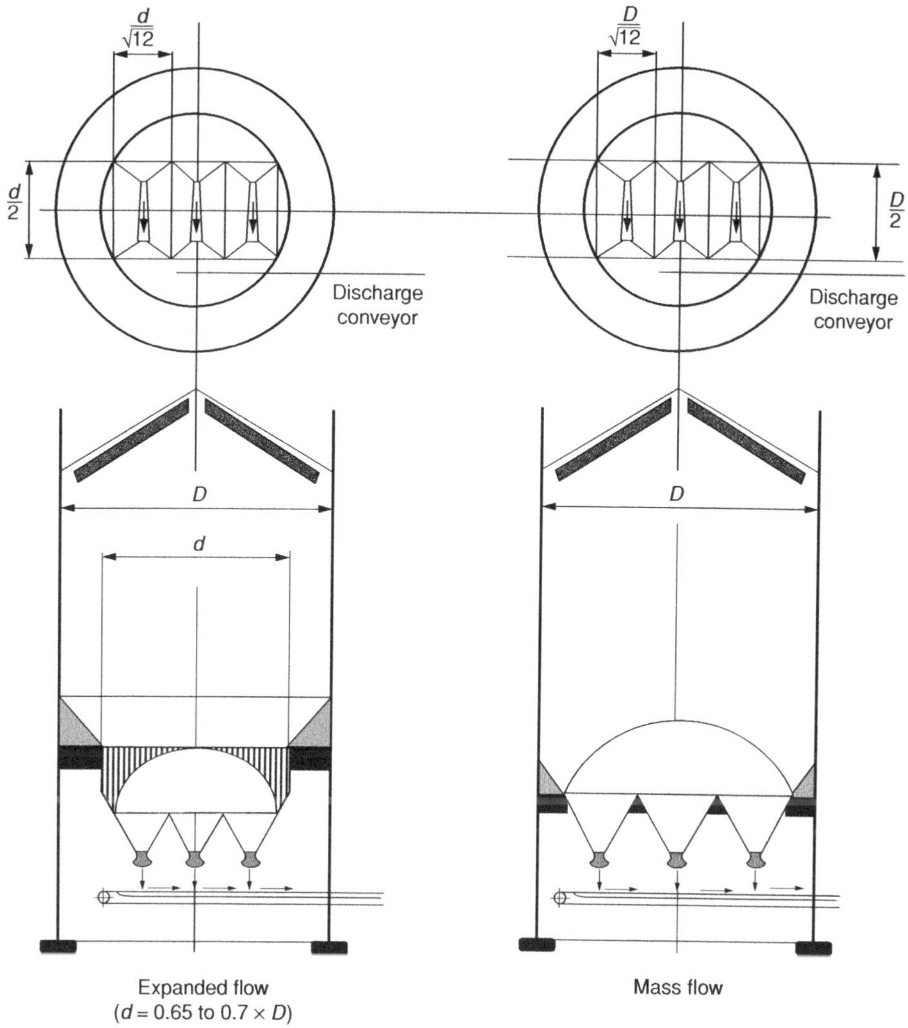

Figure 2.21 Multiple outlets in large silos with belt feeders.

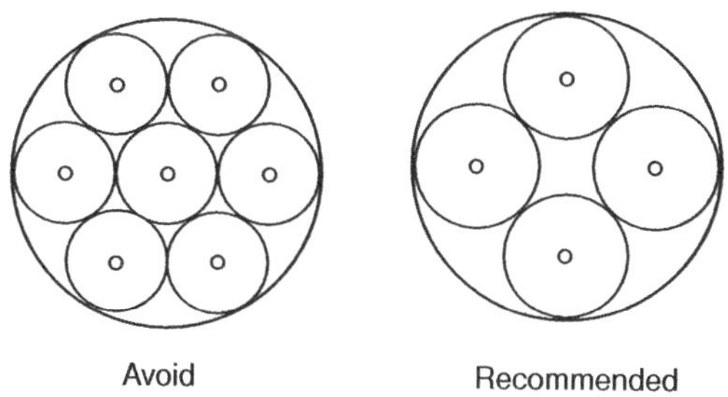

Figure 2.22 Plan views showing poor (left) and good (right) placement of outlets to minimise effects of segregation.

to increase the outlet size to match the next larger feeder size above that required by other considerations. If an elongated outlet is being used, the appropriate length of the feeder may set minimum and maximum outlet lengths. Also, the speed of the feeder may set the outlet size larger than required to overcome arching or ratholing or to accommodate the feed rate. A small feeder operating at too high a speed may not be efficient or may be subject to unacceptable abrasive wear or cause excessive particle attrition. Enlarging the feeder to reduce its speed may correct these problems.

Sometimes with rate limiting powders an air permeation system or air-assisted discharge can be used to increase flow rate and thereby limit the outlet size. Some powders are also more suitable to being handled in a fluidised state, which also limits the outlet size.

The shape of the outlet is the next consideration, in particular whether it should be circular, square or elongated. If there is a need to interface with downstream equipment such as a standpipe, this often requires the outlet to be circular because most standpipes are circular (to be able to effectively withstand explosion pressures). If the vessel is feeding into a pressurised environment, either a standpipe or a special type of feeder may be required, either of which may limit the choice of outlet configuration. Another aspect is the flexibility of location of the inlet point to downstream equipment, for example a conveyor or process vessel. If this downstream inlet is offset from the vessel centreline, a circular or square outlet will have to be followed by a conveyor or sloping chute, either of which may be unacceptable. Using an elongated outlet with a screw or belt feeder may be a better choice.

The type of feeder to be used also influences the shape of the outlet. If a rotary valve is chosen because of process considerations, the hopper outlet must generally be either circular or square. On the other hand, if a screw or belt feeder is to be used, the outlet is often elongated. It must be noted that with such an outlet the interface details become extremely important to ensure that the outlet area is fully active. There are several other feeder types besides these three, but the general considerations are similar.

If a circular or square outlet is being used, the hopper configuration must generally be either a cone or a pyramid, respectively. Associated with a pyramidal geometry are valley angles and inflowing valleys, both of which must be considered. Generally, a pyramidal geometry should be avoided if a mass flow pattern is required.

If the outlet is to be elongated, several hopper configurations are possible. Perhaps the most common is a transition hopper as shown in Figure 2.23. If mass flow is required, the end wall angle can be no less steep than the conical mass flow angle, σ_c, the sidewall angle can be no less steep than the mass flow planar angle, σ_p, and the outlet length-to-width ratio must be at least 3:1.

Another hopper configuration compatible with an elongated outlet is a chisel hopper as shown in Figure 2.24. Here the end walls are vertical, so only the slope of the sidewalls needs to be considered when designing for mass flow. The minimum length-to-width ratio for this design to achieve mass flow is 2:1.

A wedge-shaped hopper can also be used with elongated outlets, although as with pyramidal hoppers there is concern with valley angles and inflowing valleys (see Figure 2.25). One way around this is to use a series of alternating one-dimensional wedges as shown in Figure 2.26. With this arrangement, the minimum length-to-width ratio of the outlet can be as small as 2:1 since the end walls are vertical.

Figure 2.23 Example of mass flow transition hopper.

2.9.4 *Inserts*

Once the hopper geometry has been chosen, the next consideration is whether or not an insert is required within the hopper. A common insert configuration is an inverted cone or pyramid as shown in Figure 2.27. Unfortunately, studies have shown that such inserts have a limited range of effectiveness and the loads acting on them are usually extremely high. The supports required to resist these loads often provide impediment to flow greater than the beneficial effect of the insert.

An alternative to this type of insert is a Binsert® (see Figure 2.28). This system was developed for controlling and expanding flow patterns in bins by using a mass flow

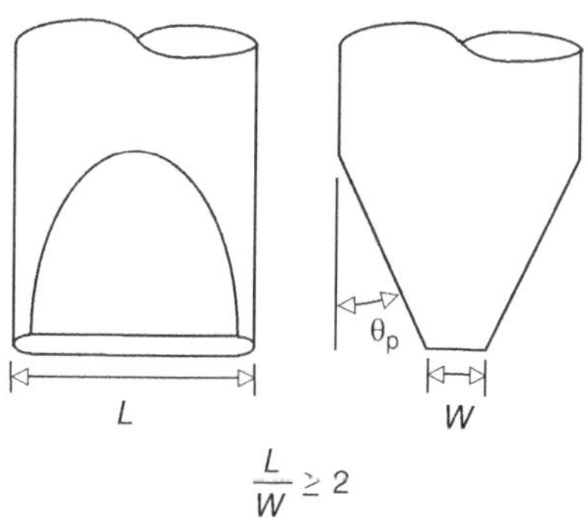

Figure 2.24 Mass flow chisel hopper.

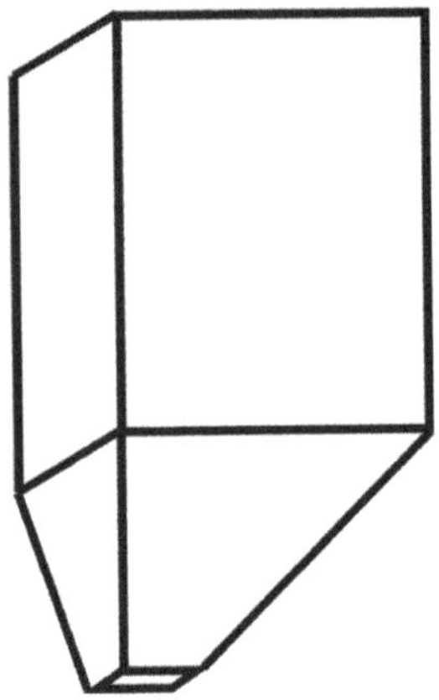

Figure 2.25 Pyramidal hopper.

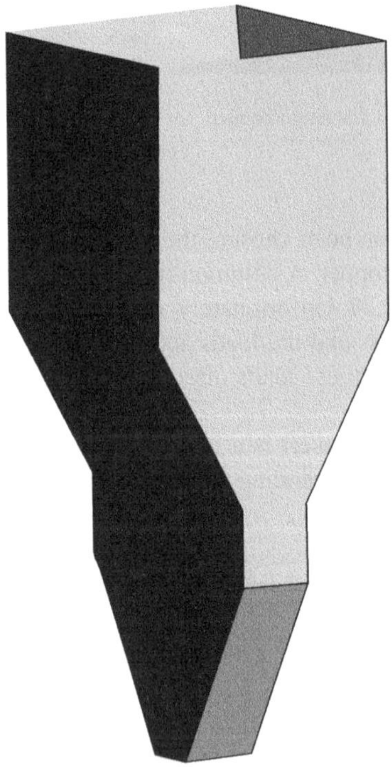

Figure 2.26 Example of alternating one-dimensional wedge hoppers.

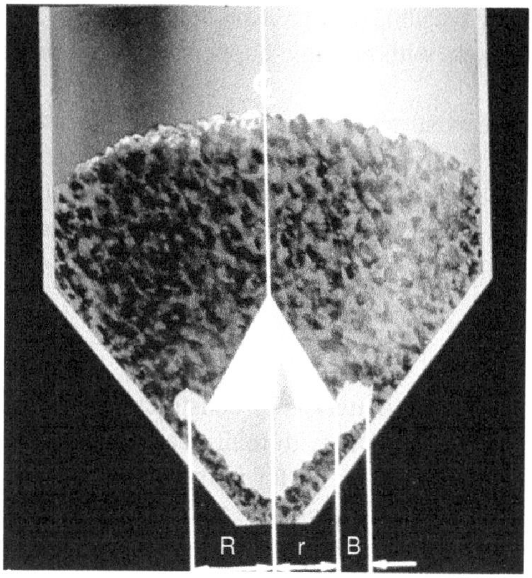

Figure 2.27 Conical insert.

hopper-within-a-hopper. The inner hopper is designed to provide mass flow through the central channel and also through the annular space between it and the outer hopper. This concept is of considerable use in minimising particle segregation, saving headroom for mass flow bin designs, eliminating ratholing problems in existing funnel flow bins by creating mass flow and blending solids by either a single pass or recycling the solids through the bin.

2.9.5 *Cylinder geometry*

It is most common to make the cylinder section of a bin circular in cross section, although there are instances in which a non-circular cross section is desirable. For example, a square cylinder has $(4/\pi - 1)$ times more cross-sectional area than a circular cylinder with diameter

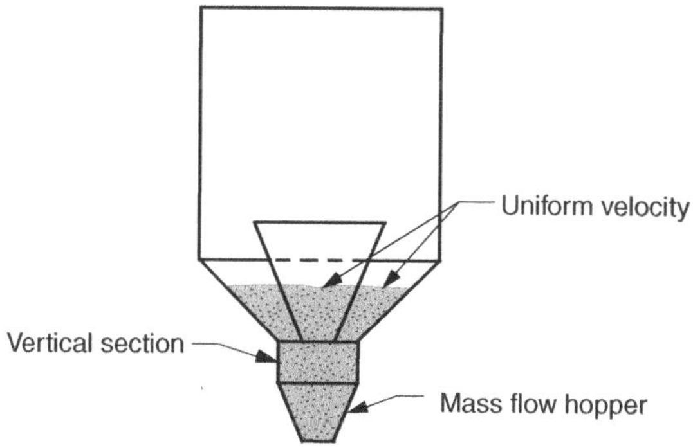

Figure 2.28 Anti-segregation Binsert®.

equal to the square cylinder's width, so it requires less height to attain the same capacity. Also, large rectangular bunkers with multiple hoppers are common in some industries, e.g. coal-fired power plants.

Designers must be aware of structural issues as well as potential flow issues when designing bins with flat walls. The structural issue is that of bending moments which must be taken into consideration. The flow issue is that flat walls are usually connected to hoppers that have inflowing valleys, which often results in a funnel flow pattern and can lead to flow stoppages.

2.9.6 *Materials of construction*

The next selection criterion is the materials of construction. The first choice to be made is probably whether or not to use metal, with reinforced concrete being the most common alternative. The most significant driving factor in this decision is generally economics. If the bin is of such a size that it must be field erected, reinforced concrete should certainly be considered. It is important to remember that a double layer of reinforcing steel (rebar) will likely be required, especially if the bin has multiple outlets or any type of asymmetry in loads applied to the vessel walls. If abrasion and/or corrosion are of concern, reinforced concrete would generally be better than metal. Whatever material of construction is chosen for the cylinder section, it is important to remember that only the hopper section (not the cylinder section) influences flow. Having a rough-walled cylinder section is, in many cases, actually a benefit in that more of the bulk solid's weight is carried by shear along the cylinder walls, so there is less pressure applied to the top portion of the hopper section.

If metal is chosen as the material of construction, the next choice is whether to make the construction bolted or welded. The primary factors to consider here are ease of fabrication, transportation and installation. Generally, a welded structure is best if it is of a size that can be shop fabricated and then lifted into place with a crane. For larger vessels that must be field welded, bolted construction is generally less expensive than welded, although it is usually not as robust because stress concentrations of bolted joints tend to weaken the structure, and there is the possibility of leakage whenever gaskets are used.

No matter the material of construction from a structural standpoint, very often some form of liner is required in the hopper section, particularly if a mass flow pattern is desired. Liners encompass a wide variety of materials, including light gage stainless steel, ultra-high molecular weight polyethylene, epoxy coatings, etc. One of the main reasons for using a liner or coating is to achieve a low wall friction angle, and, as a result, less steep hopper angles required for mass flow. When choosing a liner or coating it is important to consider both the initial condition of the surface and also whether or not this improves, or at least does not degrade, with use.

Sometimes liners or coatings are chosen to provide good abrasive wear or impact resistance, and other times chemical resistance is a factor. Whatever liner or coating is chosen, it is important to consider the ability to repair and/or replace it. This usually has to be done in the field unless the vessel is quite small and can be taken out of service for several days or longer.

Attachment of a liner must be considered in terms of both being strong enough to hold the liner in place and also not creating flow problems. The latter are due primarily to protrusions or joints that in most cases are detrimental to achieving mass flow. As to the issue of holding a liner in place, it is especially important to consider differential thermal expansion and contraction.

Before a liner or coating is selected for a given application, it is important to test it with the actual bulk solid to be stored in the bin. Not only should flow tests be run to measure wall friction, but abrasive wear tests in order to estimate the wear life of the liner and changes in wall friction with wear should also be performed.

The thickness of some liners can affect their frictional properties. For example, 2B finish is quite different in sheet thickness material as opposed to plate thickness.

2.9.7 *Type of feeder and valve*

If the hopper outlet is circular, almost any type of feeder can be used, with little detrimental effect on the flow pattern. This includes using a rotary valve, vibrating pan, screw, belt, vibrating louver, etc. If the outlet is elongated, the choice of feeder is limited and special consideration must be given to the design to ensure that the full outlet area will be active. Generally speaking, for most applications only a screw or belt feeder should be considered for an elongated outlet. For heavy-duty applications an apron feeder is often used. A vibrating pan feeder can be used if it is orientated to feed across the narrow dimension of the slot.

When considering a discharge valve, the first issue is sizing it to avoid flow impediments. This generally means that inside dimensions should be over-sized with respect to the section immediately above to which the valve is to be attached. Over-sizing the inside dimensions of bolted flange connections by 10 mm or so is sufficient for most applications.

Slide gates and butterfly valves are the most common discharge valves, while in some applications a pin gate is used. Whatever the type of valve that is chosen, it should generally be used only for maintenance purposes, i.e. operated fully open or fully closed, and not used to control the rate of discharge.

2.9.8 *Safety and environmental considerations*

The final selection criterion involves safety and environmental issues. In order to avoid over- and under-pressurising the vessel, a pressure/vacuum relief vent is usually specified. Generally, this is located on the bin roof. Depending on the material to be stored, it may be necessary to also include provisions for explosion venting, fire suppression and inert gas blanketing. Consideration should be given to the effects of contaminants, atmospheric gases, humidity and temperature on the material's flow properties, and the structure designed accordingly. Any risks associated with human exposure to the bulk solid in the vessel must also be taken into consideration.

2.10 Operational aspects

The following is a brief discussion of typical problems that can occur with storage bins, along with some common solutions.

2.10.1 *No flow or erratic flow*

Probably the most serious problem is one of *no flow*, i.e. the condition in which no material discharges when the feeder is started or the gate is opened. This is often due to arching or ratholing of the bulk solid within the vessel. If it is determined through testing that this problem is due to the cohesive strength of the bulk solid, changes should be considered to its moisture content, temperature, time of storage at rest and/or particle size distribution to make it less cohesive. Sometimes use of a chemical additive can improve the flow properties of a solid. If instead of cohesion the problem is one of particle interlocking, the solution is to decrease the particle size and/or alter the particle shape to make the particles more rounded.

If changing the properties of the bulk solid is not practical, consideration should be given to enlarging the outlet and feeder or using a mechanical flow aid such as an air cannon, external vibrator, vibrating discharger, internal agitator, flexible walls, etc.

Sometimes the problem of no flow can be caused by an outlet area that is not fully live. This could be due to a partially opened slide gate, a feeder that withdraws over only a portion of the outlet area, or protrusions into the flow channel caused by mating flanges, access doors, poke holes, etc.

Another operational problem is *erratic flow*. This is a condition of alternating mass flow and funnel flow resulting in intermittent problems of arching and perhaps ratholing. The solution to this problem is to change the design of the vessel such that a mass flow pattern occurs.

2.10.2 *Flow rate problems*

If *flooding*, i.e. uncontrolled flow of fine powder through the outlet, occurs, the solution is usually to change the flow pattern from funnel flow to mass flow. If the flow pattern is already mass flow and flooding is still occurring, consideration should be given to decreasing the fill and/or discharge rates.

If, to the other extreme, the discharge rate is insufficient for the downstream process and changing the speed of the feeder does not correct the problem, the solution is to enlarge the outlet or add an air permeation system. If the mismatch between actual and required flow rates is so great that these alternatives are not capable of solving the problem, a fluidised or air-assisted discharge should be considered.

2.10.3 *Particle segregation*

If segregation causing downstream quality control problems can be isolated to a particular bin, the first thing to consider is the type of segregation that is occurring. If it is a side-to-side distribution of particles varying in size, chemical composition or some other attribute (generally the result of segregation due to the mechanisms of sifting, dusting, trajectory or dynamic effects), a mass flow pattern should be used. One should also consider using a distributor at the bin inlet to further minimise the segregation problem. If, on the other hand, the segregation profile is top-to-bottom (due, for example, to the fluidisation mechanism), it is necessary to first break this fill/segregation pattern. One way to do this is by using tangential entry into the vessel. It will probably also be necessary to convert the vessel's flow pattern to mass flow.

2.10.4 *Excess stagnant material*

Another operational problem is that of limited live capacity. This is usually one of the results of rathole formation. To correct this it is necessary to change the flow pattern from funnel flow to mass flow or at least expanded flow.

If the operational problem is one of degradation (caking, spoilage, oxidation), again the solution is to convert from funnel flow to mass flow or at least expanded flow.

2.10.5 *Structural concerns*

Sometimes there are problems with self-induced bin vibrations. If these vibrations are high in frequency but low in amplitude, an interesting phenomenon called *silo music* (humming) or *silo honking* may be experienced. This can be a nuisance to personnel or neighbours nearby but is usually not a structural concern. If, on the other hand, the bin vibrations are low in frequency but high in amplitude, the result is what are called *silo quakes*. These apply massive dynamic loads that most vessels are not designed to withstand. Structural failures have occurred due to this mechanism.

Structural problems are more common in bins and silos than in almost any other industrial or commercial structure (see Figure 2.29). There are three major causes of such problems:

(a)

Figure 2.29 Examples of structural failure.

(b)

(c)

Figure 2.29 *(continued)*.

design errors, construction errors and operational issues (Carson 2000b). Within the category of design errors, some of the problems that often occur are:

- Material properties and flow patterns not considered
- Out-of-round bending of a circular cylinder
- Large and/or non-symmetric loads caused by inserts
- Lack of proper consideration of specific requirements for structure type, such as bolted metal or reinforced concrete
- Temperature or moisture effects resulting in thermal ratcheting or particle swelling

Construction errors include poor quality workmanship (incorrect bolts, incorrect placement of reinforcing steel, incorrect material type or thickness) and unauthorised design changes made during construction.

Operational practices that can cause structural problems include improper usage of the vessel (e.g. a change in the stored material that results in higher loads being applied to the vessel walls or adding additional outlets to the vessel wall, which changes the flow pattern). Another operational issue is improper maintenance, such as allowing material to build up on surfaces or ignoring the effects of corrosion.

Whatever the structural problem, it is essential that, at the first sign of trouble, there be an appropriate response. Continuing to empty a vessel that is already starting to fail may well result in complete collapse of the structure. Engineers trained in structural issues as well as solids flow should be consulted to ensure that property damage is kept to a minimum and personnel are kept out of harm's way.

Silos and bins should be inspected on a routine basis to anticipate potential flow and structural problems before these become major.

2.10.6 *Process problems*

If the operational problem is one of cross-contamination, this is most likely due to a funnel flow pattern. If this is the case, changing to mass flow should correct it. It should be recognised that even with mass flow there will be a velocity distribution across the vessel, so achieving perfect first-in-first-out is seldom achievable.

If there are problems with inventory control, this is usually a signal that the flow pattern needs to be changed from funnel flow to mass flow (Pittenger *et al.* 1999). Sometimes a weighing system is needed in order to provide proper inventory information.

If the bin is being used as a processing vessel such as for drying, purging or conditioning of the bulk solid, non-uniform processing may occur. This is usually an indication that a funnel flow pattern needs to be changed to mass flow. Sometimes an insert is used in such applications, but this must be done with caution (Carson *et al.* 1995).

2.10.7 *Abrasive wear and attrition*

If the bin wall is wearing abnormally, one should first determine where the wear area is occurring. If it is in an area where impact is occurring, it is important to minimise the opportunity for the bulk solid to impact the vessel walls, such as by maintaining a minimum level in the vessel as much of the time as possible. Other options include reducing the fall height of material as it enters the vessel or redirecting the incoming stream to a wear plate.

If the wear problem is due to abrasion, the first step should be to test the abrasive wear characteristics of alternate liners (Johanson & Royal 1982) and then consider changing the wall material or liner.

Another operational problem is unacceptable attrition. If this is occurring, consideration should be given to using a letdown chute, reducing the size of the bin or changing the type of feeder to one that provides fewer pinch points.

2.10.8 *Feeder problems*

Finally, if the operational problem involves the feeder such that it cannot be started, requires excessive power to operate or experiences excessive wear, these all could be indications that the flow pattern is incorrect for the material. The importance of the feeder interface cannot be over-stressed, particularly for elongated outlets.

References

ASTM D6940-03 (2003a) Standard practice for measuring sifting segregation tendencies of bulk solids. In: *Annual Book of ASTM Standards*, Vol. 04.08. American Society for Testing and Materials International, West Conshohocken, PA.

ASTM D6941-03 (2003b) Standard practice for measuring fluidization segregation tendencies of powders. In: *Annual Book of ASTM Standards*, Vol. 4.08. American Society for Testing and Materials International, West Conshohocken, PA.

Bates, L. (1997) *User Guide to Segregation*. British Materials Handling Board, Elsinore House, Marlow, England.

Beverloo, W.A., Leniger, H.A. & Van de Velde, J. (1961) The flow of granular solids through orifices. *Chem. Eng. Sci.*, **15**, 260–269.

Carson, J.W. (2000a) Step-by-step process in selecting a feeder. *Chem. Process., Powder Solids Annu.*, 38–41.

Carson, J.W. (2000b) Silo failures: case histories and lessons learned. *Proceedings of the 3rd Israeli Conference for Conveying and Handling of Particulate Solids*, Vol. 1, pp. 4.1–4.11, Dead Sea, Israel.

Carson, J.W., Purutyan, H. & Rotter, J.M. (2003) The dangers of relying on wall friction values in codes. *4th International Conference for Conveying and Handling of Particle Solids*, Budapest, Hungary.

Carson, J.W., Royal, T.A. & Goodwill, D.J. (1986) Understanding and eliminating particle segregation problems. *Bulk Solids Handling*, **6**, 139–144.

Carson, J.W., Royal, T.A. & Pittenger, B.H. (1995) Mass flow purge and conditioning vessels. *Chem. Process.*, **58**(8), 77–80.

Jenike, A.W. (1964) *Storage and Flow of Solids*. University of Utah Engineering Experiment Station, Bulletin No. 123.

Johanson, J.R. & Royal, T.A. (1982) Measuring and use of wear properties for predicting life of bulk materials handling equipment. *Bulk Solids Handling*, **2**, 517–523.

Pittenger, B.H., Carson, J.W., Prescott, J.K. & Purutyan, H. (1999) Uniform purging of resins in contact bed purge vessels. *Polym. Eng. Sci.*, **39**, 1802–1811.

Rotter, M.J. (2005) Silo design loads from stored bulk solids: the provisions of the new Eurocode EN 1991-4 (2005). *Proceedings of the IMechE Seminar on Specifying and Designing Safe Storage Silos*, IMechE.

Royal, T.A. & Carson, J.W. (1991) Fine powder flow phenomena in bins, hoppers, and processing vessels. Presented at *Bulk 2000*, London.

Troxel, T.G., Carson, J.W. & Bengston, K.E. (2005) Proven techniques for air-assisted handling of powders in bins and hoppers. *7th World Congress of Chemical Engineering* (C13-008).

3 Silo and hopper design for strength

J. MICHAEL ROTTER

3.1 Introduction

Silos and hoppers are widely used in a great many different industries for storing a huge range of different solids. The sizes of these silos may vary from capacities less that 1 tonne to the largest containing as much as 100 000 tonnes. The size of the silo has a strong bearing on the number of different considerations required: small silos generally do not produce structural problems, but in large silos many different aspects need careful attention.

The designs used for silos also vary very much (Figure 3.1). In some industries (e.g. grain storage), there is a competitive industry producing standard silo products which function extremely well and cost-effectively provided the conditions remain those anticipated. In other industries (e.g. cement and mineral ore storage) very large silos are used and every silo must be individually designed for the special conditions. It should be noted that each silo is normally designed to contain a very limited range of solids, and that the use of a silo designed for one kind of solid to store different solids can easily cause damage. Bulk solids vary very much in their properties, and a silo that is perfectly adequate to store one material may be very dangerous for another.

The terms *silo*, *bunker*, *bin* and *hopper* are often used to refer to similar containers in different industries. Here, the word 'hopper' is exclusively used with a special meaning for the converging part leading to a gravity discharge outlet. All complete storage containers are referred to as silos, irrespective of the stored solid, geometry and industrial sector. A characteristic form to describe the parts of the silo is shown in Figure 3.2. The transition, which lies at the junction between the vertical wall and the hopper, should be noted.

This chapter provides a brief outline of the development of understanding of pressures that develop in silos and their consequences for the safety of the silo structure. More structural failures occur in silos than in any other engineered structural form, considering the numbers of each, and these failures occur in all countries and all industries. Structural design considerations for silos are therefore a key aspect of bulk solids handling systems.

The chapter refers extensively to the provisions of the recently developed European standards for silo pressures (EN 1991-4 2007) and for metal silo structural design (EN 1993-4-1 2007), for which the author was the chief contributor and editor. Further useful information relating to the structural design of all silos may be found in Rotter (2001a).

3.2 Why pressures in silos matter

3.2.1 *General*

The pressures that develop in a silo are very different from those developing in a tank that contains fluid. Fluid pressures depend uniquely on the head, and in most fluid storages flow velocities are so low that dynamic effects are small. By contrast, pressures in silos

(a) 10 000 tonne steel grain storages, Australia

(b) Corrugated steel storage, Germany

(c) Rectangular concrete silo battery, Austria

(d) Older concrete and newer steel silos, France

(e) Salt storage with control room, Italy

(f) FRP farm silo, France

Figure 3.1 Different geometries and sizes of silo.

are dominated by frictional phenomena, the flow of bulk solids is controlled by frictional considerations and is largely independent of head, and there are few analogies between fluid and solid storage that are either valid or practically useful. In this context, it is worth noting that sound mechanics equations to describe fluid flow have existed for over a hundred years, but no comparable agreed set of equations yet exists to deal with bulk solids flow.

Pressures that develop in stored solids can have an important impact on their free flow from a silo if the bulk solid is prone to developing a small cohesive strength under stress

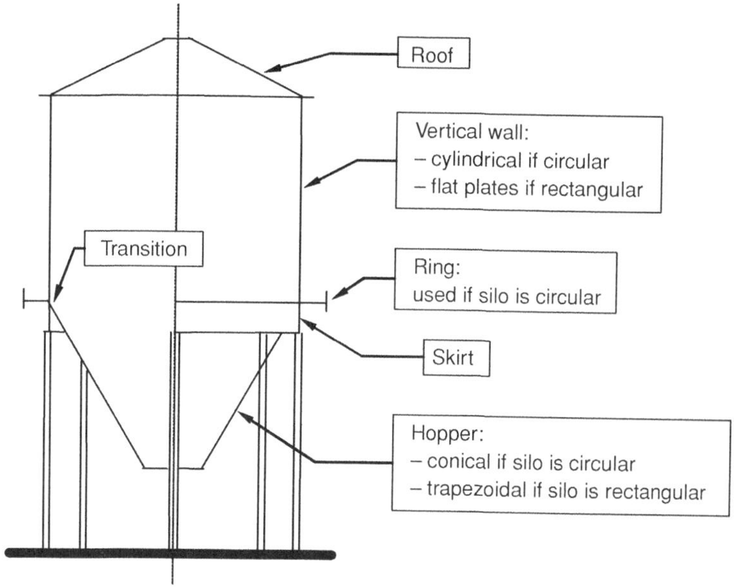

Figure 3.2 Terminology for parts of a typical silo.

(e.g. flour). These aspects are dealt with in the accompanying chapter by Dr John Carson and are not commented on further here.

The most critical aspect of pressures in silos is their effect on the structure designed to contain the solid. Because the properties of solids vary widely, the pressures can also vary very much both in magnitude, distribution and stability. Some conditions lead to very unpredictable pressure peaks that can cause serious damage, whilst other arrangements are very benign and do not cause any concern even to the unwary. This chapter tries to make some clear distinctions between these different situations.

In particular, where pressures in silos are being defined for the purposes of structural design, an understanding of the consequences for the structure is absolutely vital. Thus, it is often imagined that high pressures, wherever and whenever occurring, are the most damaging event. This is very far from the truth, and many theories of silo pressure and scientific articles on pressures are very misleading because their authors did not understand what stress conditions would be induced in the structure by the pressures, nor the conditions that lead to structural failure. This chapter sets out some pointers to that information and it is hoped that the reader will appreciate that this subject is not straightforward, but a full explanation is beyond the scope of this chapter.

3.2.2 *Classifications of silos*

Silos are commonly classified according to the cross-sectional shape in plan section. Most silos are circular, but some are rectangular and interstitial gaps between adjacent circular silos may even be star-shaped. The pressure regime is principally important in silos of larger dimensions, and the circular silo dominates these: for this reason, this chapter is chiefly concerned with the circular planform.

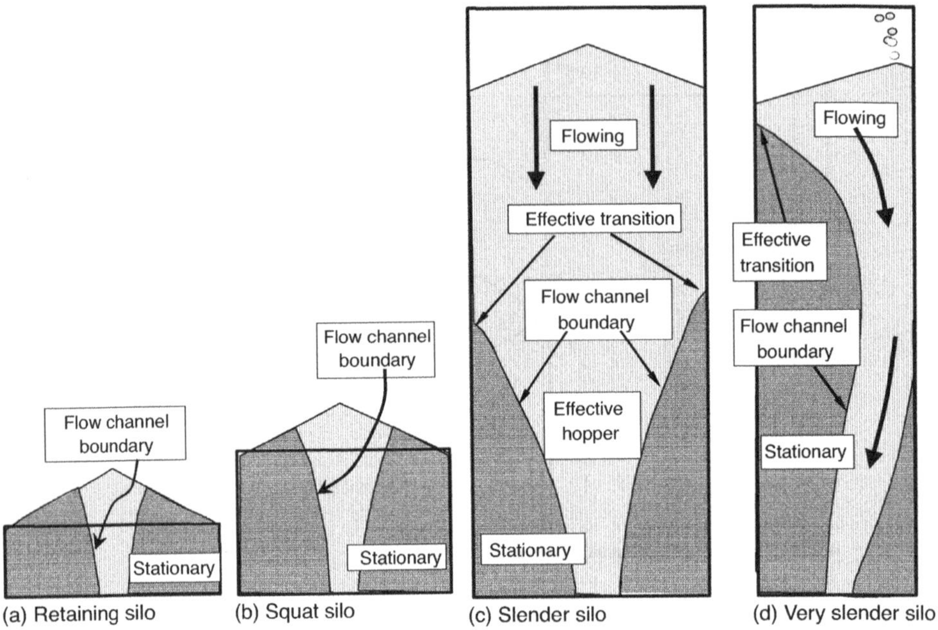

Figure 3.3 Silo conditions for different aspect ratios.

A second key distinction is the overall size of the silo. Small silos do not present structural challenges and can be designed using fairly simple calculations. Very large silos need great attention to many details. For this reason, EN 1991-4 divides silos into three categories according to the mass of solid stored, and has different design requirements for each. The break points occur at 100 tonnes, 1000 tonnes (for special cases) and 10 000 tonnes. The standard on structural design of steel silos makes similar divisions, though at different values because it is concerned with aspects of the structure, not the loading. The break points occur at 100, 200 (with eccentric discharge), 1000 (elevated) and 5000 (ground supported) tonnes, with considerable design calculation effort being demanded where the largest sizes are used.

A third key classification is necessary to define the pressure regime. This is the aspect ratio (height H divided by horizontal dimension D). Most silos research has studied slender silos ($H/D > 2$) and most of this chapter is concerned with this geometry. In squat silos ($H/D < 1$), the top surface profile plays an important role and issues of the difference between filling and discharge pressures are much reduced (Pieper & Stamou 1981). EN 1991-4 gives different rules for each aspect ratio, classing them as slender, intermediate, squat and retaining (Figure 3.3).

3.2.3 *Metal and concrete silos*

Metal and concrete silos carry their loads in very different ways, so the kinds of damage that can occur in each type are very different and the critical design considerations are different. For this reason, the later part of the chapter examines these two cases in separate sections.

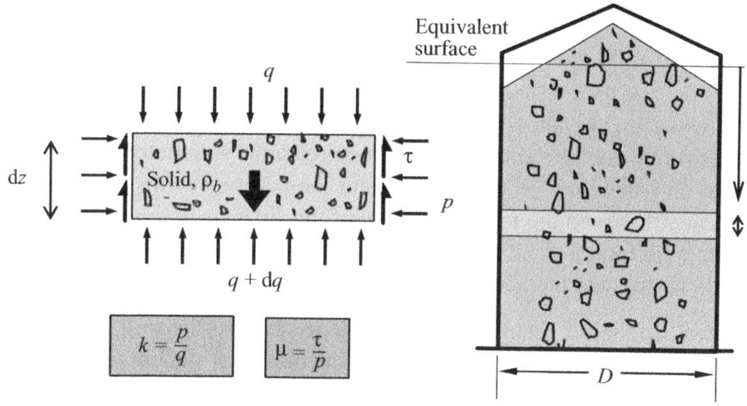

Figure 3.4 Silo contents, notation and a slice of solid.

3.3 Pressures in silos: basic theory

3.3.1 *Early studies*

A brief historical account of the developing understanding of silos may seem strange in a chapter that advises on silo design and management, but there are good reasons for it. The field of silo pressures is full of misunderstandings and misinterpretations, and many of these continue and are repeated today, so an appreciation of the reasons for some misconceptions provides a valuable background.

Although silos have been used to store solids (e.g. grains) for thousands of years, the earliest scientific studies of the pressures in silos were only undertaken at the end of the nineteenth century. Several researchers performed simple experiments and developed simple theories in this period (for a good description, see Ketchum 1907), but the most important of these was Janssen (1895) who both performed experiments on a tall square model silo and developed the theory which is almost universally used as the single reliable reference point in a sea of uncertainties concerning silo pressures.

3.3.2 *Janssen silo pressure theory for vertical walls*

This theory is so critical to understanding many aspects of silos that the derivation is set out here.

A tall silo with vertical walls, whose horizontal cross section can effectively take any shape, is shown in Figure 3.4. The equilibrium of forces on a slice of the solid with unit weight (or less formally bulk density) ρ_b at some depth z is shown, where the slice has height dz, plan area A and perimeter against the wall U. The stresses acting on it may vary across the horizontal surface above and below, and around the perimeter with the wall, so the mean values are used in this analysis. The mean vertical stress is q, the consequential mean horizontal pressure against the wall p and the frictional shear stress (termed frictional traction) on the wall τ. Vertical equilibrium of this slice of solid leads to

$$(q + dq)A + U\tau dz = qA + \rho_b A\, dz$$

or

$$\frac{\mathrm{d}q}{\mathrm{d}z}A + U\,\tau = \rho_b A \tag{3.1}$$

The vertical stress q on the slice need not be uniform: the analysis considers only the mean value. Horizontal equilibrium of the slice requires some symmetry to exist in the wall pressures p, but they need not be constant around the perimeter (this becomes a serious issue later). Shear stresses on the top and bottom of the slice are assumed to integrate to a zero resultant on each face.

Two assumptions are next made (as used by Janssen):

a The full wall friction is assumed to be developed against the wall at every point, so that the mean frictional shear τ is related to the mean normal pressure p on the wall through the wall friction coefficient μ (Figure 3.4) as

$$\tau = \mu p \tag{3.2}$$

b The normal pressure p (mean value around the perimeter) is deemed to be related to the mean vertical stress q through a lateral pressure ratio K (Figure 3.4) as

$$p = Kq \tag{3.3}$$

Inserting these into Equation (3.1) leads to

$$\frac{\mathrm{d}q}{\mathrm{d}z} + \frac{U}{A}\mu K q = \rho_b \tag{3.4}$$

which may be solved to yield

$$q = q|_{z=0} = 0 + \frac{\rho_b A}{\mu U}\left(1 - e^{-zU/(AK\mu)}\right) \tag{3.5}$$

If the *mean* vertical stress in the solid q is taken as zero at some reference height $z = 0$ (Figure 3.4) (this condition is met at the centroid of the top pile of solids), then

$$q|_{z=0} = 0 \tag{3.6}$$

and Equation (3.5) can be more neatly written as

$$q = q_0(1 - e^{-z/z_0}) \tag{3.7}$$

in which

$$q_0 = \rho_b\, z_0 \tag{3.8}$$

and

$$z_0 = \frac{1}{\mu K}\,\frac{A}{U} \tag{3.9}$$

Here, q_0 represents the mean vertical stress in the solid that is reached asymptotically at great depth. The length measure z_0 defines the rate at which the asymptote is approached and is commonly termed the *Janssen reference depth*.

The origin of the vertical coordinate z (at the centroid of the top pile of solids) is called the *equivalent surface*.

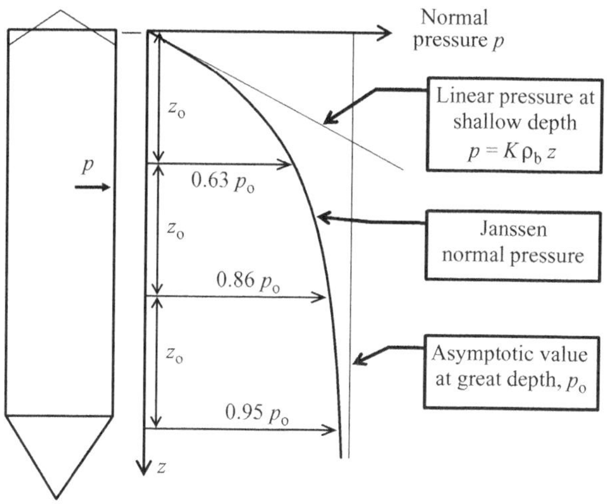

Figure 3.5 Janssen pressure pattern.

It is natural to transform Equation (3.7) into pressures normal to the wall p (Figure 3.4)

$$p = p_0(1 - e^{-z/z_0}) \tag{3.10}$$

in which the *asymptotic normal pressure* at great depth is given by

$$p_0 = \frac{\rho_b A}{\mu U} = K \rho_b z_0 \tag{3.11}$$

The typical pattern of pressure defined by this equation is shown in Figure 3.5.

Since many silos have circular cross sections, it is useful to simplify the above equations to specialise them for a silo of radius R.

$$z_0 = \frac{R}{2\mu K} \quad \text{and} \quad p_0 = \frac{\rho_b R}{2\mu} \tag{3.12}$$

The values of the wall friction coefficient μ and the lateral pressure ratio K may be measured in control tests on the particular solid being stored (see Chapter 1).

A few deductions may be made from these equations. At great depth, the mean pressure p depends only on the radius R and the wall friction coefficient μ, not on the depth below the surface. A smooth wall leads to higher pressures than a rough wall. The pressures all vary linearly with the solid bulk density ρ_b, so this is a key parameter in any silo evaluation.

The asymptotic value of pressure p_0 is actually more robust than the pressure distribution according to Janssen, because it does not need the assumption of a lateral pressure ratio. At great depth, conditions are stable, and neither the mean vertical stress q nor the mean wall pressure p changes. The equilibrium of a simple slice then simply equates the weight of the slice to the support given by wall friction, which becomes (adopting $\tau = \mu p$),

$$\mu p_0 U = \rho_b A \tag{3.13}$$

or

$$p_0 = \frac{\rho_b A}{\mu U} = \frac{\rho_b R}{2\mu} \tag{3.14}$$

Thus, every theory that assumes that the wall friction is fully developed must reach the same asymptotic value of lateral pressure p_0 at great depth. This applies whether the silo is just filled or is being emptied.

At shallow depths, the pressures vary linearly with depth and are approximated by

$$p = K\rho_b z \tag{3.15}$$

which is the 'earth pressure' against a retaining wall. However, this theory does not take proper account of the surface profile in defining wall pressures near the surface, and this matters in squat silo geometries (see EN 1991-4 2007).

The Janssen theory is the main descriptor of filling pressures in all standards.

3.3.3 *The lateral pressure ratio K*

The theory of Janssen was rapidly found to give quite a good representation of the pressures in a silo after it was filled. It is relatively easy to measure the bulk density ρ_b and wall friction coefficient μ, but the lateral pressure ratio K was less easy. Both bulk granular solids and soils (which are granular solids) were not well understood in the early twentieth century, so it was natural that the earth pressure theory of Rankine (1857), which defined two limiting values of K, should be adopted as applicable in a silo. These are limiting values because, at these values, the solid is ready to deform by shearing into a different shape. They are the Rankine active and passive limits, given by

$$\text{Active} \quad K_a = \frac{1 - \sin\phi_i}{1 + \sin\phi_i} \tag{3.16}$$

$$\text{Passive} \quad K_p = \frac{1 + \sin\phi_i}{1 - \sin\phi_i} \tag{3.17}$$

where ϕ_i is the angle of internal friction of the solid, found by shearing the solid under a compressive stress normal to the plane of shearing. For a typical solid with $\phi_i = 30°$, $K_a = 0.33$ and $K_p = 3.0$. The ratio of these two values is later found to be relevant and can be seen as $K_p/K_a = 9$.

In the first use of Janssen's theory (Koenen 1895), it was assumed that the solid in a silo after filling was in a Rankine active state, giving a low value of lateral pressure ratio K, and leading to smaller pressures. However, after extensive damage to many silos, it was widely recognised by the 1960s that this was an underestimate of K.

This situation is best explained using understandings that came much later. In Figure 3.6, a silo wall is retaining bulk solid. The pressure against the wall depends on the extent to which the wall moves inwards or outwards. In the limit, the two Rankine states are reached where the solid can deform plastically, but if the wall is rigid and does not move at all, a state referred to as K_0 exists. This is not far from the Rankine active state, but the value of K_0 is perhaps 50% larger than K_a. If the wall is flexible, the value of K may fall slightly as it moves outwards. The stored bulk solid is essentially in an elastic state, not at a plastic limit.

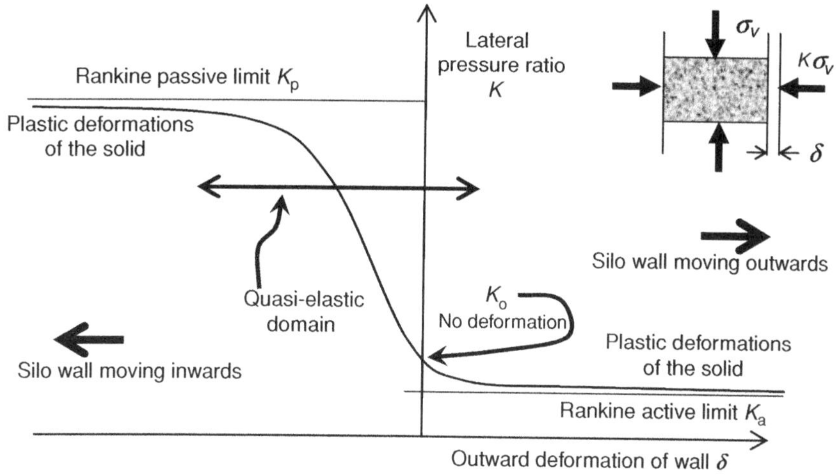

Figure 3.6 Effect of wall horizontal movement on lateral pressure ratio K.

The value for K_0 has long been approximately related to the angle of internal friction ϕ_i of the solid (Jaky 1948) as

$$K_0 = 1 - \sin\phi_i \tag{3.18}$$

The background to this equation may be read in Muir Wood (1990).

The ideal K_0 relates to conditions in which the vertical and horizontal stresses are principal stresses and both uniform. Since the state of the silo after filling has both a non-uniform vertical stress pattern and shear stresses against the wall, it is best here to assign the value K_f for the filling state, noting that $K_f > K_a$, but $K_f \approx > K_0$.

It is best to measure the lateral pressure ratio K directly (see Chapter 1), but it has long been common to estimate it from the measured angle of internal friction ϕ_i. Accounting for the above effects, the European standard EN 1991-4 (2007) defines the filling value of K_f for design purposes as

$$K_f = 1.1(1 - \sin\phi_i) \tag{3.19}$$

3.3.4 *Pressures in hoppers*

The Janssen theory describes pressures in a parallel-sided vessel. The corresponding theory for a converging channel came much later, and is normally attributed to Walker (1964, 1966), though it was first derived by Dabrowski (1957) and was probably also found by Jenike and others in the late 1950s.

The hopper height is H and the vertical coordinate is taken with its origin at the hopper apex, using coordinate x (Figure 3.7). The steepest line on the hopper is at angle β to the vertical. For a conical or pyramidal hopper, the horizontal coordinate to the closest point on the wall is $r = x \tan\beta$ and the area of a slice becomes

$$A = k_1 r^2 = k_1 x^2 \tan^2\beta \tag{3.20}$$

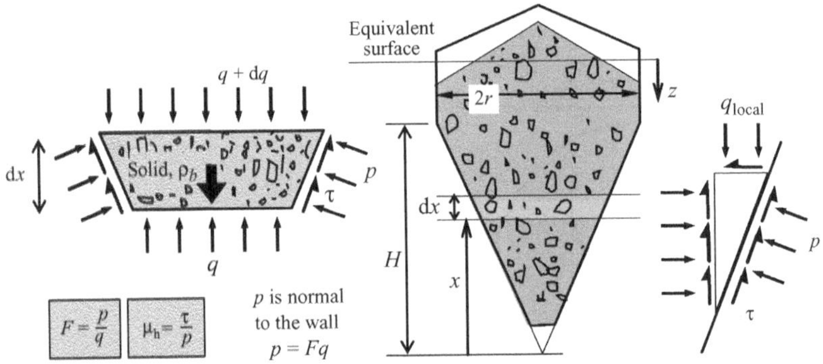

Figure 3.7 Hopper slice analysis, coordinate system and local equilibrium.

where $k_1 = \pi$ for a conical hopper and $k_1 = 4$ for a square hopper of half side r. The perimeter of the slice is given by

$$U = k_2 r = k_2 x \tan\beta \tag{3.21}$$

where $k_2 = 2\pi$ for a conical hopper and $k_2 = 8$ for a square hopper of half side r. Vertical equilibrium of the slice of solid (Figure 3.7) leads to

$$\begin{aligned}(q + dq)k_1(x + dx)^2 \tan^2\beta - qk_1x^2 \tan^2\beta + \rho_b k_1 x^2 \tan^2\beta \mathrm{d}x \\ = (p \sin\beta + \tau \cos\beta)k_2 x \tan\beta \frac{\mathrm{d}x}{\cos\beta}\end{aligned} \tag{3.22}$$

Cancelling, eliminating small terms and noting that $(k_2/k_1) = 2$ for both geometries

$$x\frac{\mathrm{d}q}{\mathrm{d}x} = 2\left(p + \frac{\tau}{\tan\beta} - q\right) - \rho_b x \tag{3.23}$$

in which p is the mean normal pressure against the hopper wall, q is the mean vertical stress in the solid, τ is the mean wall frictional traction and ρ_b the bulk density.

The two assumptions used in the Janssen analysis are next made:

a The frictional shear τ is assumed to be a fixed proportion of the local normal pressure p. This is the hopper wall friction coefficient μ_h when sliding occurs, but is some smaller value, an effective friction $\mu_{h,\mathrm{eff}}$ when there is no sliding

$$\tau = \mu_h p \tag{3.24}$$

b The mean pressure normal on the inclined wall p is deemed to be related to the mean vertical stress q (Figure 3.7) through the hopper pressure ratio F as

$$p = Fq \tag{3.25}$$

Inserting these into Equation (3.23) leads to

$$x\frac{\mathrm{d}q}{\mathrm{d}x} - 2q[F + F\mu_h \cot\beta - 1] = -\rho_b x \tag{3.26}$$

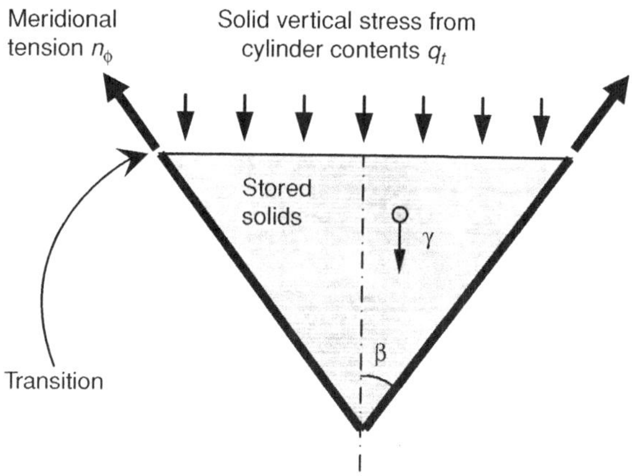

Figure 3.8 Mean vertical stress at the transition and overall hopper equilibrium.

or

$$x\frac{dq}{dx} - n\,q = -\rho_b x \tag{3.27}$$

in which

$$n = 2[F + F\mu_h \cot\beta - 1] \tag{3.28}$$

which may be solved, considering the top boundary condition $q = q_t$ at $x = H$, to yield

$$q = q_t\left(\frac{x}{H}\right)^n + \frac{\rho_b H}{(n-1)}\left\{\left(\frac{x}{H}\right) - \left(\frac{x}{H}\right)^n\right\} \tag{3.29}$$

where q_t is the mean vertical stress in the solid at the transition (Figure 3.8).

It is evident that the value of F must depend on geometry and solids properties, just as K was dependent on solids properties in the analysis of the pressures on vertical walls.

The normal pressures may be deduced from Equation (3.29) as

$$p = F\left[q_t\left(\frac{x}{H}\right)^n + \frac{\rho_b H}{(n-1)}\left\{\left(\frac{x}{H}\right) - \left(\frac{x}{H}\right)^n\right\}\right] \tag{3.30}$$

Equation (3.30) gives a variety of different forms for the hopper pressure distribution, depending on the value of F. The two components of loading are clearly separated: the weight of solids in the hopper (term involving $\rho_b H$) and the pressure derived from the cylinder (transition surcharge q_t). Equation (3.30) indicates that high local pressures can occur at the transition if the barrel has a moderate height and F is high. The distribution becomes very peaked at the transition for high n which arises if F is high and the hopper is steep and rough. This theory is used in EN 1991-4 (2007), but older standards (e.g. DIN 1055-6 1987) often gave empirical approximations to the pressure pattern which could not be guaranteed to be safe in all conditions.

These pressure patterns are illustrated in Figure 3.9, where the changing shape of the hopper wall pressures caused by transition vertical pressures q_t is illustrated for different values of F.

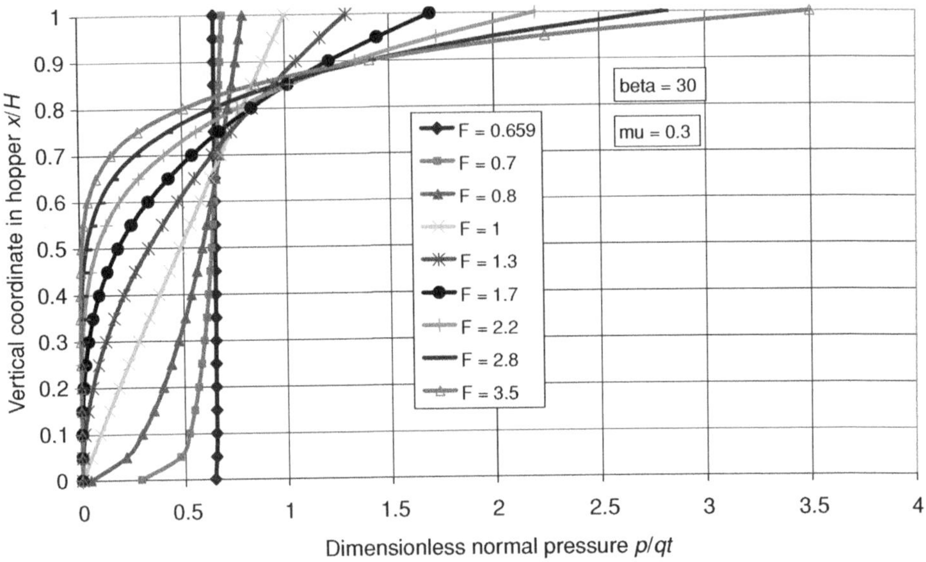

Figure 3.9 Changing pattern of pressures in hoppers as the value of F changes.

The question of whether the friction is fully mobilised in a hopper depends on its slope and the smoothness of the wall. The hopper is classed as steep if the solids slide on it, and this is met by the following test. The hopper is steep if

$$\tan \beta < \frac{1 - K}{2\mu_h} \tag{3.31}$$

where μ_h is the full wall friction coefficient on the hopper, which may have a lining. This relationship is plotted in Figure 3.10a for clarity. The effect of steepness on the pattern of pressures in hoppers during emptying of the silo is illustrated in Figure 3.10b.

However, the most critical feature of a hopper is not the wall pressure distribution but the overall equilibrium shown in Figure 3.8. Most structural failures of hoppers occur by rupture at the transition under the stress resultant n_ϕ. High values of n_ϕ are chiefly caused by an excessive vertical pressure q_t from the cylinder, probably when this is underestimated through inadequate attention to material variability (Section 3.3.6).

3.3.5 *Simple structural concepts for cylinders*

The chief goal of predicting pressures in silos is to ensure the safety of the structure. So the effect of the pressure on the structure must be a key element. All early studies of pressures assumed that the simple equilibrium between normal pressure and hoop (circumferential)

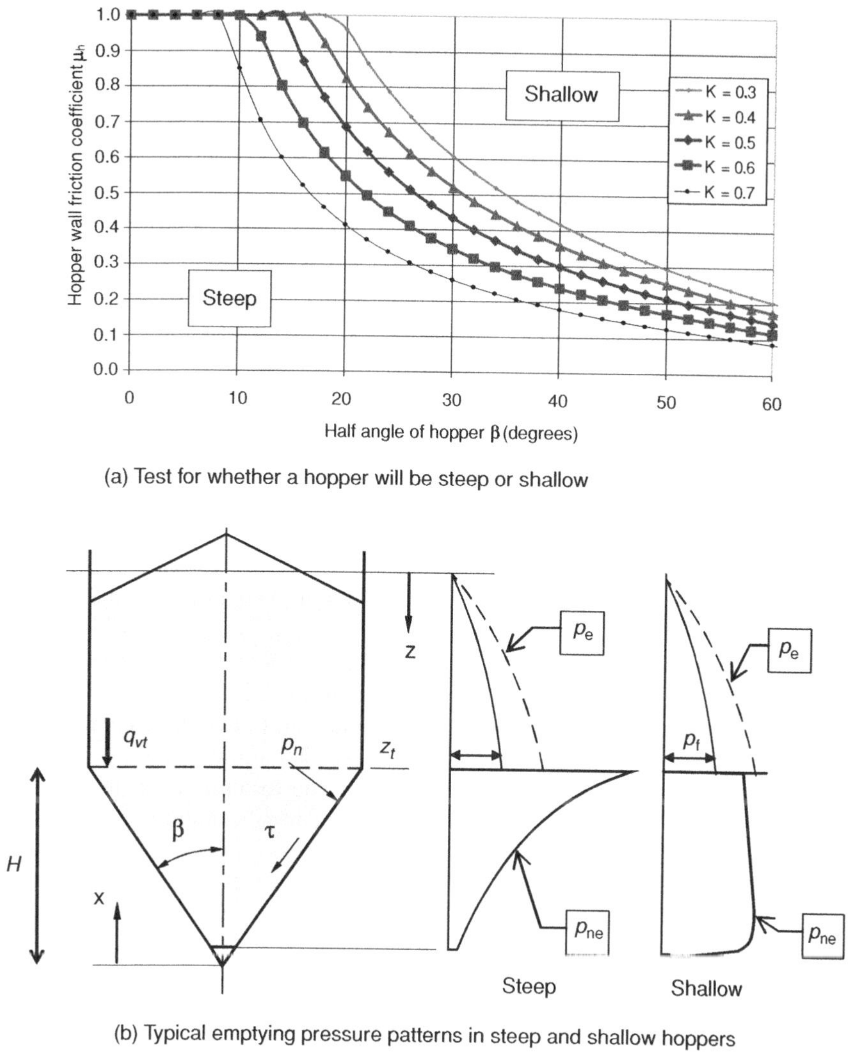

Figure 3.10 Steepness criterion and typical hopper pressures.

tension in the wall (Figure 3.11) was all that needed to be considered, leading to

$$n_\theta = pR \quad \text{for a circular silo} \tag{3.32}$$

where n_θ is the circumferential force per unit height in the wall. This equation is valid if the pressures are constant around the perimeter at any level in the silo. It indicates that higher pressures will lead to higher tensions and so presumably will be more damaging to the silo wall. This over-simplified concept has underlain much of the pressure values reported from silo research in the last century, and is certainly responsible for some failures which occurred when pressures dropped locally (see Section 3.4.5). The maximum pressure,

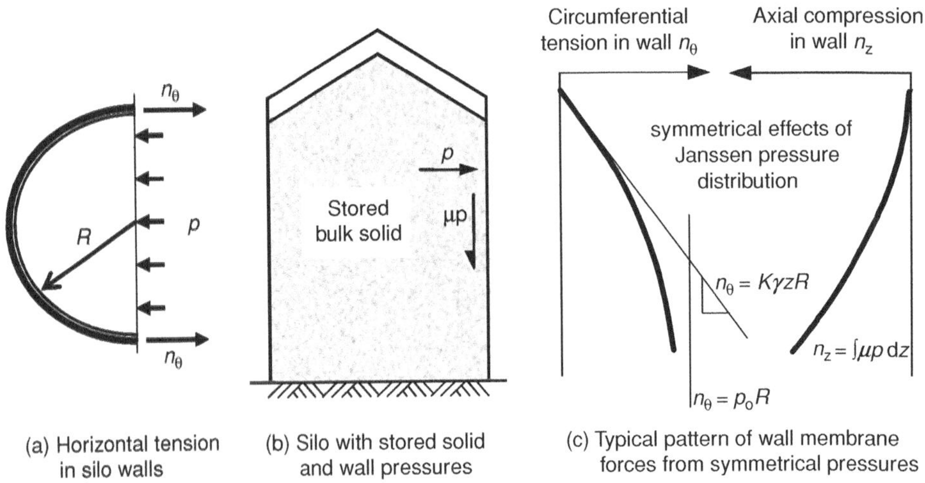

Figure 3.11 Simple structural effect of symmetrical pressures.

especially if local and of short duration, is not usually a prime cause of structural damage to silos.

Accompanying the pressure p against the wall is the frictional traction τ (Figure 3.4), which accumulates to produce vertical (axial) forces in the silo wall. Since the vertical pressure in the solid reaches an asymptotic limit (Equation (3.8)), the weight of all the additional solids must be borne instead by vertical forces in the wall (Figure 3.11c).

Adopting Janssen's theory for the pressure pattern, the resulting axial force per unit circumference n_z developing in the silo wall under symmetrical conditions is then

$$n_z = \int_0^z \tau \mathrm{d}z = \int_0^z \mu p \mathrm{d}z = \int_0^z \mu p_0 (1 - e^{-z/z_0})\, \mathrm{d}z = \mu p_0 z_0 \left(\frac{z}{z_0} - 1 + e^{-z/z_0} \right) \quad (3.33)$$

This compressive force rapidly approaches a linear increase with depth (term z/z_0) (Figure 3.11c). Thus very high forces develop in the wall towards the bottom of the silo. This force is important in thin metal silos, as it becomes the critical effect because the controlling design consideration is buckling under axial compression (see Section 3.5.2). This is the reason why metal silos must have a much greater wall thickness towards the bottom than near the top.

The above theory for cylinders is not valid for conical hoppers. For them, even the simplest stress analysis is much more complicated and is beyond the scope of this chapter. More information may be found in Rotter (2001a).

In reading what follows it should be noted that metal silos are most sensitive to vertical compression in the vertical walls, that concrete silos are most sensitive to normal pressures against the walls, and that both of these structural materials are easily damaged by unsymmetrical pressures, as noted in Sections 3.4.5 and 3.5. Finally, the hopper, which has not been discussed yet, is usually chiefly loaded by the vertical stress in the solid at the transition. These different sensitivities demand that careful attention is paid to different parts of the pressure theory, since it is not normal wall pressures alone that cause structural failures.

3.3.6 *Variability of the properties of stored solids*

The above theories are based on known properties of the stored bulk solid. However, industrial bulk solids have properties that vary considerably from time to time and from source to source. The extent of variability that a particular silo may see depends very much on its location: the solids in a silo that is part of a manufacturing process may vary rather little, whilst those at a mine or port facility are likely to vary considerably from year to year. Unfortunately, these differences cannot yet be accounted for in the design process, especially as the handling properties of solids often vary considerably when other properties (e.g. chemical composition) do not. Such changes can arise from moisture content, particle shape or surface roughness changes, traces of foreign materials and minor attrition during handling. Thus, it is wise to design all silos for the full range of properties that may arise.

In the world's first codified design rules (DIN 1055-6 1964), it was unstated, but tacitly assumed, that the silo was tall and made of concrete. Consequently, it was thought that the worst condition was normal pressures against the wall, and that a design would be safe if designed for the bulk solid that produced the highest pressures. Examining Janssen's equation (Equation (3.6)), it can be seen that these pressures are highest when the wall friction is low and the lateral pressure ratio is high. As a result, older tables of material properties, set out in standards, gave a single value of each property and tended to exaggerate the lateral pressure ratio K and underestimate the wall friction μ.

As metal silos have become much more common, the importance of vertical forces in the wall has become clear. These forces are largest when the solid has a high lateral pressure ratio K and a high wall friction μ. Thus, the single values of properties in old tables were not safe in design, and the standards were modified by adding an additional factor to the vertical force developing in the wall. In the same way, the total load on a hopper is greatest when the vertical force in the vertical wall is smallest, which occurs with a low lateral pressure ratio K and low wall friction μ. This was also accommodated in early standards by increasing the bottom force by a factor to allow a single value of each material property to be used.

Now that more potential failure modes in silos are understood, and the differing variability of different stored solids is appreciated, it is appropriate to try to define the upper and lower limits of each property value. As a result, most of the empirical additional factors can be removed from the design process, and safe design for specifically defined different extreme materials can be undertaken instead. In EN 1991-4 (2007), a central value for each property is listed, and it is then either multiplied or divided by a 'conversion factor' a to achieve upper and lower extremes. The conversion factor represents the scatter of values that particular solid may display.

The extreme values of particular properties are termed 'characteristic values' in structural design and are intended to correspond to a 10% or 90% probability of occurrence. The characteristic values that should be used in structural design calculations are shown in Table 3.1 (taken from Rotter 2001a).

Most standards for silo structural design (AS 3774 1996; DIN 1055-6 2006; EN 1991-4 2007) now acknowledge the variability of the properties of bulk solids and permit the variability of each solid in its own setting to be determined by testing. A formal methodology for establishing the variability of a given solid is given in Annex C of EN 1991-4 (2007).

Table 3.1 Values of properties for different wall loading assessments.

Purpose:	Characteristic value to be adopted: Wall friction coefficient (μ)	Lateral pressure ratio (K)	Angle of internal friction (ϕ_i)
For the vertical wall or barrel			
Maximum normal pressure on vertical wall	Lower	Upper	Lower
Maximum frictional traction on vertical wall	Upper	Upper	Lower
Maximum vertical load on hopper or silo bottom	Lower	Lower	Upper
Purpose:	**Wall friction coefficient (μ)**	**Hopper pressure ratio (F)**	**Angle of internal friction (ϕ_i)**
For the hopper wall			
Maximum hopper pressures on filling	Lower value for hopper	Lower	Lower
Maximum hopper pressures on discharge	Lower value for hopper	Upper	Upper

Note 1: It should be noted that $\phi_{wh} \leq \phi_i$ always, since the material will rupture internally if slip at the wall contact demands a greater shear stress than the internal friction can sustain. This means that, in all evaluations, the wall friction coefficient should not be taken as greater than $\tan\phi_i$ (i.e. $\mu = \tan\phi_w \leq \tan\phi_i$ always).
Note 2: Hopper normal pressure p_n is usually maximised if the hopper wall friction is low because less of the total hopper load is then carried by wall friction. Care should be taken when choosing which property extreme to use for the hopper wall friction to ensure that the structural consequences are fully explored (i.e. whether friction or normal pressures should be maximised depends on the kind of structural failure mode that is being considered).

3.4 Pressure changes during discharge of solids (emptying)

3.4.1 *First discoveries and explanations*

In some of the earliest experiments (Ketchum 1907) it was discovered that the pressures often increased when the silo was emptied. The increase was not often to a fixed value, but the pressures tended to rise and fall with time. Increases ranged from perhaps 10 to 30% as stable values, whilst very short-term local rises were seen to perhaps 2 or 3 times the Janssen value. Since the concept being used was that the Janssen theory gave the first measure of silo effects, it was natural to think that there was a 'pressure' at every level, so that this single pressure could be measured using a single pressure cell. Thus, the high pressures were imagined to occur as symmetrical high pressures at every point where they were observed.

Some effort went into trying to understand why these high pressures might occur, but the key idea came from Nanninga (1956) who suggested that the solid was in an active Rankine state after filling (higher vertical pressures than horizontal) and that during emptying it must be in a passive state (declining vertical pressures whilst the horizontal ones were retained). The transition between these two states would lead to a rapid increase in the value of K, whilst the vertical stress, in equilibrium across this change, would remain constant.

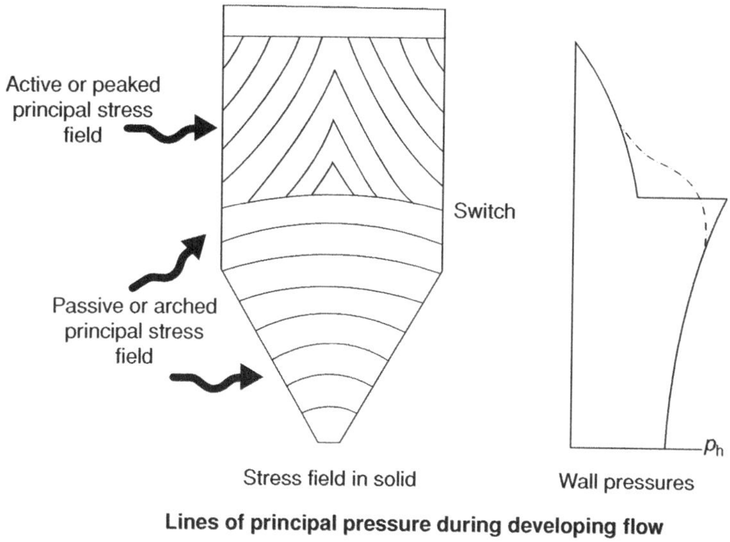

Figure 3.12 Original concept of the 'switch' during emptying. (After Gaylord & Gaylord 1984.)

Nanninga (1956) suggested that the changeover might occur over a finite depth (Figure 3.12), but later theorists who took up the idea (Arnold & McLean 1976; Jenike *et al.* 1973; Walker 1966; Walters 1973) made the change into an abrupt step. This step was termed the 'switch'.

Since the state was to pass from filling (close to an active stress state) to passive, the pressure just below an abrupt step is easily determined as the Janssen value multiplied by the ratio of passive to filling values of lateral pressure ratio (K_p/K_f). The ratio of peak symmetrical discharge pressure to symmetrical filling pressure is a very widely used variable, and its origins can be seen here to have some foundation in mechanics. This ratio is so important in silo design that it is given a symbol and defined as

$$C_e = \frac{p_e}{p_f} \tag{3.34}$$

in which p_f is the normal wall pressure after filling and during storage (taken as the Janssen pressure) and p_e is the design value of the symmetrical pressure (uniform at a given height in the silo) occurring during emptying (discharge). The above description leads to $C_e = K_p/K_f$.

It was noted above that the ratio K_p/K_a for a typical bulk solid is of the order of 9, making K_p/K_f of the order of 6. No observations of such huge increases in pressure were ever reported, so several theories were advanced which tried to explain why the switch from active to passive could produce lesser increases in pressure. The revised theories (Arnold *et al.* 1980; Jenike *et al.* 1973; Walker 1966; Walters 1973) showed that the stress pattern in the solid, involving non-uniform vertical stresses and shear stresses against the wall, could lead to rather smaller wall pressure increases. The Walker and Walters treatments relied on the solid being in a fully plastic (yielding in shear) state at all times, whilst the Jenike treatment assumed that it was elastic. Typical examples of the resulting pattern of wall

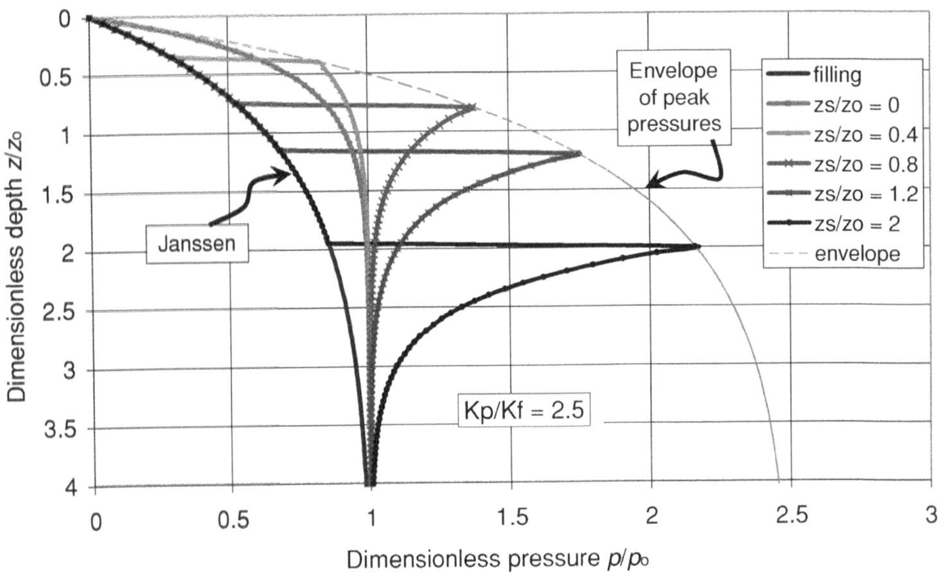

Figure 3.13 Consequences of a 'switch' in lateral pressure ratio at different levels (the switch is taken to occur at depth z_s).

pressures are shown in Figure 3.13, where it is supposed that the ratio K_p/K_f is only 2.5. Conventional wisdom, following the simple structural theory set out in Section 3.3.5, said that the design must accommodate the envelope of pressures corresponding to the maximum pressure applied at every level.

All these attempts still led to large predicted pressure increases during emptying, and for a while it was accepted that very large increases in symmetrical wall pressures must occur and should be designed for. A strange aspect of this idea was that, although many silo failures did occur, few silos failed by bursting, which is what would have been expected if the theories were accurate.

It may be noted that the pressure always returns to the Janssen asymptotic value p_0 below the 'switch'. The increase in the axial force developing in the wall is much smaller (Figure 3.14) because the switch only affects the frictional shear transfer locally.

The most widely used switch theory for vertical walls was that of Jenike *et al.* (1973), which still underlies the flow pressure rules in the Australian Standard AS 3774 (1996), leading to a high ratio of design pressures for discharge to those after filling. This type of theory is still commonly expounded (Drescher 1991) as a formal part of silo pressure behaviour.

3.4.2 *A better understanding*

The chief difficulty with the switch theory is its abrupt change from the filling pressure ratio to the discharge value. If a smoother change, based on test data in K_0 tests on solids, is used (Rotter 1999), much smaller rises in symmetrical pressure are found as the peak is

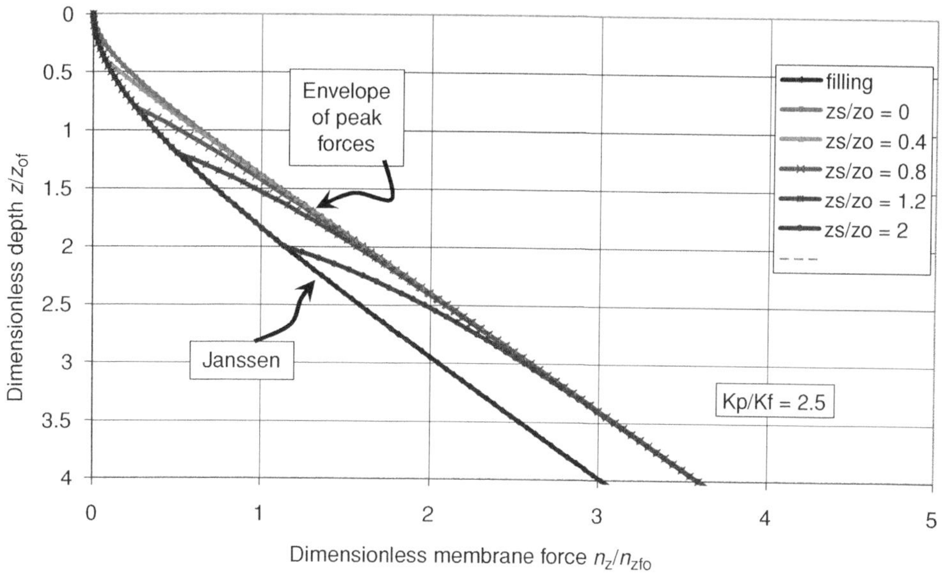

Figure 3.14 Smaller rises in the vertical force in the wall beneath a 'switch'.

rounded by the slow change (Figure 3.15). Here a progressive change in K from $K_f = 0.5$ to $K_e = 1.4$ (a ratio of $K_e/K_f = 2.8$) is assumed to occur between the heights $z/z_0 = 0.6$ and 0.75. The resulting changes in the mean vertical stress q/q_0, the mean wall pressure p/p_0 and the emptying factor C_e are shown in Figure 3.15 with the assumed ratio K/K_f at each level. Because the change is progressive (as originally suggested by Nanninga), the rise in pressure from filling to emptying is only a factor of 1.5 instead of 2.8 (i.e. the step change greatly exaggerated this phenomenon). The same analysis yields similar results for different locations of this change and thus leads to the conclusion that, although the stress field must undoubtedly change from the filling to emptying states, the magnitude of the symmetrical rise in pressure is greatly overpredicted by these simple switch theories. The European Standard (EN 1991-4 2007) consequently prescribes much smaller increases in symmetrical pressure during emptying (C_e values) than these older theories propose.

3.4.3 *Pressure observations during emptying*

Many experiments have been conducted to explore the pressures on silo walls during emptying. The data from these experiments are extremely voluminous: it is difficult for researchers to report very large quantities of data in publications. As a result, only what is judged to be the most important information is documented. A huge experimental programme on many different solids was conducted by Pieper and his team (Pieper & Wenzel 1964) in Braunschweig, and much of the following comes from their work. Unfortunately, some simplifications that they used, appropriate at that time, have been used by others for much longer than they might have wished.

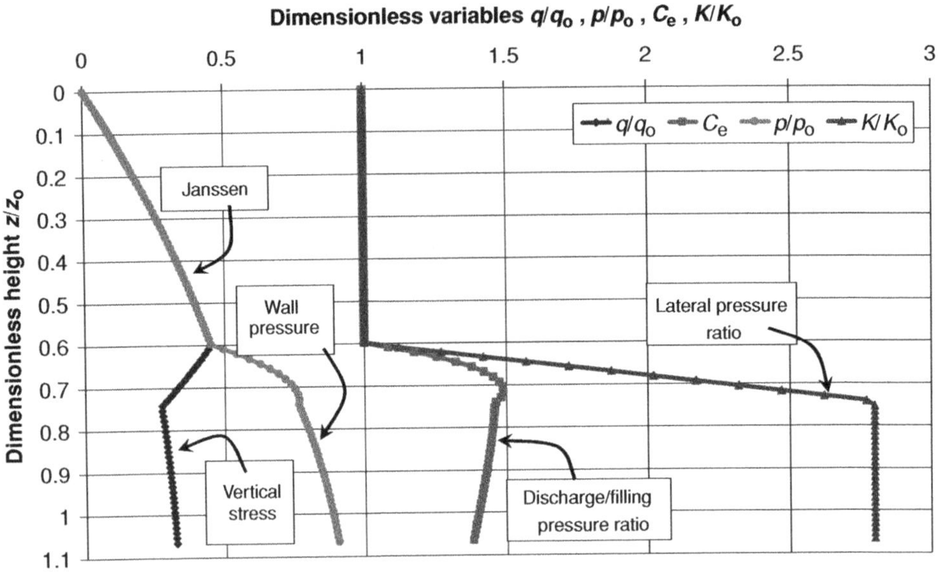

Figure 3.15 Effect on pressures of a smooth change in lateral pressure ratio.

A typical set of observations from pressure cells on the side of a tall silo subject to concentric filling and discharge and containing sand is shown in Figure 3.16, where the pressure reading is plotted against time during the test.

The lowest pressure cell, A, is the first to register pressure (at 2 min), and the pressure rises rapidly towards the Janssen asymptote. The other cells progressively start to register

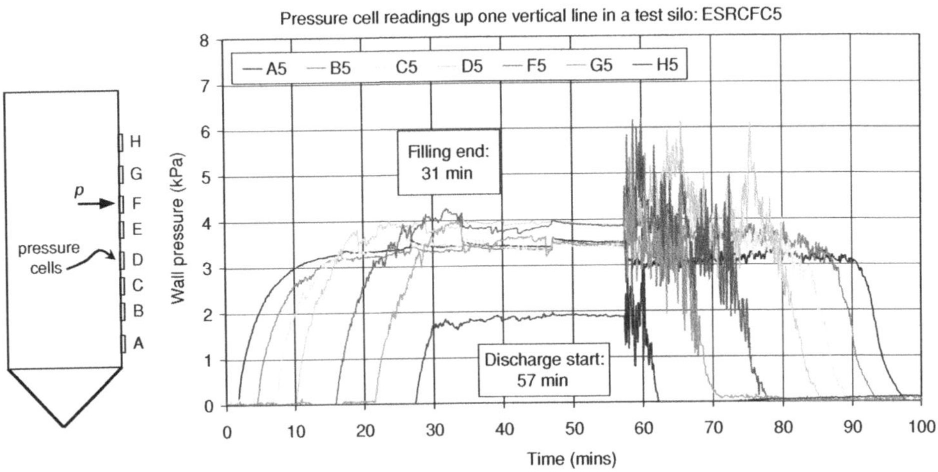

Figure 3.16 Typical pressure cell record on a vertical line of cells in a test silo.

as the silo is filled past the level they are at. It is interesting to note that at the end of filling (31 min) the pressures are not in a neat order with the lowest cell registering the highest pressure, but a little jumbled, indicating that the Janssen theory is not a precise tool like pressures in a fluid vessel, but an approximate description. After the filling process ends, the pressures are relatively stable, but not completely constant because small settlements and minor disturbances cause small increases and decreases in different places at different times.

At the instant that the discharge gate is opened (57 min) all the pressure cells begin to fluctuate quite wildly, with pressures rising for short moments to as much as twice the filling value but also falling to very low values. The largest departures from the filling state occur relatively high up the silo wall at levels D and F with the cell at F once touching 6.2 kPa from a Janssen reference value of 3.6 kPa (ratio of 1.7). But there is no evidence of a wave of high pressure passing up the silo as the stress field passes from filling to passive, and the switch theory of silo pressures on vertical walls, at least in its original form, is probably not widely believed any more.

Many silo pressure researchers, when faced with such voluminous data as this which is clearly not easily assimilated, have tried to find values that can be reported as relevant to the discharge condition, and it is quite natural that the highest pressure occurring on each pressure cell should be reported, irrespective of whether these values occurred simultaneously and whether they endured very long. Thus, the literature has many reports of major departures from the filling state, but the significance of these departures is highly questionable. The classic interpretation process is illustrated in Figure 3.17, where different cells reach peak pressures at different instants, the envelope of these peak pressures is represented as the outcome of the test, and a Janssen envelope is fitted to cover the outcome so that the result can be reduced to a single overpressure factor C_e. Alternatively, revised values of K and μ could be given to represent the emptying process (e.g. DIN 1055-6 1964). Many of the difficulties with such simplified interpretations were discussed by Rotter *et al.* (1986): in particular, the most damaging instant for the silo structure is not detected or encompassed by this process.

One must not be too unkind to the researchers who reported these experiments. The instrumentation is very expensive, so most tests were conducted with relatively few pressure cells. Faced with the challenge of where to place their few cells, most experimentalists were persuaded by the above theories that placement down a vertical line on the side of the silo would deliver the pattern of pressure to be expected, naturally a constant value at each level. Consequently, the information concerning variation of pressure at a particular level is rather sparse.

A further reason for using only one pressure cell at each level was that the simple theory used to translate pressures into forces in the structure (Equations (3.9) and (3.10)) implied that only the largest pressure needed to be found, and presumably that large pressure might well pass by every point at a particular level, even if not quite simultaneously.

The pressures recorded at different points around the circumference in the same test as in Figure 3.16 are shown for one level in Figure 3.18. First, it is clear that the pressures after filling are not at quite the same value at one level. Second, the rises and falls in pressure at different points around the circumference are not coincident, but lead to significantly unsymmetrical patterns at different instants. A detail taken from Figure 3.18 is shown in Figure 3.19.

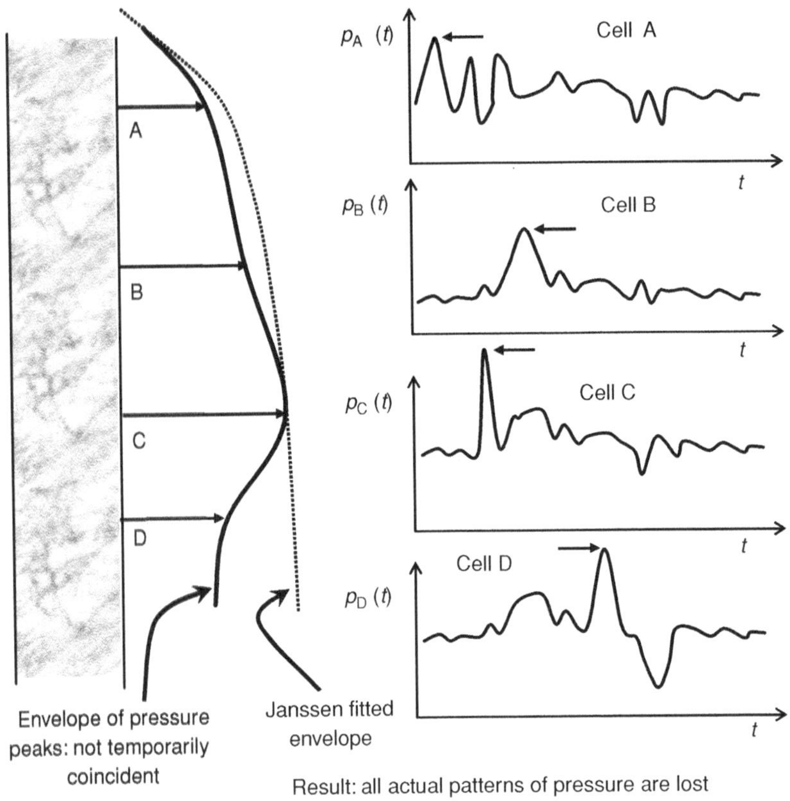

Figure 3.17 Typical interpretation process applied to pressure observations.

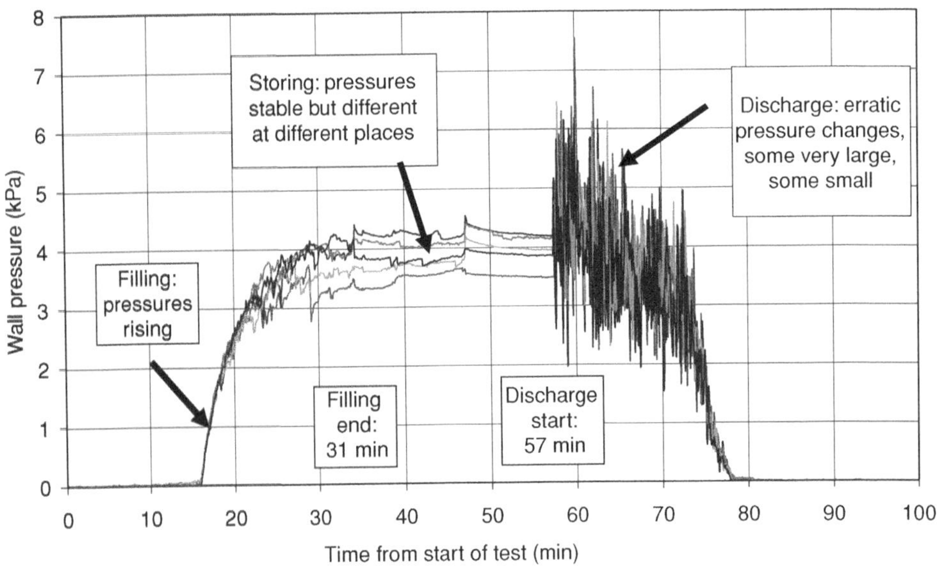

Figure 3.18 Typical pressure cell record at a single level in a test silo.

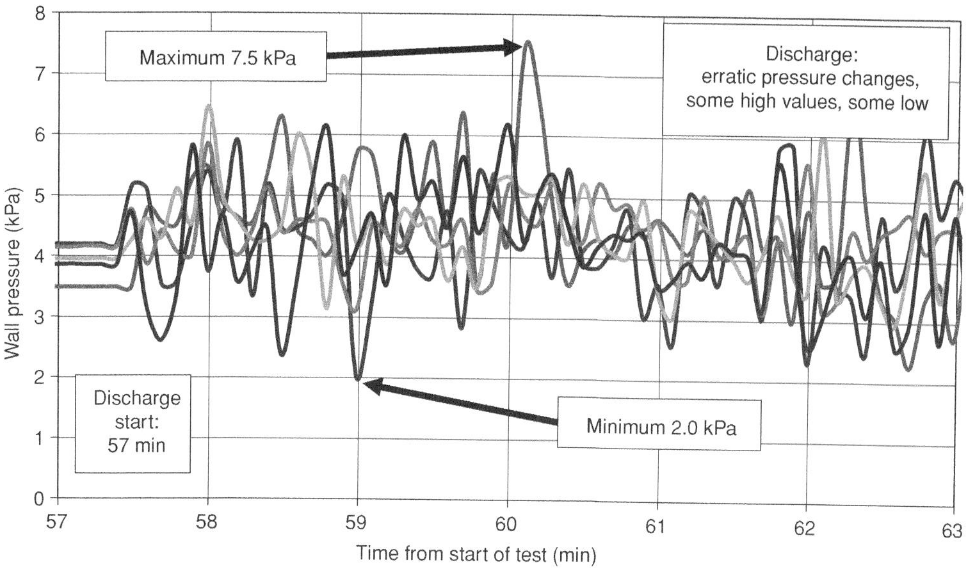

Figure 3.19 Detail of Figure 3.18 showing local rises and falls in pressure.

The key factor here is that unsymmetrical pressures are often damaging to cylindrical silo structures, whether constructed in metal or concrete, and this effect is more important than the possible peak pressure occurring at one point. In particular, if the peak pressure only occurs at one point around the circumference, then the pressures are necessarily unsymmetrical and the worst aspect of this load case is not the simple relationship between normal pressure and circumferential (hoop) tension of Figure 3.11.

In the context of the above, a key set of experiments on full-scale silos was conducted in Sweden over many years (1970–1980) by Nielsen and his co-workers (Nielsen 1998). This project arose because of the extensive cracking which had been observed in many Swedish grain silos. The experiments involved a 47 m high concrete silo of internal diameter 7 m, filled with different grains in different experiments with both concentric and eccentric filling and discharge. This huge set of experiments demonstrated many effects that are not included in any silo design, notably the progressive changes in the properties of the stored solid as it was handled, the sensitivity of pressures to anisotropic packing of the particles, the effects of imperfections in the silo walls, the difficulty of making reliable observations with pressure cells, and the fact that two pressure cells close to each other might, for limited periods, record quite different values, indicating that there can be sharp jumps in pressure on the wall.

The most critical finding for silo design was the systematic pattern of unsymmetrical pressures, both after filling and during discharge (Ooi *et al.* 1990). The ratio of the largest sustained pressure to the smallest at a single level could be as high as 2.8 under static conditions after filling and 5.6 during discharge. This kind of discovery was also made by Schmidt and Stiglat (1987) and led to the introduction of a required unsymmetrical design pressure, called a 'patch load' in the German standard (DIN 1055-6 1987). The latest version of this patch load treatment is given in EN 1991-4 (2007) where the patch load depends on the filling or discharge state, the silo aspect ratio, the eccentricities of filling and discharge and the construction medium.

The consequences of unsymmetrical pressure patterns are noted further in Section 3.5.

3.4.4 *The importance of flow patterns during discharge*

The discussion above concerning pressures during emptying has omitted a key aspect that became very clear during the 1960s and 1970s. The manner in which a solid flows within the silo has a major effect on the pressures exerted on the silo wall.

If the entire mass of solid in the silo is in motion, then it slides against the wall, producing the effect seen in Figures 3.16, 3.18 and 3.19, and the local pressure can be much influenced by variations in the straightness of the wall and its local roughness. By contrast, when the solid against the wall is at rest, the pressures generally remain close to the Janssen filling values. The work of Jenike (1961, 1964) was probably the main driver towards explicit recognition of the importance of 'flow pattern' of the solid. A modern description (EN 1991-4 2007) divides the possible flow patterns into three main categories under symmetrical conditions (Figure 3.20).

These images show an idealised version of the pattern of flow. The real boundaries of flow channels often vary a little from time to time because they depend quite sensitively on small changes in the packing of particles (Arnold 1991). Further, the idealised pattern is shown with the silo completely full, but the pattern cannot develop until some solid has come out at the bottom (unless it is being continuously replenished). However, because the critical design condition is almost always when the silo is full, this is the idealised reference shape.

Following the work of Jenike (1961, 1964), it is possible to determine with reasonable precision whether the silo will exhibit mass flow or funnel flow. The conventional diagram is similar to that for hopper steepness and shows the boundary between mass flow and funnel flow (Figure 3.21) as a function of the hopper slope and wall friction coefficient. There are similar diagrams for wedge hoppers, for which mass flow is more easily achieved (EN 1991-4 2007; Rotter 2001a). This figure marks the mass flow zone as a 'risk' because the hopper pressures may be high only in this case. The boundary distinguishes between mass flow and other types of flow: it does not distinguish pipe flow from mixed flow, and this is one of the most serious current problems in silo pressure prediction. Unfortunately, there is, as yet, no reliable method of determining the shape of a mixed flow channel, or of reliably determining when it may strike the wall at an effective transition (Figure 3.20c).

The typical patterns of symmetrical pressure against the wall for the three simple patterns of flow are shown in Figure 3.22. Under mass flow (Figure 3.22a), the high pressure that develops at the top of the hopper (sometimes referred to as 'the switch') is caused by a high F (Figure 3.9), associated with the solid below this point being in a passive stress state. Much has been made of this high local pressure, but structural research studies have shown that it is not critical to the strength of metal silos, and is indeed beneficial (Rotter 1986a; Teng & Rotter 1991).

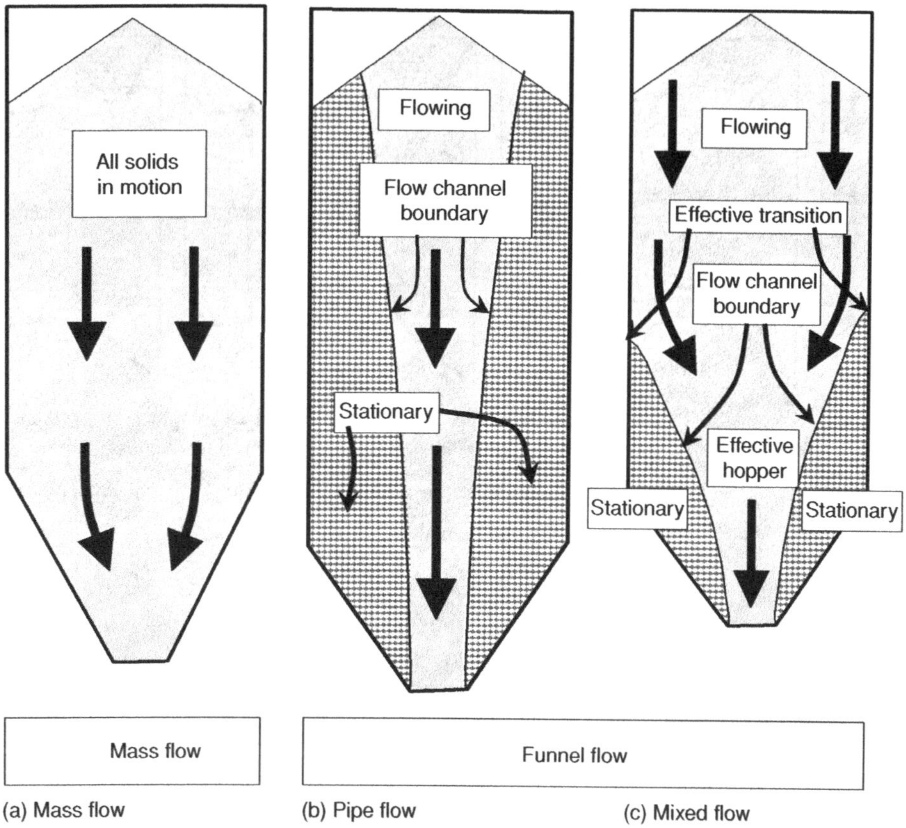

Figure 3.20 Chief categories of symmetrical flow pattern.

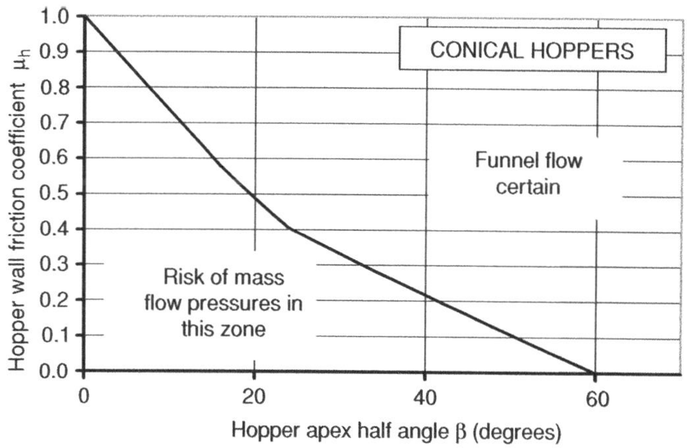

Figure 3.21 Boundary between mass flow and funnel flow in silos with conical hoppers. (After EN 1991-4 2007.)

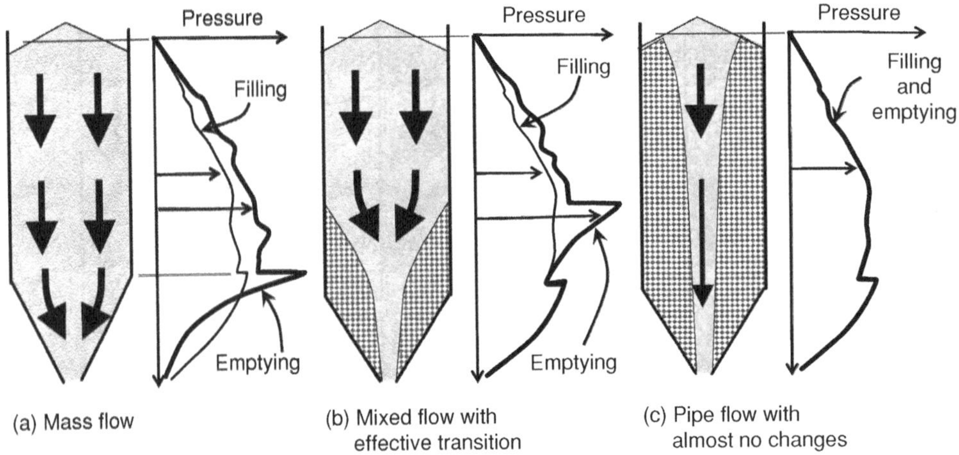

Figure 3.22 Typical patterns of average symmetric wall pressure after filling and during emptying, for different flow channel geometries.

Under pipe flow (Figure 3.22c), the pressures against the wall are largely unaffected by flow, so if the designer can be sure that no flow of solids against the wall will occur (except at the surface), lower design pressures are possible. However, under mixed flow (Figure 3.22b), the boundary of the flow channel strikes the wall and a local high pressure, comparable to that in a mass flow hopper, often develops against the wall. This pressure is somewhat unpredictable. It can vary in magnitude as the slope of the contact point changes, it can be unsymmetrical from one side to the other, it is slightly cushioned by the stored solid between the flowing solid and the wall, and in silo experiments, this is commonly the point of greatest scatter and oscillation in pressure values. Despite all of this, very few silos have ever failed by bursting at an effective transition, so this rather alarming knowledge should not be a major cause for concern.

Finally, it must be clearly repeated that it is not yet possible to predict the geometries of pipe flow and mixed flow solids flow patterns, so this rather critical distinction is not yet quantifiable. The distinction is therefore not used in the design rules of EN 1991-4 (2007).

3.4.5 *Eccentric discharge and its consequences*

The most damaging condition for most silos is the unplanned occurrence of unsymmetrical flow regimes, if the flow channel makes contact with the silo wall. This is conventionally referred to as eccentric discharge. It has caused so many silo disasters that many writers have proposed that it should never be used. But two situations arise: it may be necessary to have off-centre discharge outlets for functional reasons, and conditions in the silo (blockage of feeders, uneven thermal or moisture conditions, segregation of contents etc.) may cause unintended eccentric flow. There are numerous causes of such eccentricities.

This is a substantial subject and beyond the scope of this chapter, but EN 1991-4 now includes a simple definition of a design eccentric flow channel geometry and pressure regime which may be used to achieve a satisfactory design. The equations used adopt the theory of Rotter (1986b, 2001b). A circular silo with a part-circular flow channel in contact with the wall (Chen *et al.* 2005) is shown in Figure 3.23, together with the characteristic pressure distribution that is found in experiments. The vertical stresses induced in the wall by this unsymmetrical pattern are also shown to indicate the dramatically large effect on this silo. In particular, note that the highest compression stress occurs around the mid-height of the silo in the middle of the flowing channel.

Eccentric discharge pressures of the pattern shown in Figure 3.23 also have a very damaging effect on concrete silos, where severe bending of the wall induces substantial vertical cracks and sometimes leads to spalling.

3.5 Structural damage and its causes

3.5.1 *Introduction*

The simplest stress analysis of a cylindrical silo structure under symmetrical loads was presented above in Section 3.3.5. Unfortunately, this is often the only analysis that is applied, sometimes with unfortunate consequences for the structure. Metal and concrete silos carry their loads differently because metals are strong in tension but thin metal sections tend to buckle under compression. By contrast, concrete is very weak in tension, but can resist compression well. These aspects lead to different key design considerations.

Both metal and concrete silos are thin shell structures. Shell structures have more complex patterns of behaviour than any other structural form, they are more sensitive to small errors of geometry and they have more possible failure modes. As a result, it is common for designers to oversimplify the problem, and especially to misdiagnose the cause of structural damage. The subject is very large and only a brief outline is given here. More information may be found in Rotter (2001a) together with the Eurocodes on metal silos (EN 1993-4-1 2007) and shells (EN 1993-1-6 2007).

Shell structures tend to suffer serious effects when the pressure is not uniform at one level. A local drop of pressure can cause serious damage, of different kinds, in both metal and concrete structures. Where signs of damage are seen, possible causes of loss of symmetry should be the natural first investigation path to follow.

3.5.2 *Steel and aluminium silos*

3.5.2.1 *Bolted and welded construction* A first distinction must be made according to the form of joint that is used in metal silo construction. Many smaller steel silos have bolted joints, and where these are present, every stress developing in the wall, at every point, must be transmitted through a joint. The joints are lines of weakness, so they should be made stronger than is strictly necessary. Careful attention should be paid to edge distances, and it is most desirable that the weakest failure mode of the joint should be by bearing rather than bolt shear, since the latter is not very ductile and lack of fit in the joints may cause

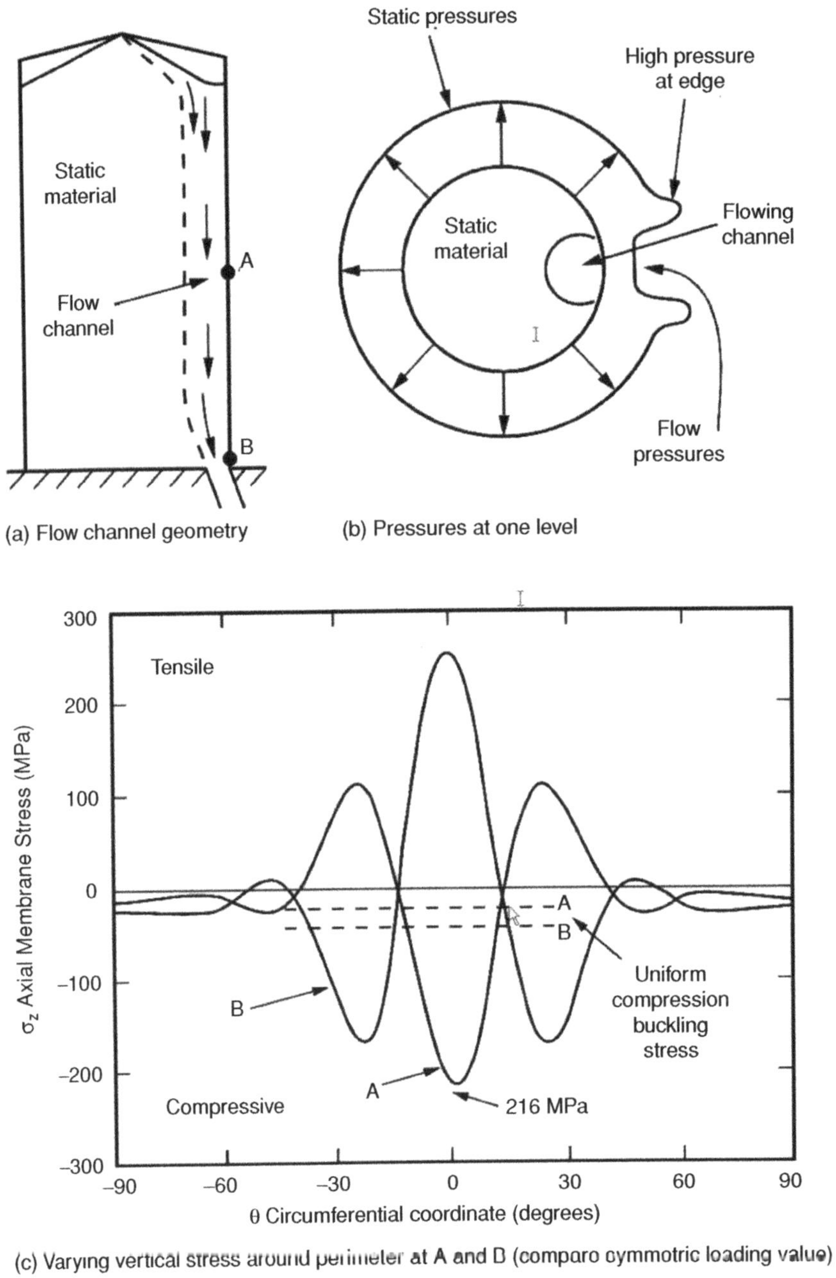

Figure 3.23 Flow channel geometry, typical pressure pattern and vertical wall stresses during eccentric discharge.

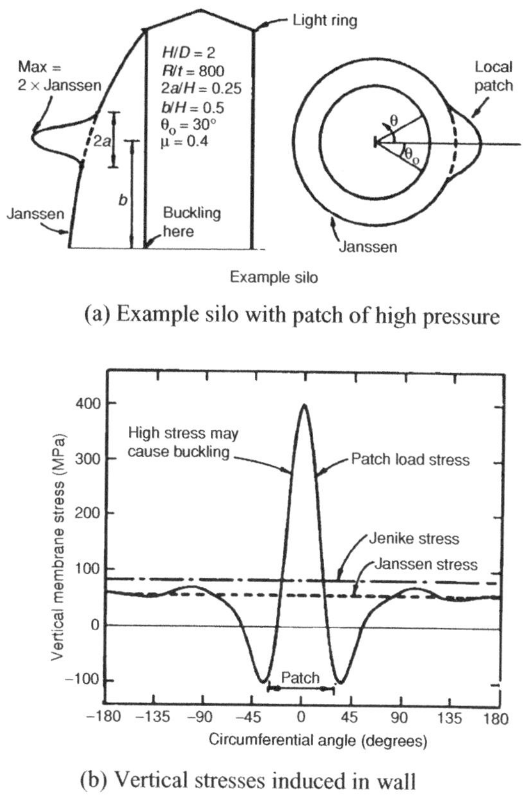

Figure 3.24 Stresses resulting from a patch of normal pressure on a thin silo.

unzipping of a complete joint from a single zone of slightly elevated pressure. Larger bolts in thin plates are more ductile than smaller bolts in thick plates. None of these problems arises in welded construction.

3.5.2.2 *Bursting of the vertical wall* Bursting failures are very uncommon and are almost all found in bolted silos where a joint detail has failed. A careful analysis of the loads and strengths in different modes shows that this failure mode is only critical near the surface, or in squat silos.

3.5.2.3 *Axial compression buckling of the vertical wall* Buckling of the vertical wall is by far the commonest failure mode in metal silos. The buckles can be huge or quite local, but all buckles should be treated as very serious because this mode of failure is often dramatically catastrophic.

Axial compression arises from the friction transmitted to the silo wall by the solids. But axial compression also develops as a result of unsymmetrical pressures against the silo wall, caused by shell bending phenomena, which cannot be explained within the space limits here. An example is shown in Figure 3.24 where a local small patch of pressure on the silo wall induces high vertical compression (not due to friction) far from the patch. In particular, a

(a) (b)

Figure 3.25 (a) Flyash silo with buckle arrested by hopper impacting the ground; (b) grain silo en route to total destruction as grain leaks from a buckled zone.

local loss of pressure can result in large increases in vertical compression stresses far above the point of pressure loss (Rotter *et al.* 2006). The location of the buckle is therefore not always a good guide to the location of the problem.

Buckling under axial compression occurs at very low stresses compared with the material strength (perhaps at 20 MPa in a metal with yield stress 250 MPa), and the strength is very sensitive to small errors of geometry. The post-buckling behaviour is also notoriously catastrophic. Two examples, where total destruction has not yet occurred, are shown in Figure 3.25. The buckles are relatively small, often with a characteristic diamond shape..

Under high internal pressures, a different form of axial compression buckle occurs, termed the 'elephant's foot' because of its smooth flat squashed shape. Also, where a buckle occurs adjacent to a support, a buckle may develop in the local high stress field, needing a more careful evaluation (the force being transmitted may not be easily determined).

3.5.2.4 *Eccentric discharge buckling of the vertical wall* A separate section is noted here for conditions of eccentric discharge. This is the commonest cause of axial compression buckles, where the low pressures against the wall in the flow channel cause high vertical compressive stresses over part of the perimeter near the mid-height of the silo (Figure 3.23). Extremely catastrophic failures are easily produced in tall silos, in which the whole silo falls over in the direction of the discharge outlet. The analysis of this problem can be found in Rotter (1986b, 2001b). The condition is often mistaken as being caused by bending moments in the wall (Jenike 1967; Wood 1983), and these moments are indeed present, but bending does not produce diamond pattern buckles. The complex behaviour of cylindrical shells under unsymmetrical loads is, unfortunately, not widely understood.

The evaluation of the buckling strength under different conditions is quite complicated and can be found in Rotter (2001a), EN 1993-4-1 (2007) or EN 1993-1-6 (2007).

3.5.2.5 *External pressure buckling of the vertical wall* When a silo is empty, the thin wall is very susceptible to buckling under extreme wind. The buckles associated with this loading tend to be much larger than those for axial compression, usually stretching either the whole height of the silo or from a plate thickness change up to the top. Similar buckles occur when a partial vacuum is induced by the discharge of solids of low permeability and the silo is inadequately vented. For advice, see EN 1993-4-1 (2007).

3.5.2.6 *Shear buckling of the vertical wall* Where a squat silo (low aspect ratio) is either eccentrically filled (unsymmetrical top pile producing different heights of solid-wall contact) or is subjected to seismic excitation, the wall can buckle in shear near the foundation. These buckles have a characteristic diagonal stripe shape, but these load cases are relatively rare.

3.5.2.7 *Rupture, plastic deformations and buckling in hoppers* Hoppers made in bolted construction are susceptible to fracture of the joint at the point where the structural stresses are most seriously mismatched with the joint strength. The pattern of stresses is not the same as the pattern of pressures, but in bolted hoppers it is important to adopt a correct pressure pattern so that these joints are well designed. Once a failure initiates, unzipping tends to occur, leading to catastrophic failure.

In welded hoppers, failure is much less likely in the hopper itself. Most failures occur near the top of the hopper, and are either rupture (the hopper is torn off, with unzipping passing around the perimeter) or plastic deformations. Both situations arise from an excessive total load on the hopper or from unsymmetrical pressures, not from a high 'switch' pressure at the transition. For design and evaluation advice, see Rotter (2001a) and EN 1993-4-1 (2007).

3.5.2.8 *Buckling and yielding in transition rings* The transition is subject to high compressions because the hopper has a sloping form. Both buckling and yielding failures can occur in these rings, but these situations are usually caused by a misunderstanding of the complex stresses in such rings especially near supports (thrust, bending, torsion and shell flexure), rather than any special event in the stored bulk solid.

3.5.3 *Concrete silos*

3.5.3.1 *General* Concrete is good in compression, but cannot resist tensile stresses at all. Unfortunately, silos are essentially structures in tension, holding in the stored solid.

When concrete is subjected to tension, it cracks at right angles to the tension. It is normal to reinforce concrete to carry the tensile forces, but this reinforcement cannot carry stresses without stretching (strains), and this same stretching causes the concrete to crack. Cracked concrete often permits ingress of moisture and may lead to degradation of the stored product. The simplest solution is to prestress the concrete with steel high strength strand, so that it is in compression before any load comes on it. Then when additional tensile stresses are induced in the wall by the stored solid, they simply reduce the pre-existing compression.

Vertical compression does not usually cause problems in concrete silos since the weight of concrete, the thickness and the good compressive strength all contribute to excellent strength.

3.5.3.2 *Cracking under bending moments* The chief problem for concrete silos is cracking under bending moments induced by unsymmetrical pressures, where a zone of low pressure occurs inside the silo, the wall bends inwards, cracking on the inside (possibly not visible without careful inspection), possibly with adjacent regions of cracking on the outside at the edges of that zone. To prevent serious cracking of this kind, all concrete silo walls must be designed with some significant bending strength, and this is arranged by using an inner and an outer layer of reinforcement and requiring the design to support unsymmetrical loads. In EN 1991-4 (2007), 'patch' loads are defined on the silo wall which are intended to produce similar bending moments in the walls to those that would be produced by the real unsymmetrical pressure patterns discussed above. However, these patch loads have not yet been properly calibrated against the outcome of tests on silo pressures, so the design magnitude is not yet very certain.

Where concrete silos are subject to eccentric discharge, the low pressures in the flowing solid cause reduced pressures against a limited part of the wall, and the primary effect of these is to induce vertical cracks associated with circumferential bending. However, the concrete silo is a shell structure, albeit thicker than the metal silos, and eccentric discharge has been shown (Rotter 2001c) also to cause cracking in the roof and severe damage to internal structures simply because the effects of the flow channel low pressure are transmitted throughout the whole structure.

3.5.3.3 *Crack observations* As noted above, cracks in concrete are at right angles to the principal tensile stress, so the orientation of cracks gives a good indication of the stress state in the wall. Since it is usually only the outside surface that can be observed, care must be taken to determine whether the cracks are caused by through-thickness tension (very serious) or external surface tension caused by bending. Diagonal cracks may, for example, indicate a flow channel of widening dimensions inside the silo.

3.5.3.4 *Ductility and delamination* Concrete is a brittle material, but most structural design relies on the assumption that the structure behaves in a ductile manner. Concrete structures achieve this by appropriate reinforcement, but where forces are applied to the structure that were not planned for in the design, brittle failures can occur. In particular, shear failures in concrete walls can cause serious cracking. Another brittle problem is that of delamination, where splitting occurs along the plane of the reinforcement. This generally occurs when the concrete is under high compressive stresses.

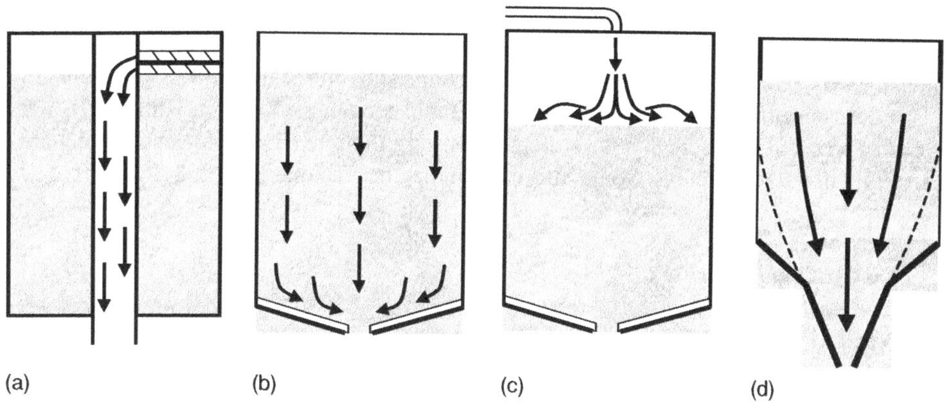

Figure 3.26 (a) Mechanical discharge with concentric pressures; (b) air injection and air slides promote mass flow; (c) pneumatic filling of powders causes almost flat top surface; (d) expanded flow hopper gives mass flow only in the bottom hopper.

3.5.3.5 *Durability considerations* Reinforcement in concrete structures must be protected from corrosion, and the conventional manner of doing this is to have a suitable thickness of concrete 'cover' over the steel. Where large cracks are able to develop in the concrete wall, the protective effect of this cover can be lost, and a significant loss of the area of reinforcement may occur. This leads to a dramatic loss of strength and has caused failures (Elghazouli & Rotter 1996).

3.6 Design situations

There are many different special circumstances that can occur in silos that need special attention. Several are specifically identified in EN 1991-4 (2007), but even these require an extensive description for a full explanation. However, a few are briefly noted here so that the reader can seek further information where it is needed.

The aspect ratio of the silo is a key determinant of conditions, as noted in Figure 3.3. Where silos have an internal system of discharging (Figure 3.26a), only filling pressures need to be considered, so simpler safe designs are possible. Where air slides are used in silos containing powders that can be fluidised (Figure 3.26b), the flow pattern will be mass flow irrespective of the indications of Figure 3.21: mass flow pressure conditions must be assumed. Where powders are filled in a condition such that they are fluidised on deposition, it should be assumed that the top surface will be flat (Figure 3.26c): this matters where the silo is relatively squat. Where an expanded flow hopper is used (Figure 3.26d), the bottom part of the hopper is subject to mass flow hopper pressures, but the upper part of the hopper may be shallow and the base of the cylinder experiences mixed flow, so proper account should be taken of this.

Internal structures within silos (tubes to assist flow, flow promotion devices such as Chinese hats and cone-within-cone structures, etc.) may be subject to large forces from stored solids. Some advice on these may be found in the Australian Standard (AS 3774 1996).

Finally, where silos may be subjected to seismic loads, much care is needed. In elevated silos, a huge mass is supported on a relatively soft spring, leading to a low natural frequency which is easily excited by seismic waves. In on-ground silos, vertical compressions and high shear forces develop in the walls due to the horizontal excitation (Rotter & Hull 1989), and care must be taken to ensure that the structure is strong enough, but also to ensure adequate connection in the base details. Some information may be found in EN 1998-4 (2006).

3.7 Concluding remarks

This chapter has given a brief outline of the key aspects of silo pressure phenomena and their implications for potential damage to silo structures. It is evident that the subject is large and requires much more detailed treatment on many issues than is possible here. However, many references to other useful sources have been given.

Our understanding of silo pressures and their consequences for storage structures is continually expanding, sometimes as a result of new catastrophes. As a result, current advice and standards are likely to be steadily improved, and better treatments should be available for many of the questions that were imperfectly answered here. The reader is invited to seek specialist advice when new problems are encountered.

Acknowledgements

The author wrote this chapter whilst in Hong Kong as a Royal Society Kan Tong Po Visiting Professor. The author is most grateful to the Royal Society, the Kan Tong Po fund and the Hong Kong Polytechnic University for their generous support. The chapter has drawn on understandings gained in extensive discussions with Dr Jørgen Nielsen and Prof J.G. Teng, and their contributions are gratefully acknowledged.

References

Arnold, P.C. (May 1991) On the influence of segregation on the flow patterns in silos. *Bulk Solids Handling*, **11**(2), 447–449.

Arnold, P.C. & McLean, A.G. (November 1976) Prediction of cylinder flow pressures in mass-flow bins using minimum strain energy. *J. Eng. Ind., Trans ASME, Ser. B*, **98**(4), 1370–1374.

Arnold, P.C., McLean, A.G. & Roberts, A.W. (1980) *Bulk Solids: Storage, Flow and Handling, Tunra Bulk Solids Handling Research Associates*, 2nd edn. University of Newcastle, NSW, Australia.

AS3774 (October 1996) *Loads on Bulk Solids Containers, Australian Standard*. Standards Association of Australia, Sydney.

Chen, J.F., Rotter, J.M., Ooi, J.Y. & Zhong, Z.J. (June 2005) Flow pattern measurement in a full scale silo containing iron ore. *Chem. Eng. Sci.*, **60**(11), 3029–3041.

Dabrowski, A. (1957) *Parcie Materialow Sypkich w Leju* (Pressures from Bulk Solids in Hoppers). Archiwum Inzynierii Ladowej, Warsawa, pp. 325–328.

DIN 1055-6 (1964, 1987, 2006) *Design Loads for Buildings: Loads in Silo Containers*. DIN 1055, Part 6, Deutsches Institut für Normung, Berlin.

Drescher, A. (1991) *Analytical Methods in Bin-Load Analysis*. Elsevier, New York.

Elghazouli, A.Y. & Rotter, J.M. (February 1996) Long-term performance and assessment of circular reinforced concrete silos. *Construction Build. Mater.*, **10**(2), 117–122.

EN 1991-4 (2007) *Eurocode 1: Basis of Design and Actions on Structures, Part 4 – Silos and Tanks*. CEN, Brussels.

EN 1993-1-6 (2007) *Eurocode 3: Design of Steel Structures, Part 1.6: General Rules – Strength and Stability of Shell Structures*. CEN, Brussels.

EN 1993-4-1 (2007) *Eurocode 3: Design of Steel Structures, Part 4.1: Silos*. CEN, Brussels.

EN 1998-4 (2006) *Eurocode 8: Design Provisions for Earthquake Resistance of Structures – Part 4: Silos, Tanks and Pipelines*. CEN, Brussels.

Gaylord, E.H. & Gaylord, C.N. (1984) *Design of Steel Bins*. Prentice Hall, New Jersey.

Jaky, J. (1948) Pressures in silos. *Proceedings of the 2nd International Conference on Soil Mechanics and Foundation Engineering*, Rotterdam, Vol. 1, pp. 103–107.

Janssen, H.A. (1895) Versuche uber Getreidedruck in Silozellen. *Z. des Vereines Dtsch Ingenieure*, **39**(35), 1045–1049.

Jenike, A.W. (1961) *Gravity Flow of Bulk Solids*. Bulletin of the University of Utah, Vol. 52, No. 29, Bulletin No. 108 of the Utah Engineering Experiment Station, Salt Lake City, Utah.

Jenike, A.W. (1964) *Storage and Flow of Solids*. Bulletin of the University of Utah, Vol. 53, No. 26, Bulletin No. 123 of the Utah Engineering Experiment Station, Salt Lake City, Utah (revised November 1976).

Jenike, A.W. (February 1967) Denting of circular bins with eccentric drawpoints. *J. Struct. Div., ASCE*, **93**(ST1), 27–35.

Jenike, A.W., Johanson, J.R. & Carson, J.W. (February 1973) Bin loads – Parts 2, 3 and 4: concepts, mass flow bins, funnel flow bins. *J. Eng. Ind., Trans. ASME*, **95**(1, Series B), 1–5, 6–12, 13–16.

Ketchum, M.S. (1907) *Design of Walls, Bins and Grain Elevators*, 1st edn. McGraw-Hill, New York (2nd edn, 1911; 3rd edn, 1919).

Koenen, M. (1895) Berechnung des Seitenund Bodendrucks in Silos. *Zentralbl. Bauverwaltung*, **16**, 446–449.

Muir Wood, D. (1990) *Soil Behaviour and Critical State Soil Mechanics*. Cambridge University Press, Cambridge, England.

Nanninga, N. (November 1956) Gibt die ubliche Berechnungsart der Drucke auf die Wande und den Boden von Silobauten Sichere Ergebnisse. *Die Ingenieur*, **68**(44).

Nielsen, J. (1998) Pressures from flowing granular solids in silos. *Phil. Trans. R. Soc., Lond. A*, **356**(1747), 2667–2684.

Ooi, J.Y., Rotter, J.M. & Pham, L. (1990) Systematic and random features of measured pressures on full-scale silo walls. *Eng. Struct.*, **12**(2), 74–87.

Pieper, K. & Stamou, K. (March 1981) Lasten in Niedrigen Silos. Lehrstuhl für Hochbaustatik, Technische Universität Braunsweig, 94 pp.

Pieper, K. & Wenzel, F. (1964) *Druckverhältnisse in Silozellen*. Wilhelm Ernst und Sohn, Berlin.

Rankine, W.J.M. (1857) On the stability of loose earth. *Phil. Trans. R. Soc., Lond.*, **147**, 9.

Rotter, J.M. (1986a) On the significance of switch pressures at the transition in elevated steel bins. *Proceedings of the Second International Conference on Bulk Materials Storage Handling and Transportation, Institution of Engineers*, Wollongong, Australia, pp. 82–88.

Rotter, J.M. (1986b) The analysis of steel bins subject to eccentric discharge. *Proceedings of the Second International Conference on Bulk Materials Storage Handling and Transportation*, Institution of Engineers, Wollongong, Australia, pp. 264–271.

Rotter, J.M. (1999) Flow and pressures in silo structural integrity assessments. *Proceedings of the International Symposium: Reliable Flow of Particulate Solids III*, Porsgrunn, Norway, August 1999, pp. 281–292.

Rotter, J.M. (2001a) *Guide for the Economic Design of Circular Metal Silos*. Spon, London.

Rotter, J.M. (2001b) *Pressures, Stresses and Buckling in Metal Silos containing Eccentrically Discharging Solids*. Festschrift Richard Greiner, TU Graz, Austria, pp. 85–104.

Rotter, J.M. (2001c) *Report on Damage to the Raw Meal Silo at Cebu*. Technical Investigation Report C01-05, School of Civil and Environmental Engineering, University of Edinburgh.

Rotter, J.M. & Hull, T.S. (1989) Wall loads in squat steel silos during earthquakes. *Eng. Struct.*, **11**(3), 139–147.

Rotter, J.M., Ooi, J.Y. & Zhong, Z. (2006) Critical pressure conditions in silos. *Proceedings of the 5th International Conference for Conveying and Handling of Particulate Solids*, Sorrento, Italy, 27–31 August, 6 pp.

Rotter, J.M., Pham, L. & Nielsen, J. (1986) On the specification of loads for the structural design of bins and silos. *Proceedings of the Second International Conference on Bulk Materials Storage Handling and Transportation*, Institution of Engineers, Australia, Wollongong, July 1986, pp. 241–247.

Schmidt, K.H. & Stiglat, K. (1987) Anmerkungen zur Bemessungslast von Silos. *Beton und Stahlbetonbau*, **9**, 239–242.

Teng, J.G. & Rotter, J.M. (1991) The strength of welded steel silo hoppers under filling and flow pressures. *J. Struct. Eng., ASCE*, **117**(9), 2567–2583.

Walker, D.M. (1964) *A Theory of Gravity Flow of Cohesive Powders*. Central Electricity Generating Board, UK, SW Region, R&D Dept, Report No 22.
Walker, D.M. (1966) An approximate theory for pressure and arching in hoppers. *Chem. Eng. Sci.*, **21**, 975–997.
Walters, J.K. (1973) A theoretical analysis of stresses in silos with vertical walls. *Chem. Eng. Sci.*, **28**, 13–21.
Wood, J.G.M. (1983) The analysis of silo structures subject to eccentric discharge. *Proceedings of the Second International Conference on Design of Silos for Strength and Flow*, Stratford-upon-Avon, pp. 132–144.

4 Pneumatic conveying

DAVID MILLS AND MARK JONES

4.1 Introduction

Pneumatic conveying systems are basically quite simple and are ideally suited for the transport of powdered and granular materials in factory, site and plant situations. The system requirements are a source of compressed gas, usually air, a feed device, a conveying pipeline and a receiver to disengage the conveyed material and carrier gas. The system is totally enclosed, and if it is required, the system can operate entirely without moving parts coming into contact with the conveyed material.

High, low or negative pressures can be used to convey materials. For hygroscopic materials dry air can be used, for toxic materials a closed loop system can be used and for potentially explosive materials an inert gas such as nitrogen can be employed. A particular advantage is that materials can be fed into reception vessels maintained at a high pressure if required.

4.1.1 *System flexibility*

With a suitable choice and arrangement of equipment, materials can be conveyed from a hopper or silo in one location to another location some distance away. Considerable flexibility in both plant layout and operation is possible, such that multiple point feeding can be made into a common line, and a single line can be discharged into a number of receiving hoppers.

With vacuum systems, materials can be picked up from open storage or stockpiles, and they are ideal for clearing dust accumulations and spillages. Pipelines can run horizontally, as well as vertically up and down, and with bends in the pipeline any combination of orientations can be accommodated in a single pipeline run. Material flow rates can be controlled easily and monitored to continuously check input and output, and most systems can be arranged for completely automatic operation.

Pneumatic conveying systems are particularly versatile. A very wide range of materials can be handled and they are totally enclosed by the system and pipeline. This means that potentially hazardous materials can be conveyed quite safely. There is minimal risk of dust generation and so these systems generally meet the requirements of any local Health and Safety legislation with little or no difficulty.

Pneumatic conveying plants take up little floor space and the pipeline can be easily routed up walls, across roofs or even underground to avoid any existing equipment or structures. Pipe bends in the conveying line provide this flexibility, but they will add to the overall resistance of the pipeline. Bends can also add to problems of particle degradation if the conveyed material is friable and suffer from erosive wear if the material is abrasive.

4.1.2 *Industries and materials*

A wide variety of materials are handled in powdered and granular form, and a large number of different industries have processes that involve their transfer and storage. Some of the industries in which bulk materials are conveyed include agriculture, mining, chemicals, pharmaceuticals, paint manufacture and metal refining and processing. In agriculture very large tonnages of harvested materials such as grain and rice are handled, as well as processed materials such as animal feed pellets. Fertilisers represent a large allied industry with a wide variety of materials.

A vast range of food products from flour to sugar and tea to coffee are conveyed pneumatically in numerous manufacturing processes. Confectionery is an industry in which many of these materials are handled. In the oil industry fine powders such as barite, cement and bentonite are used for drilling purposes. In mining and quarrying, lump coal and crushed ores and minerals are conveyed. Pulverised coal and ash are both handled in very large quantities in thermal power plants for the generation of electricity.

In the chemical industries materials include soda ash, polyethylene, PVC and polypropylene in a wide variety of forms from fine powders to pellets. Sand is used in foundries and glass manufacture, and cement and alumina are other materials that are conveyed pneumatically in large tonnages in a number of different industries.

4.1.3 *Modes of conveying*

Much confusion exists over how materials are conveyed through a pipeline and to the terminology given to the mode of flow. First, it must be recognised that materials can either be conveyed in batches through a pipeline, or they can be conveyed on a continuous basis, 24 hours a day if necessary. In batch conveying the material may be conveyed as a single plug if the batch size is relatively small. For continuous conveying and batch conveying if the batch size is large, two modes of conveying are recognised (Mills 2004).

4.1.3.1 *Dilute phase* If the material is conveyed in suspension with the carrier gas through the pipeline it is referred to as dilute phase conveying. Provided that a material can be fed reliably into a pipeline almost any material can be conveyed in dilute phase, regardless of the particle size, shape or density.

4.1.3.2 *Dense phase* If the material can be conveyed at low velocity in a non-suspension mode, through all or part of the pipeline, it is referred to as dense phase conveying. In dense phase conveying two modes of flow are recognised. One is moving bed flow, in which the material is conveyed as a moving bed on the bottom of the pipeline. The other mode is plug flow, in which the material is conveyed as full bore plugs separated by air gaps. Moving bed flow is only possible in a conventional conveying system if the material to be conveyed has good air retention characteristics. Plug-type flow is only possible in a conventional conveying system if the material has good permeability.

4.1.3.3 *Conveying air velocity* For dilute phase conveying, a relatively high value of conveying air velocity must be maintained. This is typically in the region of 11 m/s for a very fine powder, to 16 m/s for a fine granular material, and beyond for larger particles and higher density materials. For dense phase conveying, air velocities can be down to 3 m/s, and lower in certain circumstances.

4.1.3.4 *Solids loading ratio* The solids loading ratio is a useful parameter in helping to visualise the flow. This is the ratio of the mass flow rate of the material conveyed divided by the mass flow rate of the air used to convey the material. It is expressed in a dimensionless form. For dilute phase, maximum values that can be achieved are typically of the order of 15 (15 kg of material conveyed with 1 kg of air), although this can be higher if the conveying distance is short and the conveying line pressure drop is high.

With moving bed flows, solids loading ratios of well over 100 can be achieved if materials are conveyed with pressure gradients of about 20 mbar/m of horizontal pipeline. For plug-type flows the use of solids loading ratio is not as appropriate, for as the materials have to be very permeable, maximum values are only of the order of about 30. Despite the low value of solids loading ratio, materials can be reliably conveyed at velocities of 3 m/s and below in plug-type flow.

4.2 Conveying system types

A wide range of pneumatic conveying systems are available, and they are all generally suitable for the conveying of dry bulk particulate materials. The majority of systems are conventional, continuously operating, open systems, in a fixed location. To suit the material being conveyed, or the process, however, innovatory, batch operating and closed systems are commonly used. Many of these systems can be either positive or negative pressure in operation or a combination of the two. In this chapter some of the more common systems are presented. With such a wide range and choice of system types, a useful starting point is to consider the alternatives in pair groupings (Mills *et al.* 2004):

- *Open and closed systems*: Open systems are the norm for pneumatic conveying, particularly when conveying with air. Closed systems would only be employed for very specific circumstances, such as with highly toxic and potentially explosive materials.
- *Positive pressure and negative pressure systems*: Materials can be sucked as well as blown and so either pressure or vacuum can be employed for pneumatic conveying. This is often a matter of company or personal preference.
- *Fixed and mobile systems*: The majority of pneumatic conveying systems are in fixed locations and so this is not identified as a particular case. A variety of mobile systems are available for specific duties.
- *High and low pressure systems*: In pneumatic conveying, high pressure typically means any pressure above about 1 bar gauge. For systems delivering materials to reception points at atmospheric pressure, 6 bar gauge is typically the upper limit, due to the problems of air expansion. Very much higher pressures can be employed if delivering materials to reception points maintained at pressure, such as chemical reactors and fluidised bed combustion systems.
- *Conventional and innovatory systems*: Conventional systems are those in which the material is simply fed into a pipeline and either blown or sucked, and so this is not identified as a particular case since this is the norm. Innovatory systems are those in which the material to be conveyed is conditioned in some way, either at the feed point

or along the length of the pipeline. This is generally in order to convey the material at low velocity and hence in dense phase, if the material has no natural capability for low velocity conveying.

- *Batch and continuously operating systems*: Both of these types of conveying are common in industry.
- *Single and multiple systems*: The majority of conveying systems are single units. It is possible, however, to combine units for certain duties.
- *Dilute and dense phase systems*: Dilute and dense phase conveying do not relate to any particular type of system. Any bulk particulate material can be conveyed in dilute phase. It is primarily the properties of the material that determine whether the material can be conveyed in dense phase, particularly in conventional conveying systems.

4.2.1 *Open systems*

Where strict environmental control is not necessary an open system is generally used. Most pneumatic conveying systems can ensure totally enclosed material conveying, and so with suitable gas–solid separation and venting, the vast majority of materials can be handled quite safely in an open system. Many potentially combustible materials are conveyed in open systems by incorporating necessary safety features. Air is used for the conveying of most materials. Nitrogen and other gases can be used for particular materials and applications, but because of the added cost of operation closed loop systems are more commonly used in these cases.

4.2.2 *Positive pressure systems*

Although positive pressure conveying systems discharging to a reception point at atmospheric pressure are probably the most common of all pneumatic conveying systems, the feeding of a material into a pipeline in which there is air at a positive pressure does present a number of problems. A wide range of material feeding devices, however, are available that can be used with this type of system, from venturis and rotary valves to screws and blow tanks.

A sketch of a typical low positive pressure pneumatic conveying system is given in Figure 4.1. With the use of diverter valves, delivery to a number of alternative reception points can be arranged very easily with positive pressure systems. Although multiple point feeding into a common line can also be arranged, care must be taken, particularly in the case of rotary valve feeding of the pipeline, since air leakage through a number of such valves can be quite significant in relation to the total air requirements for conveying.

4.2.3 *Negative pressure (vacuum) systems*

Negative pressure systems are commonly used for drawing materials from multiple sources to a single point. There is little or no pressure difference across the feeding device and so multiple point feeding into a common line presents few problems. As a result the rotary valve and screw can also be a much cheaper item for feeding a pipeline in a negative pressure system than in a positive pressure system. The filtration plant has to be much larger, however, as a higher volume of air has to be filtered under vacuum conditions (Mills 2004; Mills *et al.* 2004). A sketch of a typical system is given in Figure 4.2.

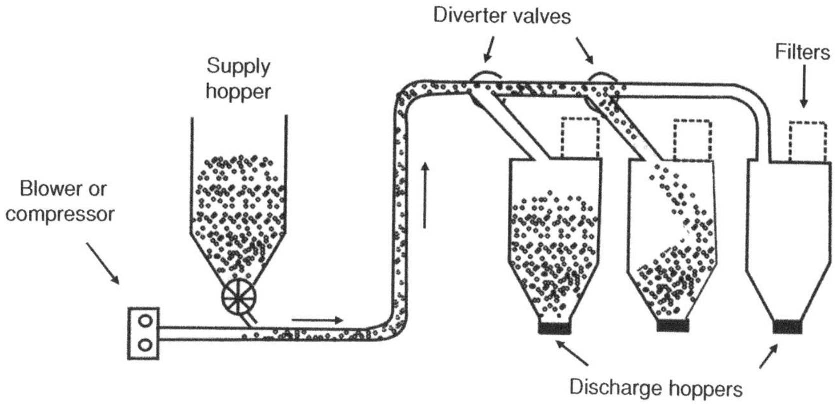

Figure 4.1 Typical positive pressure conveying system.

Negative pressure systems are also widely used for drawing materials from open storage, where the top surface of the material is accessible. This is achieved by means of suction nozzles. Vacuum systems, therefore, can be used most effectively for off-loading ships. They are also particularly useful for cleaning processes, such as the removal of material spillages and dust accumulations. Another application is in venting dust extraction hoods. Vacuum systems have the particular advantage that all gas leakage is inward, so that the injection of dust into the atmosphere is virtually eliminated. This is particularly important for the handling of toxic and explosive materials or any material where environmental considerations have to be taken into account. It is not always necessary to employ a closed system with these materials, therefore, provided that adequate safety measures are taken, particularly with regard to exhaust venting.

As a result of the conveying air being drawn through the air mover, it is essential that the exhauster should be protected from the possibility of the failure of one or more of the filter elements in the gas–solid separation system. This can be achieved by incorporating a back-up filter. A sketch of a typical negative pressure conveying system operating with a vacuum

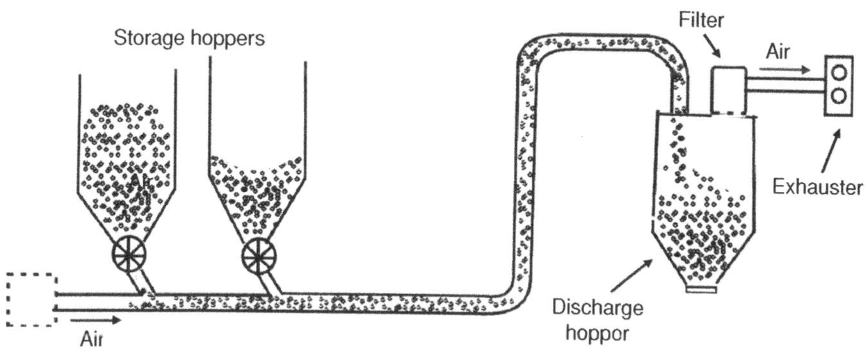

Figure 4.2 Typical negative pressure conveying system.

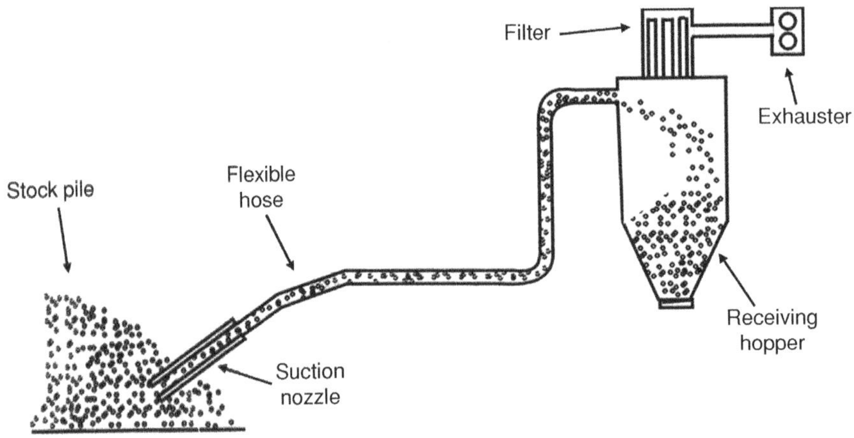

Figure 4.3 Vacuum conveying from open storage.

nozzle is given in Figure 4.3. If a high vacuum is used for the conveying of a material, consideration should be given to the stepping of the pipeline part way along its length. Air is compressible and the volume increases considerably with decrease in pressure. If the pipeline is not stepped, extremely high conveying air velocities could occur towards the end of the pipeline. The situation is the same for high positive pressure conveying systems.

4.2.4 *Staged systems*

The systems illustrated above have all been single stage systems. In hydraulic conveying, for very long distance conveying, it is usual to stage systems. At the end of one stage the material is pumped back to pressure and fed into the pipeline of the next stage. Although this is perfectly possible for pneumatic conveying it is very rare that it is ever done. Distance capability is limited with pneumatic conveying and the cost implications are probably against it. Combined systems, however, are quite common in which vacuum systems feed into positive pressure systems.

4.2.5 *Shared negative and positive pressure systems*

Combined negative and positive pressure systems that share a common air mover represent a very versatile type of conveying system, combining many of the advantageous features of both the negative and positive pressure systems. They are often referred to as suck–blow or push–pull systems, as illustrated in Figure 4.4. They can be used to transfer material from multiple sources to multiple discharge locations and can thereby extend vacuum systems over much longer distances.

Protection has to be provided for the exhauster/blower from the possible ingress of material, as with negative pressure systems. It should be noted that the available power for the system has to be shared between the two sections and that the pipelines for the two parts of the system have to be carefully sized to take account of different operating pressures.

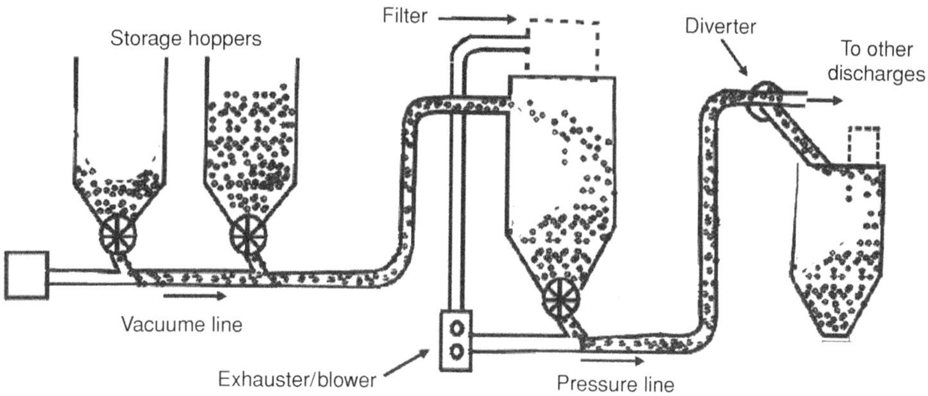

Figure 4.4 Combined negative and positive pressure system.

Some air movers, such as positive displacement blowers, operate on a given pressure ratio, and this will mean that the machine will not be capable of operating over the same pressure range with the combined duty as compared with their individual operation. This will mean that the system capability is limited in terms of both tonnage and distance. Although there is only one air mover, two filter units will be required, as shown in Figure 4.4.

4.2.6 *Dual vacuum and positive pressure systems*

If the conveying potential of a system requiring the vacuum pick-up of a material needs to be improved beyond that capable with a combined negative and positive pressure system, particularly in terms of conveying distance, then a dual system should be considered. In this combination the two conveying elements are separated and two air movers are provided. A sketch of a typical system is given in Figure 4.5. Filters and valves have been omitted from the sketch of the system for clarity.

As two air movers are provided, the most suitable exhauster can be dedicated to the vacuum system and the most appropriate positive pressure system can be used for the onward transfer of material. If the vacuum off-loading section is only a short distance, it is possible that the material could be conveyed in dense phase over the entire conveying distance.

The system shown in Figure 4.5 is typical of a ship off-loading system. With a high vacuum exhauster a material such as cement could be off-loaded at a rate of 800 tonne/h through a single pipeline. Twin vessels on the quayside would allow continuous conveying to shore-based reception vessels, which could be some 500 m distance if a high pressure compressor was to be used. For the onward conveying two pipelines would probably need to be used to achieve the 800 tonne/h.

4.2.7 *Batch conveying systems*

The systems considered so far have all been capable of continuous conveying. In many processes, however, it may be more convenient to convey one batch at a time. Although a

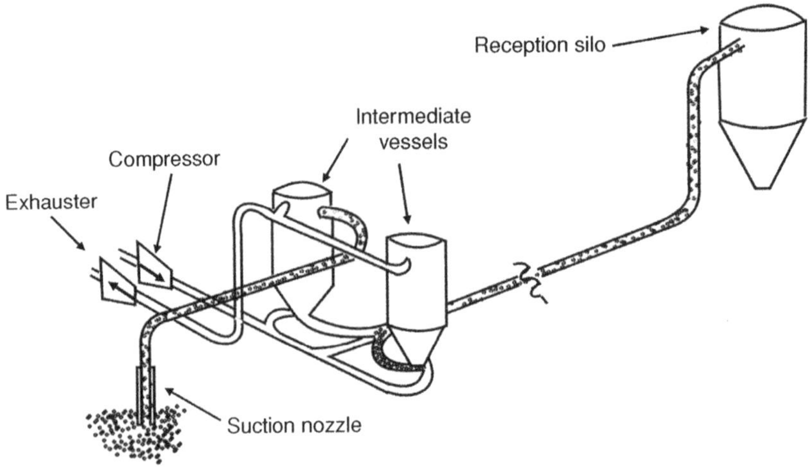

Figure 4.5 Typical dual vacuum and positive pressure system.

batch conveying system may be chosen for a specific process need, the mode of conveying is, to a large extent dictated by the choice of pipeline feeding device. The majority of batch conveying systems are based on blow tanks, and blow tanks are chosen either because of their high pressure conveying capability, or because of the abrasive nature of the material.

Two types of system are considered. In one, the batch size is relatively large, and the material is fed into the pipeline gradually, and so can be considered as a semi-continuous system. In the other, the material is fed into the pipeline as a single plug.

4.2.7.1 *Semi-continuous systems* It should be noted that when batches of material are fed into the pipeline gradually, there is essentially no difference in the nature of the gas–solids flow in the pipeline with respect to the mode of conveying through the pipeline. This is certainly the case during the steady state portion of the conveying cycle, regardless of the value of solids loading ratio.

The blow tanks used vary in size up to 50 m^3 or more, generally depending upon the material flow rate required as well as a need to maintain a reasonable frequency of blow tank cycling. The material can be conveyed in dilute or dense phase, depending upon the capability of the material, the pressure available and the conveying distance, as with continuously operating systems.

With a single blow tank it is not possible to utilise the pipeline while the blow tank is being filled with material or when the system is being pressurised. Since batch conveying is discontinuous, steady state values of material flow rate, achieved during conveying, have to be higher than those for continuously operating systems in order to achieve the same time averaged mean value of material flow rate. This means that air requirements and pipeline sizes have to be based on the maximum, or steady state, conveying rate. The intermittent nature of the conveying cycle is illustrated in Figure 4.6.

In comparison with a continuously operating system, therefore, the batch operating system would appear to be at a disadvantage. Blow tank systems, however, can operate at very much higher pressures to compensate, and twin arrangements in series can be configured

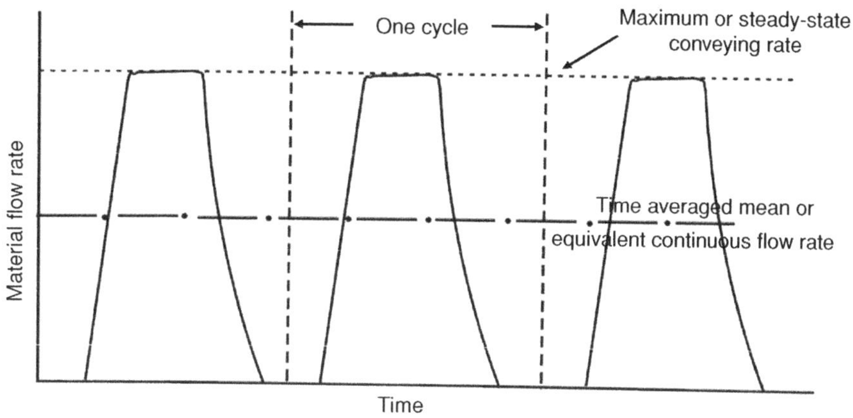

Figure 4.6 Sketch showing the transient nature of batch conveying.

to operate continuously. It should be emphasised that blow tanks can be operated at low as well as high pressure, depending upon the system requirements.

If a material needs to be fed into a chemical reactor or a boiler plant that is maintained at a pressure of 25 bar for example, the blow tank can be designed to operate at 30 bar for the duty. When delivering material to a reception point at atmospheric pressure, however, air supply pressures greater than about 6 bar are rarely used. This is mainly because of the problem of air expansion and the need for a stepped pipeline to prevent excessively high values of conveying air velocity. A typical batch conveying system based on a single blow tank is illustrated in Figure 4.7.

4.2.7.2 *Single plug systems* In the single plug conveying system the material is effectively extruded into the pipeline as a single plug, although the material is generally well aerated. It is typically about 10 m long. This plug is then blown through the pipeline as a coherent plug. A certain amount of material will tail off the end of the plug as it is conveyed, but the

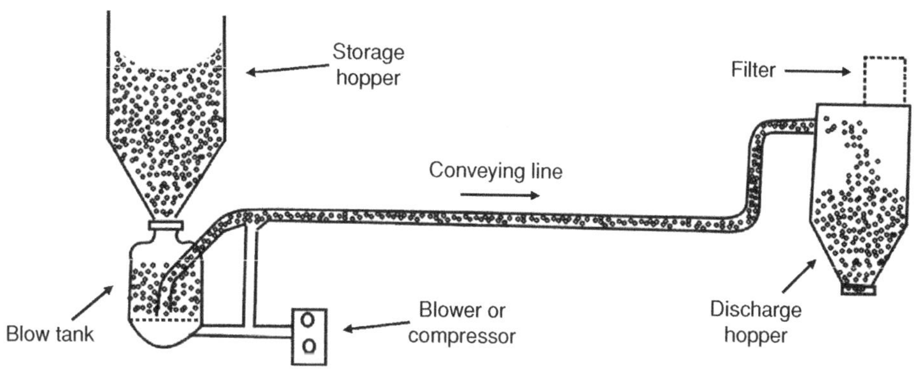

Figure 4.7 Typical single blow tank conveying system.

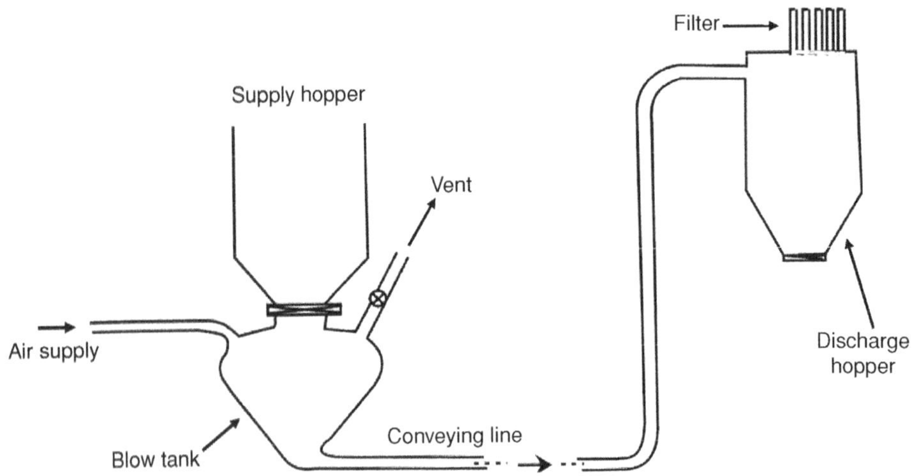

Figure 4.8 Typical single plug conveying system.

front of the plug will sweep up material deposited in the pipeline by the previous plug. Blow tanks are generally used as the feeding device and a sketch of a typical system is given in Figure 4.8.

The air pressure has to overcome the frictional resistance of the plug of material in the pipeline. As a result blow tank sizes are rarely larger than about 5 m^3, unless very large diameter pipelines are employed. In terms of system design, a cycling frequency is selected to achieve the required material flow rate, which determines the batch size. The pipe diameter is then selected such that the frictional resistance of the plug results in a reasonable air supply pressure to propel the plug at the given velocity. The material will be conveyed at a low velocity, in what may be regarded as dense phase, but solids loading ratios have no significance here, and steady state conveying, as depicted in Figure 4.6, does not apply either. Single plug systems are capable of conveying a wide range of materials, and generally at much lower velocities than can be achieved in continuously operating systems.

Many coarse, granular materials are either friable or abrasive and can only be conveyed in dilute phase with conventional conveying systems, and so single plug systems can represent a viable alternative, although it would always be recommended that tests be carried out to confirm this. Material discharge often represents a problem with this type of system. Although the plugs of material are conveyed at a relatively low velocity, once they are discharged from the pipeline the high pressure air released behind the plug can cause severe erosion of the pipeline on venting.

4.2.8 *Mobile systems*

All of the systems described so far have been essentially fixed systems. The only real flexibility in any of the systems has been the capability of moving vacuum nozzles in negative pressure systems. By the use of flexible hoses these can be moved, and they find wide application in ship off-loading systems, and the clearing of material from stockpiles or spillages. Many road sweeping vehicles employ vacuum conveying for their operation.

Many bulk particulate materials are transported from one location to another by road, rail and sea. Many materials, of course, are transported in a pre-packaged form, or in bulk containers, and can be transported by road, rail, sea or air, in a similar manner to any other commodity. Many transport systems, however, are specifically designed for bulk particulate materials and have a capability of self loading, self off-loading, or both. These are generally mobile versions of the above static conveying systems, depending upon the application and duty.

4.2.8.1 *Road and rail vehicles* Road vehicles are widely used for the transport of a multitude of bulk particulate materials, such as cement, flour, sugar and polyethylene. Road vehicles often have their own positive displacement blower mounted behind the cab and so can off-load their materials independently of delivery depot facilities. The material containing element on the truck can generally be tipped to facilitate discharge, which can be via a rotary valve, or the container might double as a blow tank that can be pressurised.

Rail cars or wagons generally rely on delivery depot facilities for off-loading. Because of their length tilting is not an option and so multiple point off-loading is often employed. They may be off-loaded by rotary valve, or the rail car may be capable of being pressurised so that it can be off-loaded as a blow tank. Whereas road vehicles are typically designed to operate with air at 1 bar gauge for this purpose, rail vehicles are usually designed to 2 bar gauge and can generally be off-loaded in about 1 hour. The base of the rail car is usually angled at about 5° in herringbone fashion around each discharge point and fluidised to facilitate removal of as much material as possible.

4.2.8.2 *Ships* Large bulk carriers usually rely on port facilities for off-loading and these are generally similar to that depicted in Figure 4.5. Intermediate bulk carriers, however, often have on-board facilities for off-loading. Such vessels are often used for the transfer of materials such as cement to storage depots at ports for local supply, or to off-shore oil and gas rigs. Materials are typically transferred from storage holds in the ship by a combination of air-assisted gravity conveyors and vacuum conveying systems, into twin blow tanks. High pressure air is supplied by on-board diesel driven compressors and materials are conveyed to dock-side storage facilities through flexible rubber hose, which solves the problems of both location and tidal movements.

4.2.9 *Closed systems*

The systems illustrated above have all been open systems in which air is usually the conveying gas and this is simply drawn from the atmosphere and returned back to it, after being filtered. For certain conveying duties it is necessary to convey the material in a controlled environment. If a dust cloud of the material is potentially explosive, nitrogen or some other gas can be used to convey the material. In an open system such environmental control can be very expensive, but in a closed system the gas can be re-circulated and so the operating costs, in terms of inert gas, are significantly reduced.

If the material to be handled is toxic or radioactive, it may be possible to use air for conveying, but very strict control would have to be maintained. A closed system would be essential in this case. Continuous conveying systems are probably the easiest to arrange in the form of a closed loop. A sketch of a typical system is given in Figure 4.9.

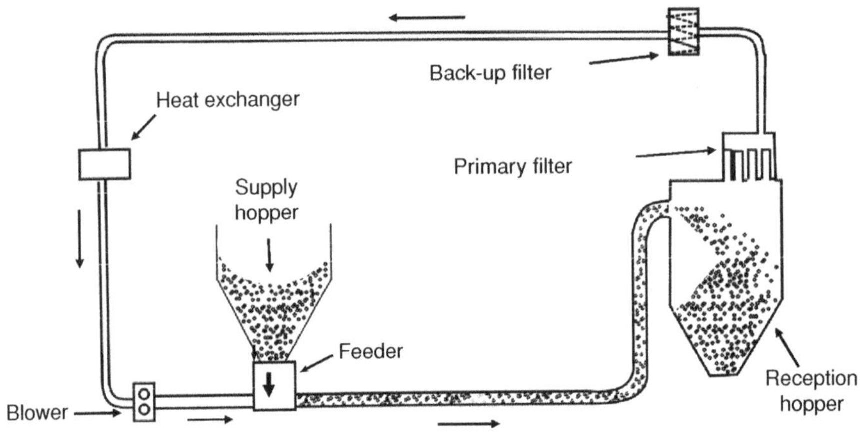

Figure 4.9 Closed loop pneumatic conveying system.

A null point needs to be established in the system where the pressure is effectively atmospheric and provision for make up of conveying gas can be established there. If this is positioned after the blower the conveying system can operate entirely under vacuum. If the null point is located before the blower it will operate as a positive pressure system.

A back-up filter would always be recommended, because positive displacement blowers and compressors are very vulnerable to damage by dust. This is simply a precaution against an element in the filter unit failing. There will generally be an increase in temperature across an air mover and so in a closed loop system it may be necessary to include a heat exchanger, otherwise there could be a gradual build up in temperature. The heat exchanger can be placed either before or after the air mover, depending upon the material being conveyed.

4.2.10 *Innovatory systems*

The systems illustrated above have all been conventional systems in which the material is simply fed into a pipeline and either blown or sucked to its destination. Unless the material to be conveyed has natural bulk characteristics such as good air retention or very good permeability, it is unlikely that it will be possible to convey the material at low velocity, and in dense phase, in a conventional continuous or semi-continuous conveying system such as those described above. Even if a high pressure system is employed it is unlikely that such a material will convey in dense phase, unless the pipeline is relatively short. Dense phase conveying is not synonymous with high pressure, it is material property dependent.

For materials that are abrasive, alternatives to conventional systems may have to be considered in order to reduce damage to the conveying system, particularly if the materials are not naturally capable of being conveyed in the dense phase mode, and hence at low velocity. A similar situation applies to materials that are friable, for considerable damage can occur to the conveyed particles. Particle degradation can occur in high velocity suspension flow, and erosion of bends in the pipeline and other plant surfaces subject to particle impact will occur if an abrasive material is conveyed in dilute phase.

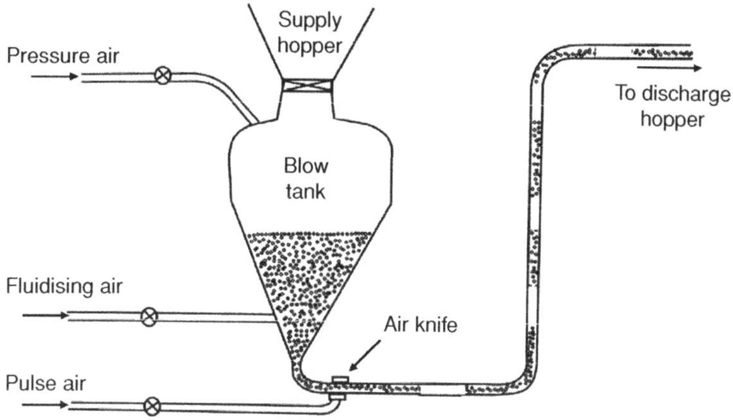

Figure 4.10 Pulse phase conveying system.

With a need to convey many materials at low velocity, much development work has been undertaken to find means of conveying materials at low velocity that have with no natural dense phase conveying capability. The innovatory systems produced as a result of these developments have centred around some form of conditioning of the conveyed material, either at the feed point into the pipeline or along the length of the pipeline. Since the modifications are essentially based on the pipeline, types of conveying system have not changed significantly.

4.2.10.1 *Plug forming systems* The pulse phase system was developed in the late 1960s. A typical pulse phase system is shown in Figure 4.10. An air knife, positioned at the start of the pipeline, intermittently pulses air into the pipeline to divide the discharging material into discrete short plugs. Blow tanks are commonly used for the feeding of materials in this type of system.

No further conditioning of the material occurs along the length of the pipeline. The pulse phase system was initially proposed as a solution to the problem of conveying cohesive bulk solids, but subsequent developments have shown that a wider range of materials can be conveyed successfully.

4.2.10.2 *By-pass systems* The most common by-pass systems employ a small pipe running inside the conveying line, having fixed ports, or flutes, at regular intervals along its length. This inner pipe is not supplied with an external source of air, but air within the conveying line can enter freely through the regular openings provided. In an alternative design the by-pass pipe runs externally to the pipeline and is interconnected at regular intervals. By this means pipeline bends can also be conveniently incorporated.

If the material is impermeable the air will be forced to flow through the by-pass pipe if the pipeline blocks. Because the by-pass pipe has a much smaller diameter than the pipeline, the air will be forced back into the pipeline through the next and subsequent flutes because of the extremely high pressure gradient, and this will affect a break up of the plug of material causing the blockage.

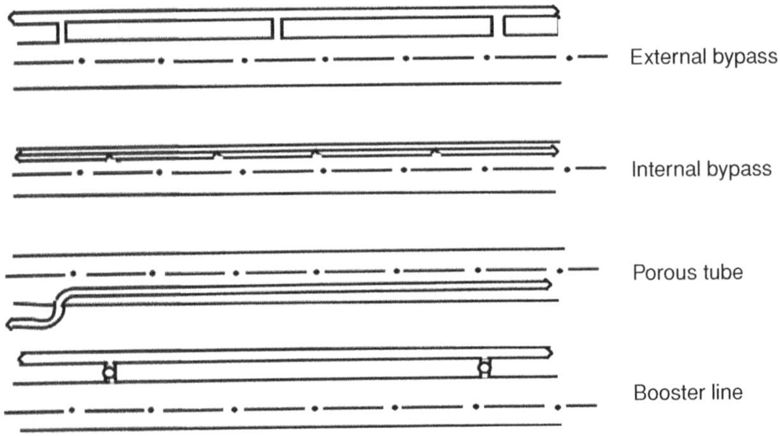

Figure 4.11 Sketch of various plug control systems.

4.2.10.3 *Air injection systems* A number of systems have been developed that inject air into the pipeline at regular points along its length. While by-pass pipe systems artificially create permeability in the bulk material, air injection will help to maintain a degree of air retention within the material. Continuous injection of air into the pipeline, however, does mean that conveying air velocities towards the end of the pipeline will be much higher as a result. In some systems, sensors are positioned between the parallel air line and the conveying pipeline and air is only injected when required. If a change in pressure between the two lines is detected, which would indicate that a plug is forming in the conveying pipeline, air is injected close to that point in order to break up the plug and so facilitate its movement. A sketch of various plug control systems, including both by-pass pipe and air addition methods, is given in Figure 4.11.

Many of the innovatory systems are capable of being stopped and re-started during operation. With most conventional systems this is not possible, and would result in considerable inconvenience in clearing pipelines if a blockage should occur as a consequence. Since they are capable of conveying materials in dense phase, operating costs for power are likely to be lower than those for a conventional dilute phase system. Capital costs for the innovatory systems are likely to be higher, however.

4.3 System components

All pneumatic conveying systems, whether they are of the positive or negative pressure type, conveying continuously or in a batch-wise mode, can be considered to consist of the basic elements depicted in Figure 4.12.

Numerous devices have been developed to feed materials into pipelines, as well as to disengage materials from the conveying air at the reception point. In vacuum systems the material feed is invariably at atmospheric pressure and so the pipeline can either be fed directly from a supply hopper or by means of suction nozzles from a storage vessel or stockpile. Pressure capability, control and air leakage are important points to consider

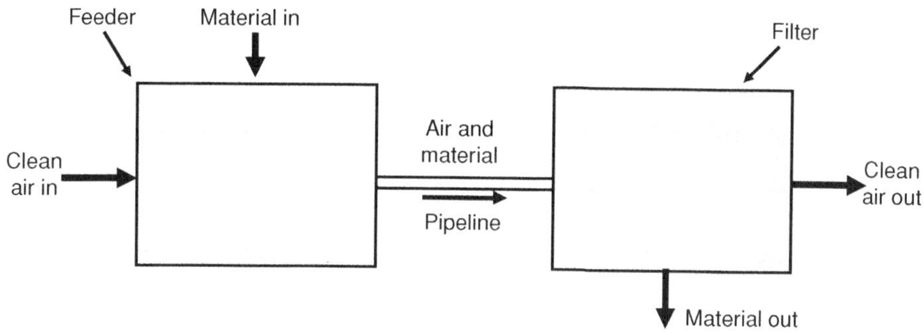

Figure 4.12 Basic elements of a pneumatic conveying system.

here. A wide range of compressors, blowers, fans and exhausters are also available, but consideration on choice must be given to the operating characteristics of the machine.

4.3.1 *Pipeline feeding devices*

Many diverse devices have been developed for feeding pipelines. Some are specifically appropriate to a single type of system, such as suction nozzles for vacuum systems. Others, such as rotary valves, screws and gate valves, can be used for vacuum and positive pressure systems. The approximate operating pressure ranges for various pipeline feeding devices are illustrated in Figure 4.13.

The air mover can be positioned at either end of the system shown in Figure 4.12. In positive pressure systems the material has to be fed into the pipeline with air maintained at pressure. As a consequence of this there may be a loss of conveying air across the feeding device. In certain cases this air flow can interfere with the feeding process. In negative

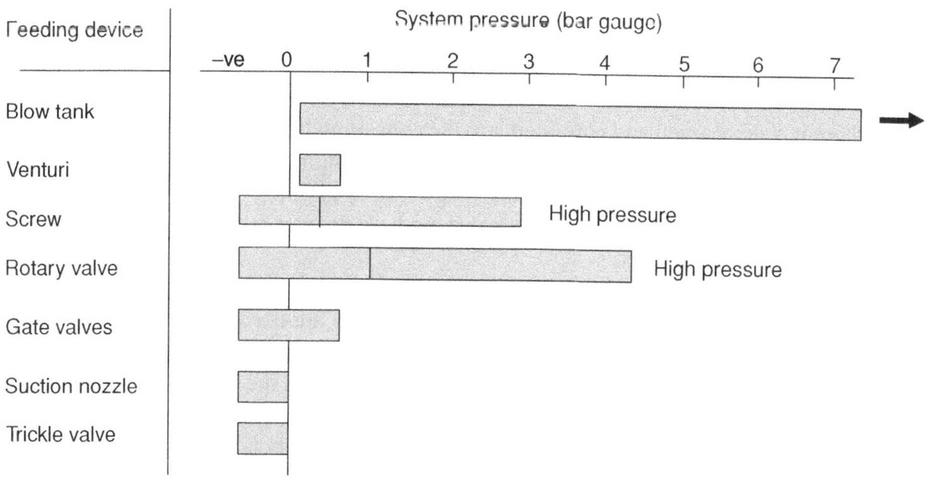

Figure 4.13 Approximate operating pressure ranges for various feeding devices.

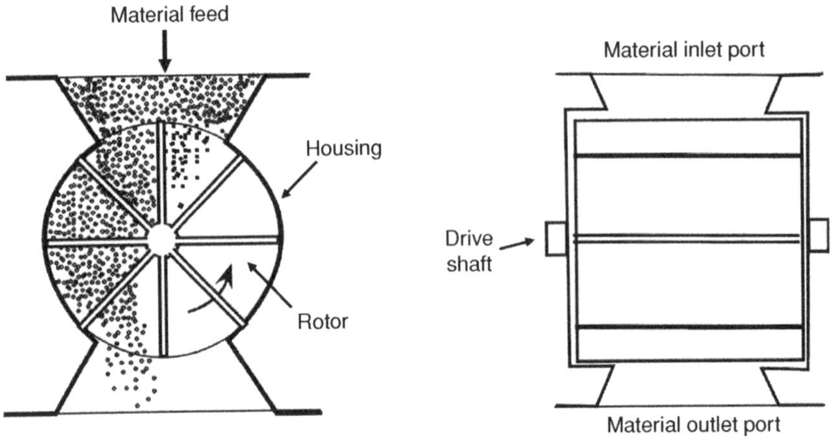

Figure 4.14 Drop-through rotary valve.

pressure systems there will be no adverse pressure drop across the feeding device, but a similar problem in terms of air ingress into the conveying system may occur with the receiving hopper since this will have to be maintained under vacuum. If theses losses or ingress are significant, the air supply will have to be increased to compensate, for the correct air flow rate to the pipeline must be maintained for conveying the material.

4.3.1.1 *Rotary valves* The rotary valve is one of the most commonly used devices for feeding material into pipelines. This type of feeder consists of a bladed rotor working in a fixed housing. In many applications in which it is used its primary function is as an air lock, and so is often referred to as a rotary air lock. It is a positive displacement device and so material flow rate can be controlled by varying the speed of the bladed rotor. This type of valve, usually referred to as a 'drop-through' feeder, is depicted in Figure 4.14 and is generally suitable for free-flowing materials.

Material from the supply hopper continuously fills the rotor pockets at the inlet port which is situated above the rotor. It is then transferred by the motor-driven rotor to the outlet where it is discharged and entrained into the conveying line. For cohesive materials it is generally necessary to either profile the rotor pockets, to minimise 'hang up' of material, or to physically purge the material from the discharge side by aligning this with the conveying pipeline. For pelletised materials an off-set valve, with a side inlet, is generally used in order to prevent shearing of particles.

By the nature of the feeding mechanism, rotary valves are more suited to relatively non-abrasive materials. This is particularly the case where they are used to feed materials into positive pressure conveying systems. By virtue of the pressure difference across the valve, and the need to maintain a rotor tip clearance, air will leak across the valve. Wear, therefore, will not only occur by conventional abrasive mechanisms, but by erosive wear also. Air leakage through the blade tip clearances can generate high velocity flows, which will entrain fine particles, and the resulting erosive wear can be far more serious than the abrasive wear. Wear resistant materials can be used in the construction of rotary valves, and removable

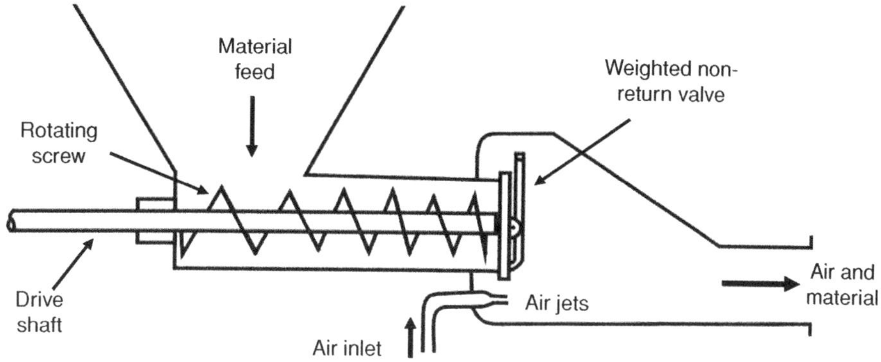

Figure 4.15 Variable pitch type of screw feeder.

lining plates can be incorporated to help with maintenance, but wear can only be minimised, it cannot be eliminated if an abrasive material is to be handled.

For positive pressure conveying systems it is usual to have a small opening in the side of the rotary valve so that the air leaking across the valve can be vented. The vented air is usually directed to the supply hopper above, where it can be filtered, for this air can carry fine material with it. To increase the pressure capability of rotary valves, end plates are usually fitted to the rotors and the radial surfaces are provided with seals in order to minimise the air leakage.

4.3.1.2 *Screw feeders* Much of what has been said about rotary valves applies equally to screw feeders. They are positive displacement devices, feed rate control can be achieved by varying the speed, they can be used for either positive pressure or vacuum feeding duties, air leakage is a problem when feeding into positive pressure systems, and they are prone to wear by abrasive materials. For high pressure duties, in order to minimise air leakage, a variable pitch screw is employed, as illustrated in Figure 4.15.

This high pressure type of feeder, however, is only suitable for fine materials, such as cement and fine grades of fly ash, that are capable of being 'compressed' to a higher bulk density in the reducing pitch section of the screw, since this is how the 'air seal' is achieved. As a consequence of this the power required to drive the screw is quite high.

4.3.1.3 *Venturi feeders* Since the basic problem with feeding positive pressure systems is that the air leakage arising from the adverse pressure gradient can interfere with the flow of the material into the pipeline, this situation can be improved, to a certain extent, by using venturi feeders. These basically consist of a reduction in pipeline cross-section in the region where the material is fed from the supply hopper, as shown in Figure 4.16. It will be seen that there are no moving parts with this type of feeding device, which has certain advantages with regard to wear problems, but there is no inherent means of flow control either, and so this has to be provided additionally.

A consequence of the reduction in flow area is an increase in air velocity and a decrease in pressure in the region of the throat. With a correctly designed venturi the pressure at the throat

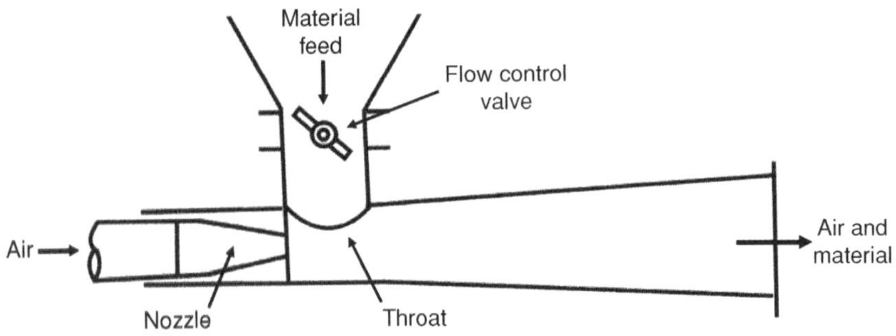

Figure 4.16 Venturi feeder.

should be the same as that in the supply hopper which, for the majority of applications, is atmospheric pressure. This then encourages the material to flow more readily under gravity into the pipeline, since under these conditions there is no leakage of air in opposition to the material feed. In order to keep the throat at atmospheric pressure, and also of a practical size that will allow the passage of material and for it to be conveyed, a relatively low limit has to be imposed on the air supply pressure.

Since there are no moving parts, these feeders are potentially suitable for abrasive and friable materials. Care must be exercised in using venturis to feed such materials into the conveying line, however, for the high air velocity in the throat may lead to considerable erosion and particle degradation in this region. There is no inherent means of flow control and so this means that the venturi would need to be fed from a belt, screw or vibratory feeder. A rotary valve could also be used, since there would be no adverse pressure drop across the valve.

4.3.1.4 *Gate lock valves* These are probably the least used of all devices for feeding pneumatic conveying system pipelines. They are variously known as double flap valves, double dump valves and double door discharge gates. They basically consist of two doors or gates which alternately open and close to permit the passage of the material from the supply hopper into the conveying line, as illustrated in Figure 4.17.

To a certain extent the gate lock might be termed an intermittent feeder, since it discharges material between 5 and 10 times a minute. In contrast, the rotary valve has approximately 150–200 discharges per minute from its pocketed rotor. This reduction in the number of discharges means that the air supply, in terms of flow rate, and particularly pressure, must be correctly evaluated to prevent the possibility of line blockage. With few moving parts this type of feeder can be used to feed friable materials, and with appropriate materials of construction it is also suited to the handling of abrasive materials.

4.3.1.5 *Suction or vacuum nozzles* A specific application of vacuum conveying systems is the pneumatic conveying of bulk particulate materials from open storage and stockpiles, where the top surface of the material is accessible. Vacuum systems can be used most effectively for the off-loading of ships and for the transfer of materials from open piles to storage hoppers. They are particularly useful for cleaning processes such as the removal of material spillage and dust accumulations. In this role they are very similar to the domestic

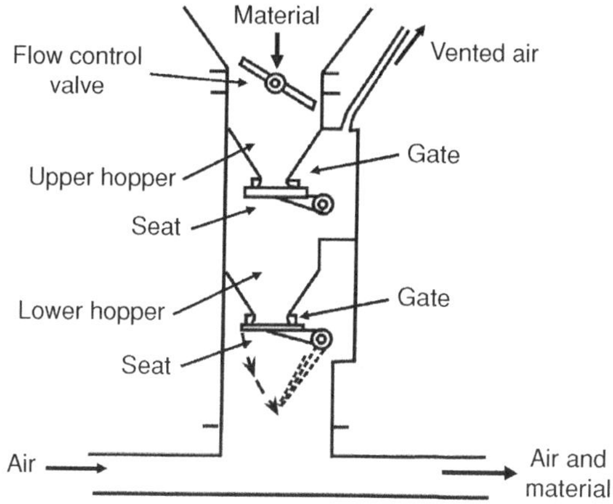

Figure 4.17 Gate valve feeder.

vacuum cleaner. For industrial applications with powdered and granular materials, however, the suction nozzles are rather more complex. A sketch of a typical suction nozzle for vacuum pick-up systems is shown in Figure 4.18. It will be seen that the device has no moving parts and so a means of controlling material flow rate is critical.

It is essential with suction nozzles to avoid filling the inlet tube solidly with material, and to maintain an adequate flow of air through the conveying line at all times. To avoid blocking the inlet pipe, sufficient air must be available at the material feed point, even if the suction nozzle is buried in the bulk solid material. Indeed, the vacuum off-loading system must be able to operate continuously with the nozzle buried in the material in order to maximise the material flow rate. Sufficient air must also be available for conveying the material through

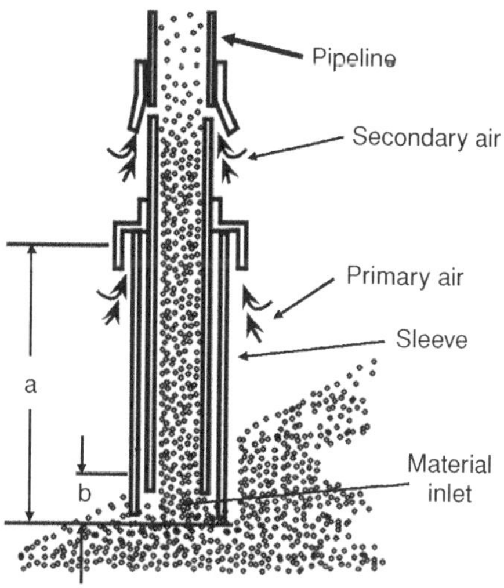

Figure 4.18 Suction nozzle for vacuum pick-up systems.

the pipeline once it is drawn into the inlet pipe. In order to achieve maximum conveying rate it is necessary to maintain as uniform a feed to the line as possible. To achieve this two air inlets are required, one at the material pick-up point and another at a point downstream.

The conveying pipeline is provided with an outer sleeve at its end, and primary air for material feed is directed to the conveying line inlet in the annular space created. The length '*a*' of this sleeve has to be long enough to ensure that it is not buried by the movement of the material and so prevent the flow of primary air. This sleeve may be many metres long for ship off-loading applications. The position of the end of the sleeve relative to the end of the pipeline, '*b*', is partly material dependent and could be positive or negative. This dictates the efficiency with which the material is drawn into the conveying line. Secondary air for conveying the material is generally introduced via a series of holes in the pipeline. Some form of regulation of both the primary and secondary air is necessary, and the proportion of the total which is directed to the material inlet is particularly important.

4.3.1.6 *Trickle valves* Trickle valves, as a device on their own, are only suitable for vacuum conveying systems, since there is no adverse pressure drop against which to feed. The greatest problem with this class of feeder is that of flow rate control. A typical device was shown in relation to the control of the venturi feeder in Figure 4.16. This type of device is often used because of the cost advantage over almost any alternative method of feeding vacuum conveying systems. A flow restriction is often fitted into the inlet section of pipeline to assist with material discharge. This resistance has the effect of slightly lowering the pressure in the pipeline at the material feed point and so helps to promote flow. There is, however, no inherent means of flow control and so changes in material properties, such as particle size and moisture content, could well result in a change in material flow rate.

4.3.1.7 *Blow tanks* Blow tanks are often employed in pneumatic conveying systems because of their capability of using high pressure air. Blow tanks are neither restricted to dense phase conveying not to high pressure use, however. Materials not capable of being conveyed in dense phase can be conveyed equally well in dilute phase suspension flow from a blow tank. Low pressure blow tanks are often used as an alternative to screw feeders and rotary valves for feeding pipelines, particularly if abrasive materials have to be conveyed. The blow tank has no moving parts and so both wear of the feeder and degradation of the material are significantly reduced. Another advantage of these systems is that the blow tank also serves as the feeder, and so the problems associated with feeding against an adverse pressure gradient, such as air leakage, do not arise.

4.3.1.7.1 *Blow tank control* In most blow tank systems the air supply to the blow tank is split into two streams. One air stream pressurises the blow tank and may also fluidise or aerate the material in the blow tank. This air stream serves to discharge the material from the blow tank. The other air stream is fed directly into the discharge line just downstream of the blow tank. This is generally referred to as supplementary air and it provides the necessary control over the material flow in the conveying line. This is very similar to the means by which vacuum nozzles are controlled.

4.3.1.7.2 *Top and bottom discharge* There are numerous different types of blow tank, and for each type alternative configurations are possible. The basic features of different blow tanks are essentially similar, but different arrangements can result in very different

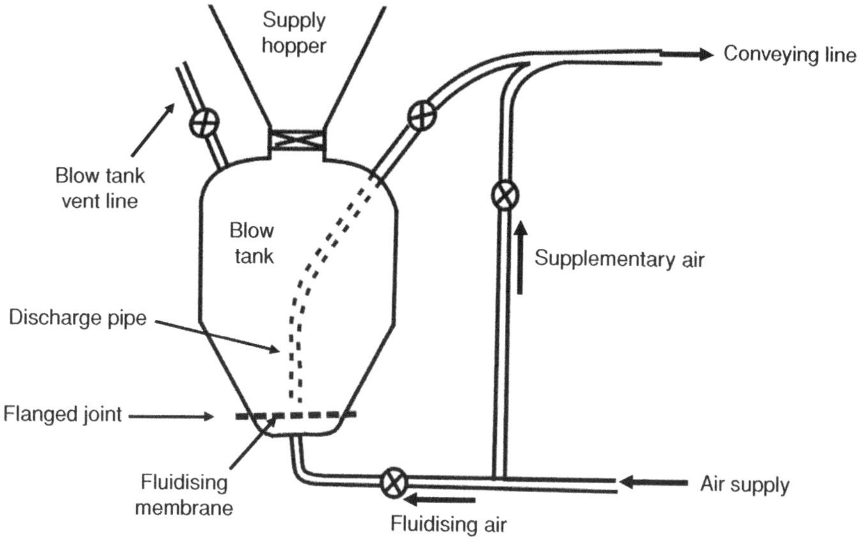

Figure 4.19 Top discharge blow tank with fluidising membrane.

conveying capabilities and control characteristics. The blow tank shown in Figure 4.19 is typical of a top discharge type.

Discharge is arranged through an off-take pipe which is positioned above the fluidising membrane. With this type of blow tank, however, it is not possible to completely discharge the contents, although with a conical membrane very little material will remain. In a bottom discharge blow tank there is no membrane and material is gravity fed into the pipeline, and so the contents can be completely discharged. A sketch of such a blow tank is given in Figure 4.20.

4.3.1.7.3 *Blow tank pressure drop* The pressure drop across the blow tank represents a potential source of energy loss to the system and so should be kept as low as possible. In the case of top discharge blow tanks this is particularly important. The discharge pipe must be kept as short as possible because the pressure drop in this line will be high owing to the very high material concentration. Supplementary air should be introduced as close to the top of the blow tank as possible. With very large blow tanks the discharge pipe should be turned through 90° just above the membrane and be taken through the side of the vessel. If the discharge pipe is kept to about 2 m, the pressure drop across the blow tank will be about 0.2 bar, which includes the membrane resistance. In the case of bottom discharge blow tanks, very short discharge lines can usually be arranged and so the pressure drop is generally no more than 0.1 bar.

4.3.1.7.4 *Twin blow tanks in parallel* If two blow tanks are used, rather than one, a significant improvement in performance can be achieved. The ratio of the mean flow rate to the steady state material flow rate can be brought close to unity if two blow tanks in parallel are used. While one is being discharged into the conveying pipeline, the other can be de-pressurised, filled and pressurised, ready for discharging when the other one is empty.

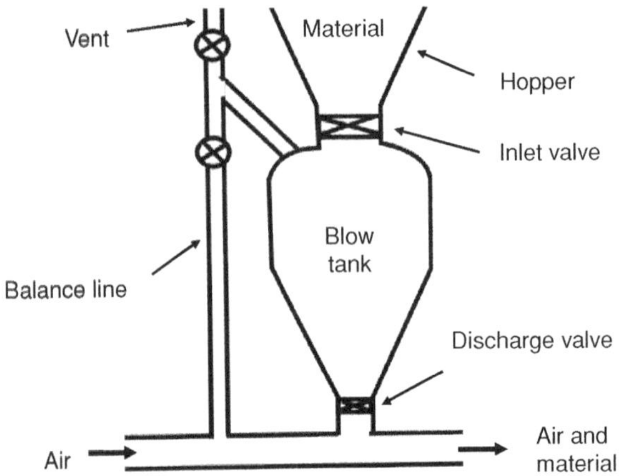

Figure 4.20 Bottom discharge blow tank.

By this means almost continuous conveying can be achieved through a common pipeline. A sketch of a typical twin blow tank arrangement is given in Figure 4.21.

4.3.1.7.5 *Twin blow tanks in series* If two pressure tanks are placed vertically in line beneath a hopper it is possible to use a high pressure air supply for the continuous conveying of a material. A sketch of such an arrangement is given in Figure 4.22. The vessel between the hopper and the blow tank transfers the material between these two, and is effectively a lock hopper. The vent line is used to release the pressure in the transfer vessel, in addition to venting on filling.

The lock hopper is filled from the hopper above. The lock hopper is then pressurised to the same pressure as the blow tank. With the transfer vessel at the same pressure as the

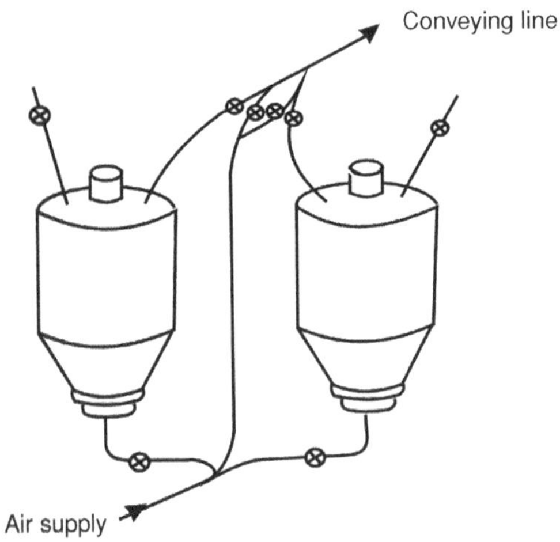

Figure 4.21 Typical twin blow tank arrangement.

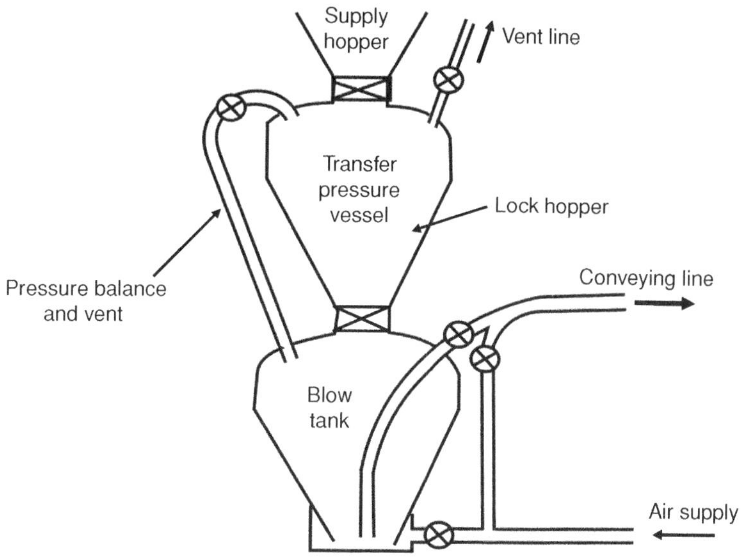

Figure 4.22 Blow tank system capable of continuous operation.

blow tank, the blow tank can be topped up to maintain a continuous flow of material. The blow tank in Figure 4.22 is shown in a top discharge configuration, but without a fluidising membrane. The air enters a plenum chamber at the base, to pressurise the blow tank and fluidise the material, and is discharged into the conveying line. Twin blow tanks, with one positioned above the other, do require a lot of headroom, and so the blow tank arrangement shown in Figure 4.22 is sometimes employed to minimise the head required.

If a lock hopper arrangement is used, as shown in Figure 4.22, the pipeline feeding device need not be a blow tank at all, despite the use of high pressure air. With the transfer pressure vessel separating the hopper and the pipeline feeding device, the feeding device can equally be a rotary valve or a screw feeder, for there is little pressure drop across the feeder. The pressure drop is, in fact, in the direction of material flow and so there are no problems of air leakage across the device, as there are with conventional feeders of this type. A rotary valve or screw may be used in this situation to guarantee the feed of a steady flow of material into a pipeline.

4.3.2 *Gas–solid separation*

Gas–solid separation devices associated with pneumatic conveying systems have two functions. The first is to recover as much as possible of the conveyed material for the next stage of the handling or treatment process. The second is to minimise pollution of the working environment by the material. The first of these functions is principally a matter of economics, in that the more valuable the material, the more trouble should be taken to ensure total recovery. However, the avoidance of environmental pollution is potentially more important, particularly since the introduction of more stringent Health and Safety at Work legislation. Where the material is known to be potentially dangerous, of course, extreme

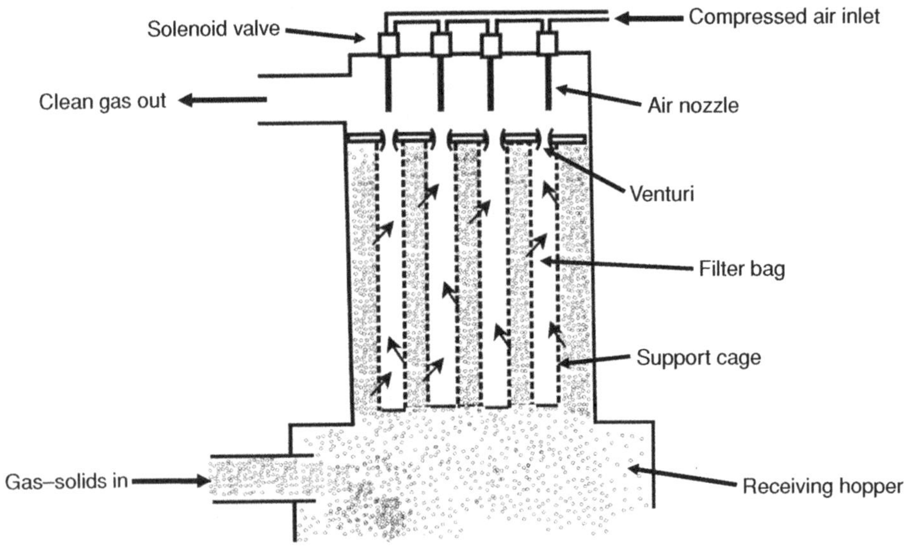

Figure 4.23 Bag filter unit with high pressure pulsed air jets.

measures must be taken to prevent its escape into the atmosphere from the handling plant. This is particularly the case with toxic and explosive materials. The choice of gas–solid disengaging system to be used on any given application will be influenced by a number of factors, notably the particle size range of the material.

Where a bulk material consists of relatively large and heavy particles, with no fine dust, it may be sufficient to collect the material in a simple bin, the solid material falling under gravity to the bottom of the bin, whilst the gas is taken off through a suitable vent. However, with a bulk solid of slightly smaller particle size it may be advisable to enhance the gravitational effect, and the most common method of achieving this is to impart spin to the gas–solid stream so that the solid particles are thrown outwards while the gas is drawn off from the centre of the vortex. This is basically the principle on which the cyclone separator operates. Where fine particles are involved, especially if they are also of low density, separation in a cyclone may not be fully effective, and in this case the gas–solid stream may be vented through a fabric filter.

4.3.2.1 *Filters* The fabric filter is now the industry standard for gas–solid separation duties in pneumatic conveying systems. With pulsed reverse air jets, on-line cleaning of these filters is possible, and very compact units are available. A sketch of such a device is given in Figure 4.23. For batch conveying it may be possible to use a shaking mechanism to clean the filter bags, rather than the reverse air jets.

The size of a filter, in terms of fabric surface area required, is specified approximately in terms of the volumetric flow rate of gas to be handled. Filtration plant is designed for steady state conditions and so if there are likely to be surges and transient effects in performance due to batch conveying or pipeline purging, for example, these effects must be taken into account. Filter fabrics will not last forever, and it is important that maintenance schedules are kept.

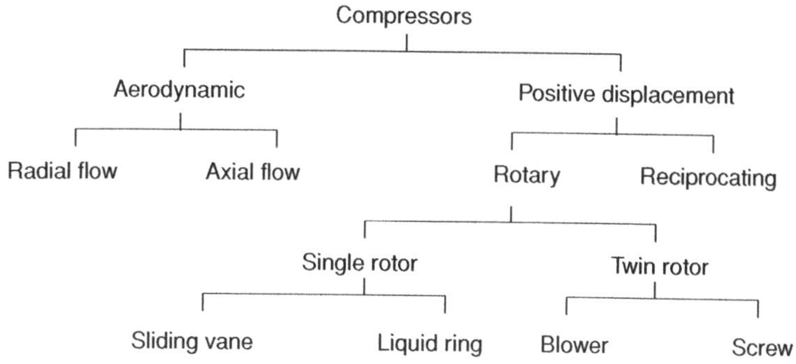

Figure 4.24 Classification of compressors.

Particular care must be taken with negative pressure systems. Firstly, it must be realised that with a system operating under vacuum, the volumetric flow rate of the air to be handled will be significantly greater than with an equivalent positive pressure system, because of the compressibility effect. Then consideration must be given to the air mover, because it has to exhaust the clean air drawn through it, and if this contains any dust, it could cause considerable damage to the machine if the dust is abrasive.

4.3.3 *Air supply*

The air mover is probably the most critical of all the equipment to be specified for a pneumatic conveying system. A wide range of air movers is available, from fans to blowers and screw to reciprocating compressors, and it is important that the correct choice is made. Not all air movers are suited to pneumatic conveying, however, and so the operating characteristics must be understood and interpreted. Plant air may be available, but it may not be economical to use. Some air movers have limitations and some are more suited as exhausters than compressors. It must also be appreciated that the air is likely to be delivered at a high temperature and may contain both oil and water vapour.

Air movers available for pneumatic conveying applications range from fans and blowers producing high volumetric flow rates at relatively low pressures to positive displacement compressors, usually reciprocating or rotary screw machines, capable of producing the higher pressures required for long distance or dense phase conveying. The basic types of air mover are categorised in Figure 4.24.

4.3.3.1 *Aerodynamic compressors* For high pressure duties centrifugal (turbo) compressors, and especially the multiple stage axial flow type, are normally manufactured only in large sizes, handling very high volumetric flow rates, and so they rarely find application to pneumatic conveying installations. The main problem with fans is that they suffer from the disadvantage that the air flow rate is very dependent upon the line pressure drop. This is a fundamental operating characteristic for pneumatic conveying, and so this class of compressor cannot be used reliably.

With a compressor having this type of operating characteristic, it means that if the solids feed rate to the system should become excessive for any reason, causing the pressure drop

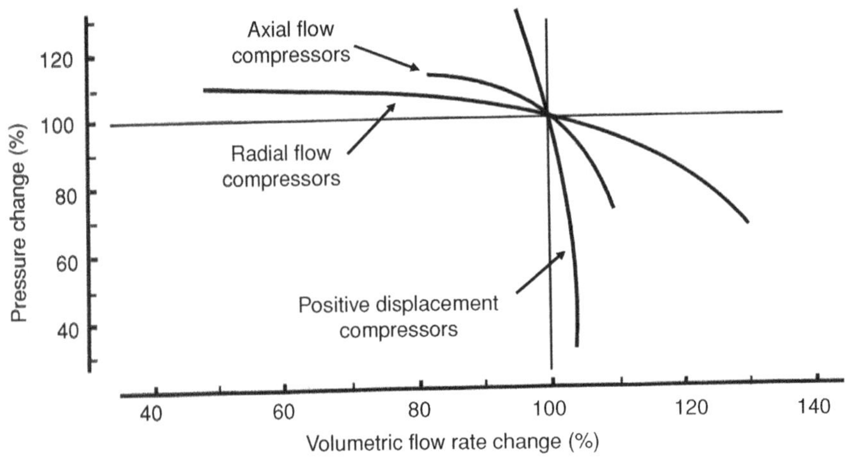

Figure 4.25 Constant speed characteristics of aerodynamic and positive displacement machines.

to increase significantly, the air flow rate may become so low that the material falls out of suspension, with the risk of blocking the line. Positive displacement machines, for which the volumetric flow rate is largely independent of the discharge pressure, are less likely to cause this type of system failure. This point is illustrated in Figure 4.25.

In order to convey materials reliably in pneumatic conveying systems a minimum value of conveying air velocity must be maintained. For dilute phase conveying this minimum velocity is typically of the order of 15 m/s, and if it drops by more than about 10 or 20% the pipeline is likely to block. A small surge in the feed rate into a pipeline of only 10% would cause a corresponding increase in pressure demand, and with either an axial flow or a radial flow machine, the reduction in the volumetric flow rate of the air would probably result in pipeline blockage. On vacuum duties they are often used for cleaning operations. With waste materials, such as paper, the blades of the fan can be sharpened so that they cut or chop the material into smaller pieces as it passes through the fan.

4.3.3.2 *Positive displacement compressors* The constant speed operating characteristic for positive displacement machines, shown in Figure 4.25, provides a basis on which the design of heavy duty conveying systems can be reliably based. A pressure surge in the conveying system will result in only a small decrease in the air flow rate delivered by the compressor, and this can be incorporated into the safety margins for the system. A pressure surge, of course, will cause a reduction in air velocity because of the compressibility effects, which must be catered for in such safety margins. With a positive displacement compressor the percentage reduction in conveying air velocity due to the constant speed characteristic will be no more than that caused by the compressibility effect.

In the classification of compressors presented in Figure 4.24, five different types of positive displacement compressor are included. The constant speed operating characteristic of each of these is similar to that shown in Figure 4.25. A particular feature of most of these machines is that very fine operating clearances are maintained between rotating parts. As a result there is no possibility of the conveyed material being conveyed through the

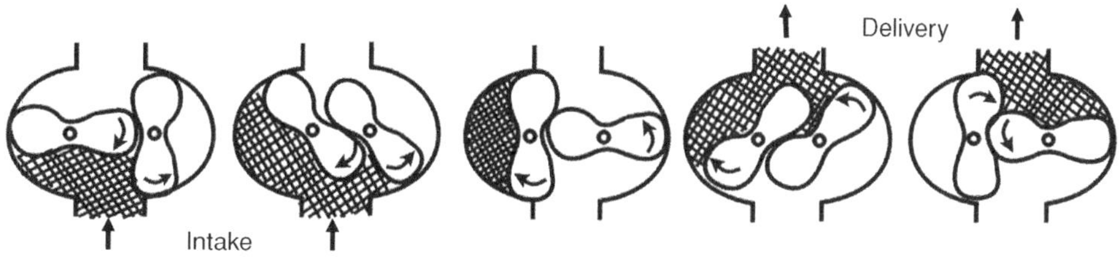

Figure 4.26 Operating principle of positive displacement blower.

compressor, as it can with a fan. Indeed, if the material being conveyed is abrasive, even dust must be prevented from entering the machine or it will suffer severe damage.

4.3.3.3 *Positive displacement blowers* Positive displacement blowers are probably the most commonly used type of compressor for dilute phase conveying systems. They provide an ideal match, in terms of pressure capability, with the conventional low pressure rotary valve, and are a typical working combination on many plants. They can be used as vacuum pumps, or exhausters, as well as blowers. The principle of operation is illustrated in Figure 4.26. Twin rotors are mounted on parallel shafts within a casing, and they rotate in opposite directions. As the rotors turn, air is drawn into the spaces between the rotors and the casing wall, and is transported from the inlet to the outlet without compression. As the outlet port is reached, compression takes place when the air in the delivery pressure pipe flows back and meets the trapped air.

Because of this shock compression the thermodynamic efficiency of the machine is relatively low and this is one of the reasons why these simple compressors are only used for low pressure applications. In order to reduce the pulsation level, and the noise, three lobed rotors, as well as twisted rotors, have been introduced. The performance of blowers would be enhanced with lubrication, but oil-free air is a general requirement of these machines. The maximum value of compression ratio with these machines is about 2.3:1. This means that at sea level (1.013 bar absolute) for blowing the maximum delivery pressure will be about 1.3 bar gauge, and for exhausting the maximum vacuum will be about 0.55 bar.

4.3.3.4 *Sliding vane rotary compressors* For medium and high pressure conveying systems the sliding vane type of rotary compressor is well suited. Figure 4.27 illustrates the operating principle of a simple single stage sliding vane compressor. It is a single rotor device, with the rotor eccentric to the casing. Compression occurs within the machine, unlike the blower, and so the air is delivered without such marked pulsations. It will be seen from Figure 4.27 that the machine will operate equally well as an exhauster for vacuum conveying duties.

4.3.3.5 *Other types* With the other machines highlighted in Figure 4.24, compression also occurs within the machine. The liquid ring compressor is another single rotor device, with the rotor eccentric to the casing. This finds widest use as a vacuum pump, partly because it is able to tolerate low concentrations of dust in the air, unlike any of the other machines. The liquid ring vacuum pump can reach 600 mm Hg in a single stage and over 700 mm Hg in two stages. For high pressure air the rotary screw compressor is now in wide use. The

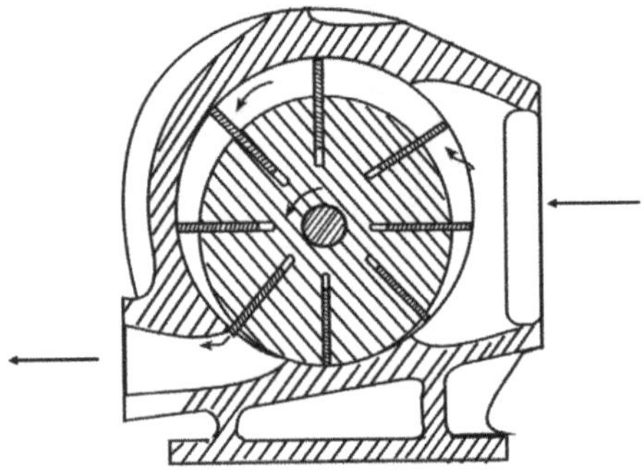

Figure 4.27 Sliding vane rotary compressor.

machine consists essentially of male and female intermeshing rotors mounted on parallel shafts. Inlet and outlet ports are at opposite ends of the compressor. Oil-free machines can reach about 4 bar with a single stage, and double this with oil injection. As these machines are generally free from pressure pulsations it is not usually necessary to operate with an air receiver, and they do not require special foundations for mounting.

4.3.4 *Air compression effects*

When compressed air is delivered into a pipeline for use, the air will almost certainly be very hot, and it may contain a lot of water and oil. The power required to provide the compressed air, and hence the operating cost, may also be very high.

4.3.4.1 *Delivery temperature* Much of the work energy that goes into compressing air manifests itself in increasing the temperature of the air. Thermodynamic equations are available that will allow this temperature to be evaluated. The compression process is adiabatic, partly because of the speed of the process, and it is far from being a reversible process. As a result, the temperature of the air leaving a compressor can be very high. If, for example, air at a temperature of 20°C is compressed to 1 bar gauge in a positive displacement blower, the minimum temperature after compression, for a reversible process, would be about 84°C, and with an isentropic efficiency of 80% it would be 100°C. If air at 20°C is compressed to 3 bar gauge in a screw compressor it will be delivered at a temperature of about 200°C.

Whether or not the air can be used without being cooled will depend to a large extent on the properties of the material to be conveyed. The most efficient form of compression is to carry out the process isothermally, and so cylinders of reciprocating compressors, for example, would be water-cooled, and if staging was employed for achieving high pressures, inter-cooling would be employed here as well.

4.3.4.2 *Oil-free air* Oil-free air is generally recommended for most pneumatic conveying systems and not just those where the material must not be contaminated, such as food products, pharmaceuticals and chemicals. Lubricating oil, if used in an air compressor, can be carried over with the air and can be trapped at bends in the pipeline or obstructions. Most lubricating oils eventually break down into more carbonaceous matter which is prone to spontaneous combustion, particularly in an oxygen-rich environment, and where frictional heating may be generated by moving particulate matter.

Although conventional coalescing after-filters can be fitted, which are highly efficient at removing aerosol oil drops, oil in the super-heated phase will pass straight through them. Super-heated oil vapour will turn back to liquid further down the pipeline if the air cools. Ultimately precipitation may occur, followed by oil breakdown, and eventually a compressed air fire. The only safe solution, where oil-injected compressors are used, is to use chemical after-filters such as the carbon absorber type which are capable of removing oil in both liquid droplet and super-heated phases.

4.3.4.3 *Water removal* As the pressure of air is increased, its capability for holding moisture in suspension decreases. As the temperature of air increases, however, it is able to absorb more moisture. If the air is not initially saturated it may well reach the saturation point during the compression process, or in a following after-cooler. Where air is compressed isothermally quite large quantities of water vapour can be condensed, and in many cases the air leaving the compressor will be saturated. In adiabatic compression the temperature of the air will rise, and because of the marked ability of warmer air to support moisture, it is unlikely that any condensation will take place during the compression process.

As compression occurs very rapidly, it is quite possible that droplets of water will be carried through pipelines with the compressed air. Also, if additional cooling of saturated air occurs in the outlet line, further condensation will occur. The removal of droplets of water in suspension is a relatively simple process. It is important, however, that water filters and drains should be carefully maintained, and be protected from frost.

4.3.4.4 *Air drying* If dry air is required for conveying, a reduction in specific humidity can be obtained by cooling the air. When air is cooled its relative humidity will increase, and when it reaches 100% further cooling will cause condensation. Beyond this point the specific humidity will decrease. If the condensate is drained away and the air is then heated, its specific humidity will remain constant, but the relative humidity will decrease. This process is adopted in most refrigerant types of air dryers. These are usually limited to a dew point of about 2°C since drying down to this level of moisture avoids problems of ice formation and freezing. The driest possible air is obtained from a desiccant dryer. These are capable of reducing the moisture level to an equivalent dew point temperature of −40°C.

4.3.5 *Power requirements*

Delivery pressure and volumetric flow rate are the two main factors that influence the power requirements of a compressor, blower or fan. For an accurate assessment of the power requirements, it will clearly be necessary to consult manufacturer's literature. By this means different machines capable of meeting a given duty can be compared. For a quick, approximate assessment, to allow a comparison to be made of different operating variables, a simple model based on isothermal compression can be used (Mills 2004):

$$\text{Power} = 202\ \dot{V}_0 \ln\left(\frac{p_1}{p_2}\right)\ \text{kW} \tag{4.1}$$

Table 4.1 Pipe diameters and wall thicknesses for 4 inch nominal bore pipeline.

	Schedule number			
Dimensions	10	40	80	160
Wall thickness (in)	0.162	0.237	0.337	0.531
Pipe bore (in)	4.176	4.026	3.826	3.438
Pipe bore (mm)	106.1	102.3	97.2	87.3
Outside diameter (in)	4.5	4.5	4.5	4.5

This will give an approximate value of the actual drive power required. If this is multiplied by the unit cost of electricity it will give the cost of operating the system. Since power requirements for pneumatic conveying can be very high, particularly if it is required to convey a material at a high flow rate over a long distance, this basic model will allow an estimation of the operating cost per tonne of material conveyed to be made. It must be emphasised that the model presented in Equation (4.1) is only for first approximation purposes.

4.3.6 *Pipelines and valves*

Decisions with regard to the specification of components for pneumatic conveying systems do not end with the feeder, air mover and filter. There are likely to be numerous valves on the plant, and the pipeline is just as important. This importance is significantly magnified if the material to be conveyed is abrasive.

4.3.6.1 *Pipelines* Decisions do have to be made with regard to the pipeline. Material, wall thickness, surface finish and steps, all have to be given due consideration, as well as bends in the pipeline.

4.3.6.2 *Wall thickness* Pipeline is available in a wide range of wall thicknesses. This is partly for operating pressure applications and for wear resistance with abrasive materials. In evaluating conveying air velocity it must be realised that the diameter of a 4 inch nominal bore pipeline is rarely 4 inches. If an abrasive material is to be conveyed, wear of the pipeline must be expected. To give the pipeline a longer life, pipe having a greater wall thickness should be used. Schedule numbers are often used to specify wall thickness. Typical dimensions for 4 inch nominal bore pipeline, for example, are given in Table 4.1.

If the material to be conveyed is not abrasive at all, a thin-walled pipeline could be suitable for the duty. Pipeline weight in kg per metre could be added to Table 4.1 and this would show a marked difference. Lighter pipe sections will certainly make construction of the pipeline easier, particularly if there are vertical sections to erect.

4.3.6.3 *Pipeline rotation* If a pipeline is to convey an abrasive material having a large particle size, the particles will tend to 'skip' along the pipeline and so wear a groove on the bottom of the pipeline. In this case a thick-walled pipeline would be essential, but if the pipeline was to be rotated periodically, this would extend the life of the pipeline considerably. For this purpose the pipeline must be located in a place where convenient access can be gained for the necessary changes to be made. With fine particles there is little tendency for the bottom section of the pipeline to wear preferentially.

4.3.6.4 *Pipeline material* Although steel is the most commonly used pipeline material, many other materials are available to suit the conveyed material and duty. It was mentioned above that thin-walled pipe would be easier to handle and erect because it is lighter. Aluminium pipe is often used for this purpose. Because of problems of moisture and condensation in pipelines there is always the possibility of steel rusting and contaminating the conveyed material. In cases where hygiene is important, such as with many food and pharmaceutical products, the pipeline will need to be made from stainless steel.

Where flexibility is required in a pipeline and this cannot be conveniently achieved with a combination of straight pipe and bends, flexible hose can be used. Where a single line needs to feed into a number of alternative lines, and a flow diverter is not wanted to be used, a section of flexible hose of the steel braided type can be used to provide the link.

Where road and rail vehicles and boats need to be off-loaded, flexible rubber hose is ideal. It is available in natural rubber and a variety of synthetic materials, and comes in a wide range of sizes. Flexibility is generally needed in ship off-loading applications with vacuum systems, and hoses provide the necessary flexibility. Care must be taken if the material is abrasive and has a large particle size, because the wear rate of rubbers can be excessive with such materials. Unfortunately, the pressure drop for gas–solid flows through rubber hose increases with increase in velocity, and more so than for steel pipeline (Mills 2004; Mills *et al.* 2004). If a very abrasive material is to be conveyed in a pipeline, consideration must be given to the use of schedule 80 pipeline or higher. An alternative to this, which is commonly adopted, is to line the pipeline with basalt. If a better material is required, then alumina ceramics can be used, but this is likely to be more expensive. A usual combination is to line the straight pipeline with basalt and use alumina for the bends. Alloy cast iron pipe can also be used.

4.3.6.5 *Surface finish* Most pipelines are supplied having a satisfactory surface finish with regard to frictional resistance to flow. For some materials, such as polyethylene, however, a particular surface finish is required for the specific purpose of reducing the problem of 'angel hairs', or particle melting, with these materials. An artificially roughened surface is often specified.

4.3.6.6 *Bends* Bends provide a pneumatic conveying system pipeline with considerable flexibility in routing, but are the cause of many problems. Each bend will add to the overall resistance of the pipeline, and hence to the conveying air pressure required. If the conveyed material is abrasive an ordinary steel bend could fail within 2 hours. An abrupt change in direction will add to the problem of fines generation with friable materials, and angel hairs will be generated in long radius bends with many synthetic materials.

Numerous different bends are available to minimise each of the above problems. Many of these are made of, or lined with, basalt, cast iron, rubber, etc., and some have a constant bore and a constant radius, as with conventional bends. Another group of bends that have been developed have neither constant bore nor constant radius. Some of these bends are shown in Figure 4.28. Care must be taken in selecting such bends, for account must be taken

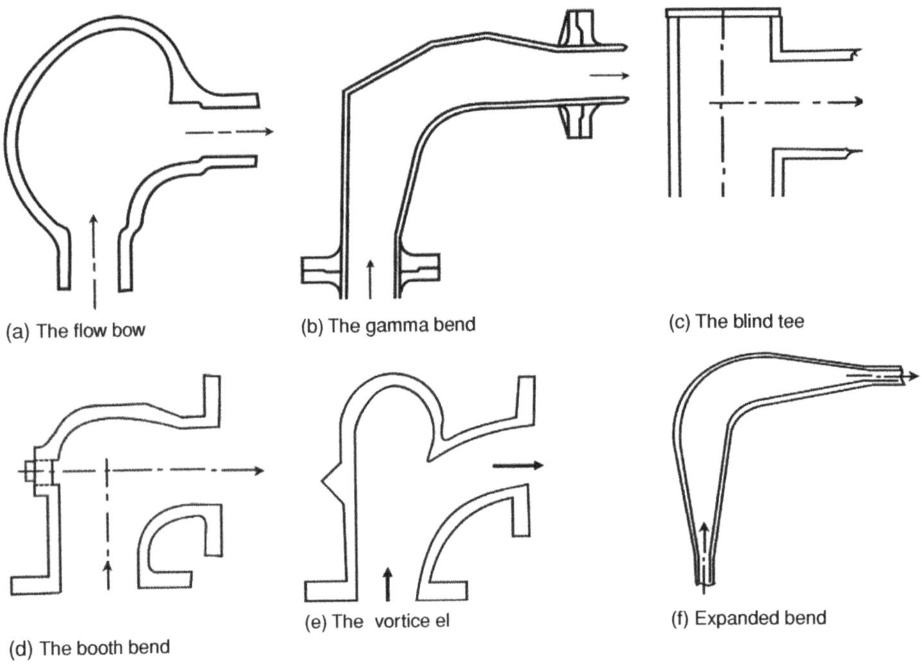

Figure 4.28 Some special bends developed for pneumatic conveying systems.

of their suitability for the material being conveyed and the pressure drop across the bend with that material.

4.3.6.6.1 *Blind tee* With an abrasive material, the simple blind tee bend shown in Figure 4.28 will probably last 50 times longer than an equivalent radiused bend. It will ultimately fail around the inside corner due to turbulence. For abrasive materials, therefore, it is extremely effective, and can even be made out of scrap material. The blind end of the bend traps the conveyed material and so the oncoming material impacts against other material, instead of the bend, and thereby protects it. The penalty is in the increased pressure drop that can result. Another problem with this type of bend is that the material that is trapped in the dead end of the bend may take a long time to be purged from the bend at the end of a conveying run. It could not, therefore, be used in pipelines required for the conveying of perishable and time-limited materials.

4.3.6.6.2 *Pressure drop* Because of the change in direction, impact of particles against bend walls, and general turbulence, there will be a pressure drop across every bend in any pipeline. The major element of the pressure drop, however, is that due to the re-acceleration of the particles back to their terminal velocity after exiting the bend. The situation can best be explained by means of a pressure profile in the region of a bend, such as that in Figure 4.29.

The pressure drop that might be recorded across the bend is quite small, and although this technique might be appropriate for single-phase flows around bends, it is inappropriate for gas–solids flow. The particles leaving the bend will be at a lower velocity than that at

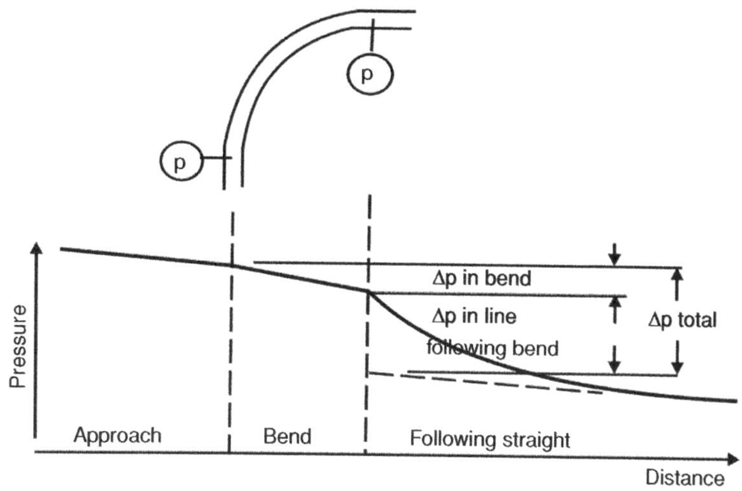

Figure 4.29 Pressure drop elements and evaluation for bends in a pipeline.

entry and so they will have to be re-accelerated. The bend was the cause of the problem but the re-acceleration occurs in the straight length of pipeline following the bend, and so it is here that the associated pressure drop occurs, and not in the bend itself.

4.3.6.7 *Steps* If high pressure air, or a high vacuum, is used for conveying a material, it would generally be recommended that the pipeline should be stepped to a larger bore part way along its length. This is to cater for the expansion of the air that occurs with decrease in pressure, and so prevents excessively high conveying air velocities towards the end of the pipeline.

Figure 4.30 illustrates the case of a high pressure dilute phase conveying system. The minimum value of conveying air velocity that must be maintained is about 15 m/s, and 60 m^3/s of free air is available to convey the material. The conveying line inlet air pressure is 4 bar gauge. From Figure 4.30 it will be seen that a 125 mm bore pipeline will be required for these conditions, and the resulting conveying line inlet air velocity will be 16.5 m/s. If a single bore pipeline is used, however, the conveying line exit air velocity will be about 81.5 m/s.

A velocity of 81.5 m/s will cause considerable damage to any conveyed material and very serious wear to the plant if the material is only slightly abrasive. By stepping the pipeline twice, as shown in Figure 4.30, it will be seen that the velocity profile can be kept within reasonably low limits. The stepping of a pipeline to a larger bore would also be recommended for high vacuum conveying systems and high pressure dense phase conveying. The stepping of a pipeline is only dependent upon conveying air pressure and should be undertaken for any length of pipeline. The stepping of a pipeline is also likely to lead to a significant improvement in performance of the conveying line (Mills 2004; Mills *et al.* 2004; Mills & Agarwal 2001).

4.3.6.8 *Valves* A number of different valves may need to be used on pneumatic conveying plant, and a wide variety of different valves are available in the market place. Rotary valves

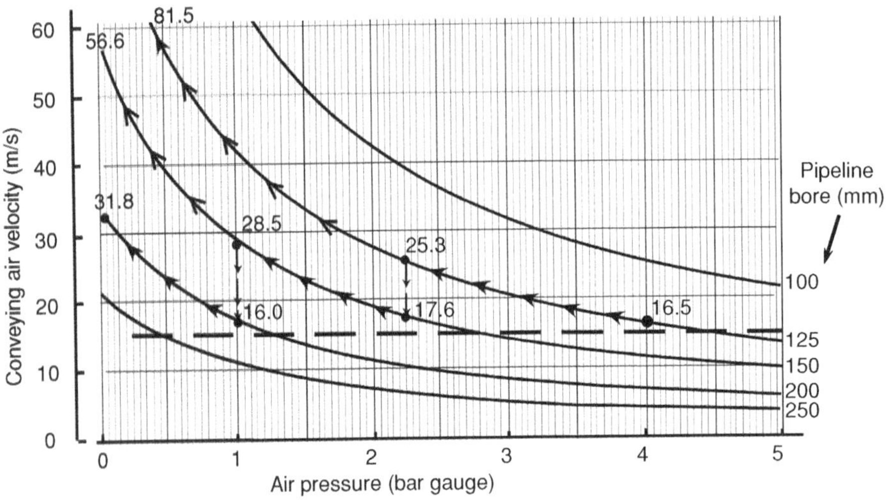

Figure 4.30 Stepped pipeline velocity profile for high pressure dilute phase system.

have been considered at length, and are ideal for controlling the feed of material into or out of a system at a controlled rate. There is, however, a requirement for many other types of valve, generally to be used for the purpose of isolating the flow. Many of these have been included on sketches of conveying systems earlier and include discharge valves, vent line valves and diverter valves.

4.3.6.9 *Discharge valves* A valve in a conveying line that is required to stop and start the flow has an onerous duty. Although the valve is only used in either the open or closed position, and is not used for flow control purposes, particulate material must be able to pass freely through when it is open. If the control surfaces of the valve remain in the flow path, as they will with pinch valves and ball valves, they must provide a perfectly smooth passage for the flow of material through the valve when open. Any small protuberances or surface irregularities that could promote turbulence in the area would result in a rapid deterioration in performance. This is particularly the case when the material to be conveyed is abrasive. Any valve for this duty is also very vulnerable during the opening and closing sequences, and so these operations should be completed as quickly as possible.

The dome valve is a more recent addition to the list of valves available, but it has been specifically designed for this type of duty, and is now being widely used in industry. The valve has moving parts, but these move completely out of the path of the conveyed material when the valve is open. On closing, the valve first cuts through the material and then becomes airtight by means of an inflatable seal. The valve can be water-cooled and so it is capable of handling hot materials.

4.3.6.10 *Isolating valves* There are many instances where material has to be transferred, usually under gravity, in batches. The valve is either open or closed and often has to provide an airtight seal. In the gate lock feeder (Figure 4.17), for example, a pair of valves is required to operate in sequence to feed small batches of material into a pipeline. Where batches of material have to be fed into blow tanks, the valve has to be capable of withstanding the pressure subsequently applied to the blow tank. Of the valves considered above only the dome valve would be appropriate for this type of duty. It finds wide use in this application, particularly with the more difficult granular and abrasive materials.

If the material to be handled is not abrasive, the butterfly valve is ideal. They are reasonably priced, require very little headroom, are not too heavy, and are reasonably airtight. They are widely used in the food and related industries and in gate lock feeders. They are, however, much too vulnerable for use with abrasive materials, since the valve remains in the flow when it is open. Disc valves, like butterfly valves, require very little headroom, but like dome valves, they swing completely out of the way of the flow of material. They cut though the material on closing, but generally rely on the subsequent pressure in the vessel below to provide the necessary seal. Their suitability for use will depend very much upon the material to be handled and the application.

Slide valves are the oldest valves in the business, and although they have been improved over the years, the disc valve is a specific development from it. They take up little space and are cheap. A particular application is in terms of back-up. If any of the other more expensive and sophisticated valves fail, and need to be replaced, this can be a very difficult and time-consuming task if the valve is holding several hundred tonnes of material in a hopper, and this must be drained out before the valve can be removed for repair or replacement.

4.3.6.11 *Vent line valves* This is a deceptively easy duty, but if it is on a high pressure blow tank handling fly ash or cement, the valve will have to operate in a very harsh environment. The air velocity will be very high, albeit for a very short period of time, but a lot of abrasive dust is likely to be carried with the air. If the material is abrasive then the choice is between a pinch valve and a dome valve. If the material is non-abrasive, a diaphragm valve could be used.

4.3.6.12 *Flow diversion* Flow diverting is a very common requirement with pneumatic conveying systems and can be achieved very easily. Many companies manufacture specific flow diverting valves for the purpose. Alternatively flow diversion can be achieved by using a set of isolating valves. The most common requirement is to divert the flow to one of two alternative routes, typically where material needs to be discharged into a number of alternative hoppers or silos. In this case the main delivery line would be provided with a diversion branch to each outlet in turn.

There are two main types of diverter valve. In one a hinged flap is located at the discharge point of the two outlet pipes. This flap provides a seal against the inlet to either pipe. The pipe walls in the area are lined with urethane, or similar material, to give an airtight seal, and this provides a very compact and light-weight unit. The other main design operates with a tunnel section of pipe between the supply and the two outlet lines. This unit would not be recommended for abrasive materials however. This design should provide a more positive seal for the line not operating, which would probably make it a more suitable valve for vacuum conveying duties. A sketch of a parallel tunnel-type diverter valve is presented in Figure 4.31 to illustrate the method of operation.

Flow diversion can equally be achieved by using a pair of isolating valves, with one placed in the branch, close to the supply pipe, and the other in the supply pipe, just downstream of the branch. This can be repeated at any number of points along the pipeline. The main

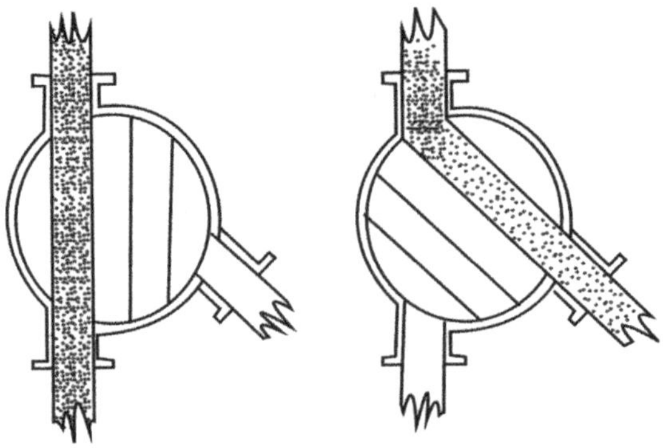

Figure 4.31 Sketch of a parallel tunnel-type diverter valve.

disadvantage with this arrangement is that a plug of material will be trapped in the short section of pipeline not in use, which will have to be blown through when the flow direction changes. If the conveyed material is abrasive, this method of flow diversion would be recommended. Either pinch valves or dome valves would need to be employed for the purpose. With two separate valves, instead of one to operate, care would have to be exercised with the sequencing when changing flow direction.

4.3.6.13 *Flow splitting* The splitting of a single line into multiple flow lines generally requires a specific pipeline layout and so there are few devices available. Flow splitting is often required on boiler plant, where coal dust might need to be sent to the four corners, and on blast furnaces, where coal or limestone powder might need injecting at a dozen or more points around the perimeter. The main requirement is generally that each outlet should be supplied with material at a uniform rate, despite the fact that the distance to each will be different. The splitting is best achieved in the vertical plane, with the line sizes carefully evaluated to provide a uniform balance for each.

4.3.6.14 *Non-return valves* Non-return valves need to be considered in terms of system protection, particularly in the event of a pipeline blockage, for blowers and compressors with positive pressure conveying systems. Should a pipeline block there is always a risk of particulate material flowing back along the pipeline when the compressor is switched off.

4.4 Conveying capability

There is essentially no limit to the capability or a pneumatic conveying system for the conveying of dry bulk particulate materials. Almost any material can be conveyed and high material flow rates can be achieved over long distances. There are, however, practical limitations and these are mainly imposed by the fact that the conveying medium, being a gas, is compressible. The limiting parameters are then mainly the economic ones of scale and power requirements. Conveying capability depends mainly upon five parameters. These are pipe bore, conveying distance, pressure available, conveying air velocity and material properties. The influence of many of these variables is reasonably predictable but that of the conveyed material is not fully understood at present.

4.4.1 *Pipeline bore*

The major influence on material flow rate is that of pipeline bore. If a greater material flow rate is required it can always be achieved by increasing the pipeline bore, generally regardless of the other parameters. In a larger bore pipeline a larger cross-sectional area is available and this usually equates to the capability of conveying more material.

4.4.2 *Conveying distance*

In common with the single phase flow of liquids and gases, conveying line pressure drop is approximately directly proportional to distance. Long distance conveying, therefore, tends to equate to high pressure, particularly if a high material flow rate is required. For the majority of conveying applications, however, it is not convenient to use high pressures. As a consequence, long distance, with respect to pneumatic conveying, typically means about 1½ km, and dilute phase conveying.

4.4.3 *Pressure*

Although air, and other gases, can be compressed to very high pressures, it is not generally convenient to use air at very high pressure. The reason for this is that air is compressible and so its volumetric flow rate constantly increases as the pressure decreases. In hydraulic conveying, pressures in excess of 100 bar can be used so that materials can be conveyed over distances of 100 km with a single stage. With water being essentially incompressible, changes in the velocity of the water over this distance are not very significant. In pneumatic conveying, air at a pressure above about 1 bar gauge is generally considered to be 'high pressure'. Although the air expansion can be accommodated to a certain extent by stepping the pipeline to a larger bore, as illustrated in Figure 4.30, this is a complex design procedure. As a consequence, air pressures above about 5 bar are rarely used for pneumatic conveying systems that deliver materials to reception points at atmospheric pressure.

4.4.4 *Conveying air velocity*

The main design parameter with respect to pneumatic conveying is conveying air velocity, and more particularly, conveying line inlet air or 'pick-up' velocity. Since the air expands along the length of the pipeline it will always be a minimum at the material feed point at the start of the pipeline, in a single bore pipeline, regardless of whether it is a positive pressure or a vacuum conveying system. In a single bore pipeline the velocity will be a maximum at the end of the pipeline. It is the value of the minimum velocity of the air that is critical to the successful operation of a pneumatic conveying system. In dilute phase conveying the particles are conveyed in suspension in the air and this relatively high value of velocity is due, in part, to the large difference in density between the particles and the air. In hydraulic conveying typical velocities for suspension flow are only about 1½ m/s, but the difference in density between water and particles is very little in comparison.

4.4.5 *Particle velocity*

In dilute phase conveying, with particles in suspension in the air, the mechanism of conveying is one of drag force. The velocity of the particles, therefore, will be lower than that of the conveying air. It is a difficult and complex process to measure particle velocity, and apart from research purposes, particle velocity is rarely measured. As a consequence it is generally only the velocity of the air that is ever referred to in pneumatic conveying. This is effectively the 'superficial air velocity', because the presence of the particles is disregarded in evaluating the value of velocity. The velocities quoted on Figure 4.30, therefore, are conveying air velocities.

In a horizontal pipeline the velocity of the particles will typically be about 80% of that of the air. This is usually expressed in terms of a slip ratio, defined in terms of the velocity of the particles divided by the velocity of the air transporting the particles, and in this case it would be 0.8. The value depends upon the particle size, shape and density, and so the value can vary over an extremely wide range. In vertically upward flow in a pipeline a typical value of the slip ratio will be about 0.7 in comparison.

These values relate to steady flow conditions in pipelines remote from the point at which the material is fed into the pipeline, bends in the pipeline and other possible flow disturbances. At the point at which the material is fed into the pipeline, the material will essentially have zero velocity. The material will then be accelerated by the conveying air to its slip velocity value. This process will require a pipeline length of many metres and this distance is referred to as the 'acceleration length'. The actual distance will depend once again on particle size, shape and density. The process was illustrated earlier in Figure 4.29 in relation to the pressure drop across a bend.

4.4.6 *Material properties*

The properties of the conveyed material have a major influence on the conveying capability of a pneumatic conveying system. It is the properties of the material that dictate whether the material can be conveyed in dense phase in a conventional conveying system, and the minimum value of conveying air velocity required. For this reason the conveying characteristics of several different materials are presented in order to illustrate the importance and significance of material properties.

Although it is the properties of the bulk material, such as particle size and size distribution, particle shape and particle density that are important in this respect, at this point in time it is the measurable properties of materials in bulk that are more fully understood. These include air–material interactions, such as air retention and permeability, and are more convenient to use. In general, materials that have either good air retention or good permeability will be capable of being conveyed in dense phase and at low velocity in a conventional conveying system. Materials that have neither good air retention nor good permeability will be limited to dilute phase suspension flow.

4.4.7 *Dense phase conveying*

As mentioned earlier, there are two main mechanisms of low velocity dense phase flow. For materials that have good air retention, the material tends to be conveyed as a fluidised mass.

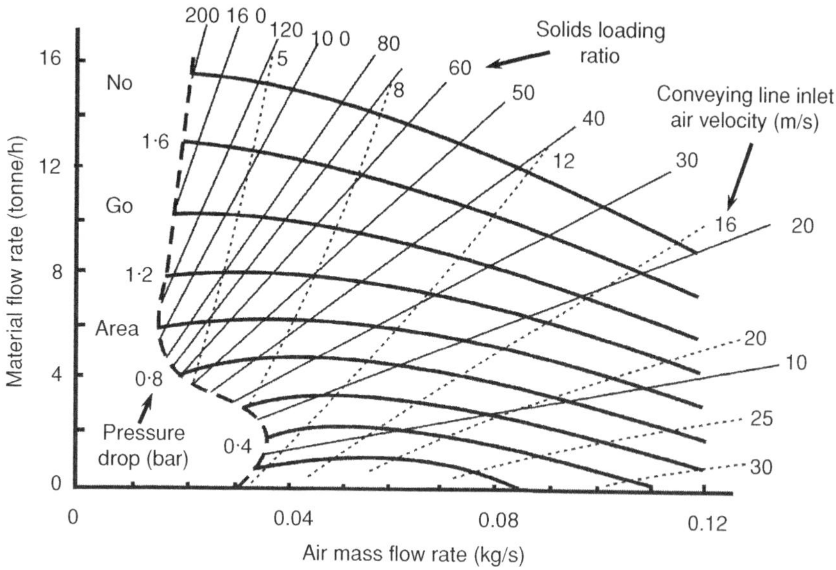

Figure 4.32 Conveying data for ordinary portland cement.

In a horizontal pipeline the vast majority of the material will flow along the bottom of the pipeline, rather like water, with air above but carrying very little material. For materials that have good permeability the material tends to be conveyed in plugs through the pipeline. The plugs fill the full bore of the pipeline and are separated by short air gaps. As the conveying air velocity is reduced, the air gap between the plugs gradually fills with material along the bottom of the pipeline and the plug ultimately moves as a ripple along the top of an almost static bed of material. As the air flow rate reduces, to give very low conveying air velocities, the material flow rate also reduces.

Materials composed entirely of large mono-sized particles, such as polyethylene and nylon pellets, peanuts and certain grains and seeds, convey very well in plug flow. In dilute phase conveying, nylons and polymers can suffer damage in the formation of angel hairs, and grains and seeds may not germinate as a consequence of damage caused at the high velocities necessary for conveying. Because of the very high permeability necessary, air will readily permeate through the material while it is being conveyed and so maximum values of solids loading ratio will typically be about 30.

4.4.8 *Sliding bed flow*

Pneumatic conveying data for cement is presented in Figure 4.32. This is essentially a performance map for the material in a given pipeline and is a graph of material flow rate against air mass flow rate, with lines of constant conveying line pressure drop plotted. The pressure drop lines are derived from experimental data obtained from conveying the material in the given pipeline, which was 53 mm bore, of 50 m length, almost entirely in the horizontal plane and included nine 90° bends having a bend diameter to bore ratio of 24:1. Since solids loading ratio is the ratio of the material to air flow rates, these can simply

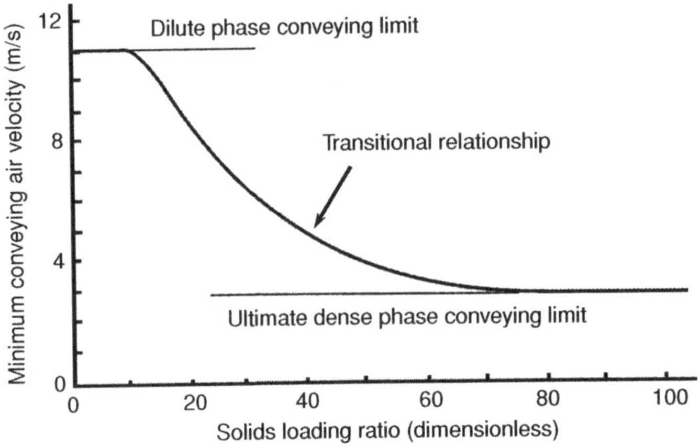

Figure 4.33 The influence of solids loading ratio on the minimum conveying air velocity for the pneumatic conveying of ordinary portland cement.

be added as straight lines through the origin. The lines of constant conveying line inlet air velocity can be plotted from basic thermodynamic relationships.

It will be seen from Figure 4.32 that the cement could be conveyed at high values of solids loading ratio and with low values of conveying line inlet air velocity, and hence in dense phase. These 'conveying characteristics' are typical of those for most fine powdered materials that have very good air retention. At very low values of conveying line pressure drop it will be seen that the minimum value of conveying line inlet air velocity for the cement is about 11 m/s. Everything to the right of this line, at higher velocities, is entirely dilute phase suspension flow. There is a natural transition between dilute and dense phase flow, and because of the air expansion the flow can change from dense to dilute phase along the length of a pipeline when velocities are in the transition zone.

The locus of the line defining the 'no go area' for the cement is given in Figure 4.33. The minimum value of conveying air velocity required for the reliable conveying of the cement is dependent upon the solids loading ratio, or the concentration of the cement in the pipeline. As the solids loading ratio of the material increases, the conveying line pressure required also increases, and so much higher pressures are required to convey a material in dense phase than in dilute phase. At low velocities, however, problems of erosive wear with abrasive materials, and particle degradation with friable materials, are significantly lower. For materials such as cement, power requirements are also very much lower.

With a limit on air pressure used for conveying, because of the problems associated with air expansion, dense phase conveying, as a consequence, is additionally limited by conveying distance. If conveying distance is doubled, for example, for the same air supply pressure, the value of solids loading ratio will fall by about half. With the same pressure there will be no need for an increase in air flow rate, but the material flow rate will have to drop by about half to compensate, since there is no increase in energy to the conveying system. Since conveying capability is dependent upon both parameters, pressure gradient is a more convenient term to use and the effect can be shown in Figure 4.34.

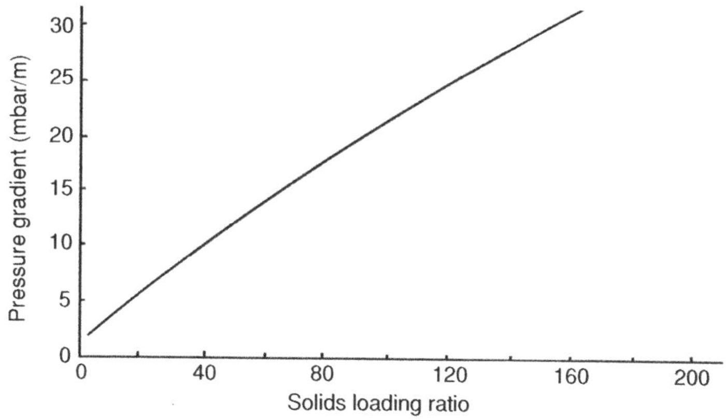

Figure 4.34 Approximate influence of solids loading ratio on conveying line pressure gradient for the horizontal conveying of cement.

For materials that are capable of being conveyed in dense phase, therefore, this capability will only be possible if the air supply pressure is relatively high or the conveying distance is relatively short. For low pressure or long distance conveying it will only be possible to convey a material in dilute phase, even if the material has dense phase conveying capability. Pressure gradient is simply the ratio of the pressure drop available for conveying, divided by the equivalent length of the pipeline. For convenience it is expressed in mbar per metre of horizontal pipeline.

Conveying distance is in terms of an equivalent length in order to take account of vertical sections and the number and geometry of bends in the pipeline. The reference for equivalent length is that for straight horizontal pipeline. For flow vertically up a scaling parameter of two approximately holds, such that the equivalent length is double the actual vertical lift in terms of straight horizontal pipeline. The equivalent length of bends depends very much upon the value of the conveying air velocity. A correlation derived for cement and fine fly ash is presented in Figure 4.35 (Mills 2004; Mills *et al.* 2004).

4.4.9 *Plug flow*

Pneumatic conveying data for polyethylene pellets is presented in Figure 4.36. The pellets were conveyed through the same pipeline as the cement in Figure 4.32 and so a direct comparison of performance is possible, as the data was obtained over the same range of air flow rates and air supply pressures. Although conveying is possible at low values of conveying air velocity, there is a marked change in material flow rate for low velocity conveying. An optimum maximum value of material flow rate occurs at a conveying line inlet air velocity of about 15 m/s, which corresponds closely with the minimum value of conveying air velocity for the dilute phase conveying of the material.

As with the cement, there is a natural transition from dilute to dense phase conveying for the material. With the lines of constant pressure drop being so close together in the dense phase conveying region, however, care must be exercised with controlling material flow rate in the region close to the conveying limit. Solids loading ratio values are significantly

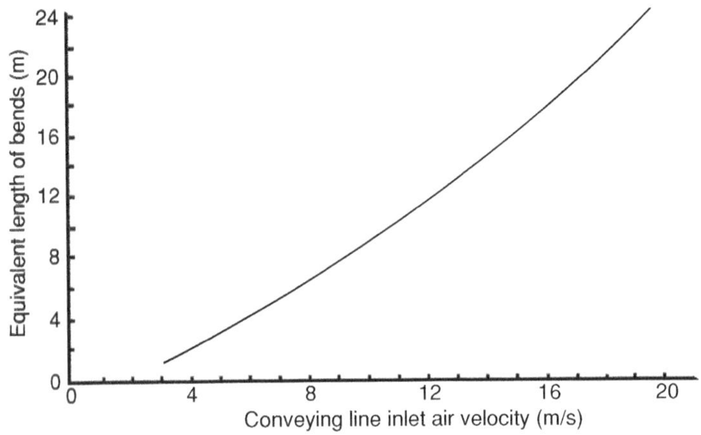

Figure 4.35 Equivalent length of long radius 90° bends for the conveying of cement.

lower because of the high permeability of the material and the very different form of the conveying characteristics with low velocity dense phase conveying of the material. For the dense phase conveying of this type of material, therefore, solids loading ratio does not have the same significance as it does for sliding bed flow.

4.4.10 *Dilute phase conveying*

Pneumatic conveying data for granulated sugar is presented in Figure 4.37. Once again this material was also conveyed through the same pipeline as that for the cement and polyethylene pellets and so further direct comparison of conveying performance is possible. Granulated sugar has very poor air retention capability, and poor permeability, and so can

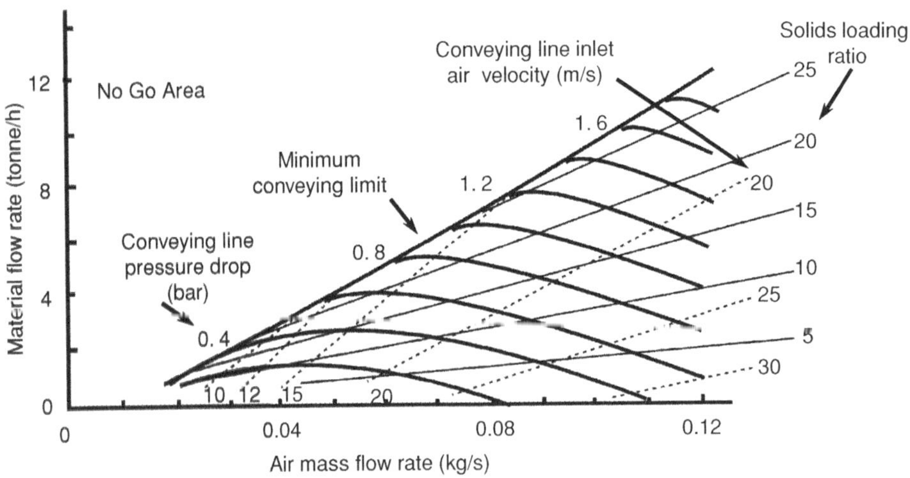

Figure 4.36 Conveying data for polyethylene pellets.

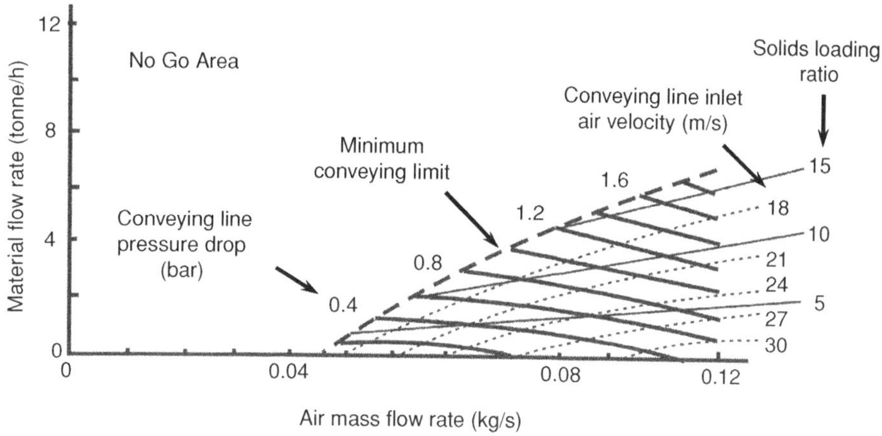

Figure 4.37 Conveying data for granulated sugar.

only be conveyed in dilute phase suspension flow in a conventional conveying system. The pipeline was relatively short and the air supply pressure was quite high, but despite the fact that the pressure gradient was very high, the sugar could not be conveyed in dense phase and hence at low velocity. High pressure, therefore, is not synonymous with dense phase conveying.

It will be seen from Figure 4.37 that the minimum conveying air velocity for the sugar was about 16 m/s. The maximum value of solids loading ratio achieved was just over 15 and so this is quite clearly dilute phase suspension flow conveying. From Figure 4.31, it will be seen that the cement could be conveyed at a solids loading ratio of about 40 in dilute phase flow. This is because the minimum conveying air velocity was only 11 m/s and the cement could be conveyed at a very much higher material flow rate.

With no dense phase conveying capability the operating envelope for the sugar is significantly smaller than that for either the cement or the polyethylene pellets. For any given pressure drop and air flow rate the material flow rate achieved for the sugar was also significantly lower than that for either the cement or the polyethylene pellets. A comparison of the performance of a number of different materials conveyed through this same pipeline is presented in Figure 4.38.

It will be seen from Figure 4.38 that the conveying capability of different materials can vary widely, and not only in terms of dilute and dense phase conveying. Every material was conveyed in dilute phase suspension flow and even here there was a 2:1 variation. Differences are clearly evident between sliding bed and plug modes of dense phase flow but particularly wide differences exist between powdered materials, having good air retention, at low air flow rates, and hence low velocities. Copper concentrate might be classified as 'medium phase' but this capability was due to the fact that the air retention capability was not as good as the other finer materials. The coke fines had a lower value of minimum conveying air velocity than the granulated sugar due to the fact that the material had a very much wider particle size distribution. It can also be recorded that particle density does not correlate very well with these materials.

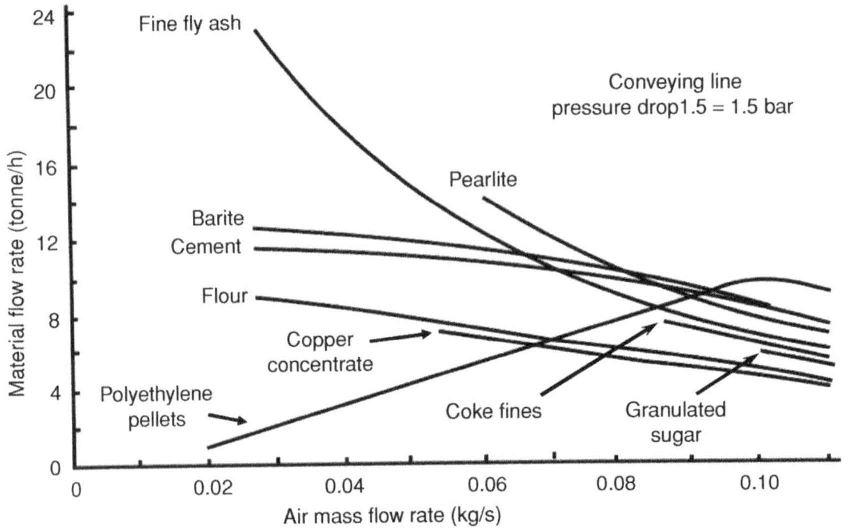

Figure 4.38 Comparison of the pneumatic conveying performance of different materials conveyed under identical conditions.

4.5 Conveying system design

With pneumatic conveying performance of materials varying as widely as the sample shown in Figure 4.38, it is not surprising that the scope and accuracy of computer-aided design programs is limited. Many manufacturing companies that serve a wide range of industries generally make a point of listing in their advertising material, a vast number of different materials that they have experience of conveying. Most reputable manufacturing companies will have a test facility, specifically for the purpose of testing clients' materials. This will generally be offered as a 'free' service and the client will be invited to witness the conveying trials to show that their material can be conveyed reliably.

It is unlikely that the geometry of the test facility will match that of the plant pipeline to be built, but with the use of appropriate scaling parameters such differences can be accounted for. With regard to the pipeline these differences include: pipeline bore; horizontal and vertical lengths; number, location and geometry of bends in the pipeline; and pipeline material. With regard to conveying conditions, conveying line pressure drop, conveying air velocity and solids loading ratio of the conveyed material can all have an influence on the conveying performance of the pipeline. With regard to the conveyed material there is mean particle size and size distribution, particle shape and particle density. If tests are carried out with a specific material, it is possible that the computer program will not have to take particle properties into account, but such a program could not possibly be used for another material, or even a different grade of the same material, with any degree of reliability.

The potential influence of material grade is illustrated in Figure 4.39, where conveying data for both a coarse and a fine grade of fly ash are compared (Mills & Agarwal 2001).

Fly ash comes from the combustion of pulverised coal in a boiler. The resulting ash is mostly carried over with the combustion gases. The coarse ash soon drops out of suspension

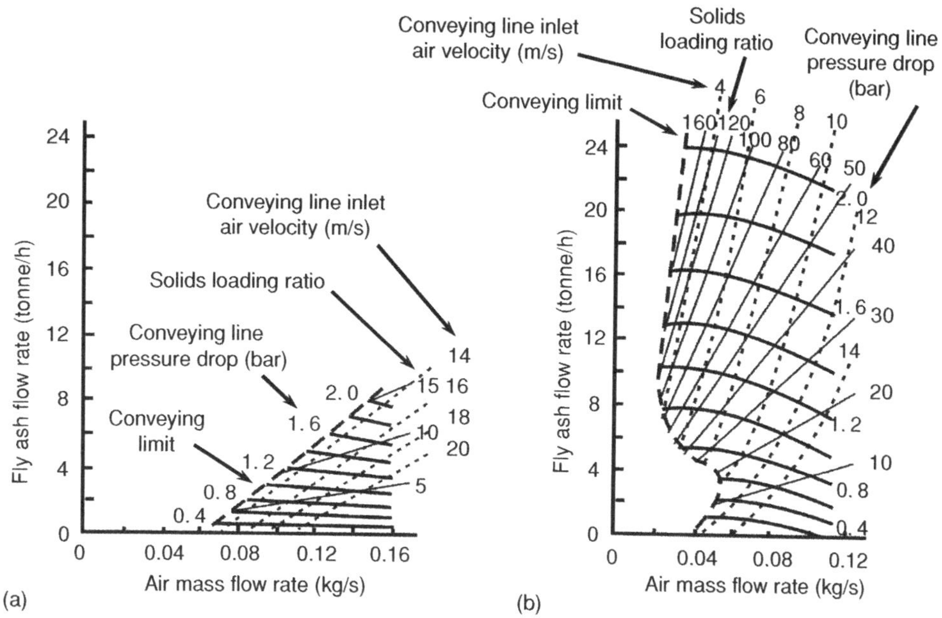

Figure 4.39 Conveying data for fly ash. (a) Coarse grade; (b) fine grade.

in the economiser and air pre-heater hoppers but the fine ash has to be physically removed by electrostatic precipitators. Both grades of ash were conveyed through the same pipeline in Figure 4.38 and it will be seen that the coarse grade had no dense phase conveying potential at all. There were also very marked differences in material flow rates achieved.

4.5.1 *Conveying air velocity*

Conveying air velocity is clearly important and that at the material feed point into the pipeline is critical. A problem comes in evaluation of this velocity, for compressors are specified in terms of a given quantity of air being delivered at a given pressure, and the reference point for each is different. The volumetric flow rate of the air delivered, $\dot{V}_0$, is that at 'free air conditions' (standard atmospheric pressure and temperature) and this will generally be close to that at the pipeline exit. The delivery pressure will be close to that at the pipeline inlet. Compressibility, therefore, must be taken into account.

The basic equation here is the Ideal Gas Law:

$$p\dot{V} = \dot{m}_a RT \tag{4.2}$$

The inclusion of the characteristic gas constant, R, in this equation means that this equation can be used for any gas. Nitrogen, carbon dioxide, superheated steam and many other gases are often used for pneumatic conveying. The following equations, however, are derived in terms of air only. If any other gas is employed the equations will have to be re-worked with the appropriate value of R.

From this equation is derived:

$$\frac{p_1 \dot{V}_1}{T_1} = \frac{p_2 \dot{V}_2}{T_2} \tag{4.3}$$

This applies for any gas and is essentially a continuity equation for the conveying system and pipeline.

The volumetric flow rate of the air (or any other gas) at any point can be obtained from:

$$\dot{V} = \frac{\pi d^2}{4} \times C \ \text{m}^3/\text{s} \tag{4.4}$$

4.5.2 *Compressor specification*

A conveying line inlet air velocity, C_1, will need to be specified for the given material and conveying conditions, and a pipeline bore, d, will also need to be evaluated. By re-arranging the above equations and substituting for constants (including R) and free air conditions it can be shown that:

$$\dot{V}_0 = 2.23 \times \frac{p_1 d^2 C_1}{T_1} \ \text{m}^3/\text{s} \tag{4.5}$$

This is the volumetric flow rate of air required for conveying the material through the pipeline. To this may need to be added an allowance for any leakage of air across the material feeding device.

The air supply pressure to be specified will be p_1 plus allowances for any pressure drops, such as that between the compressor and the material feed point into the pipeline.

It should be emphasised that absolute values of both pressure and temperature must always be used in all of the above equations.

4.5.3 *Solids loading ratio*

Solids loading ratio, ϕ, as mentioned earlier, is the dimensionless ratio of the mass flow rate of the material conveyed divided by the mass flow rate of the air used to convey the material. Air flow rate is almost exclusively expressed in terms of a volumetric flow rate and so air mass flow rate is most conveniently derived from the conveying line inlet air conditions. A further re-arrangement of the above equations gives:

$$\dot{m}_a = 2.74 \times \frac{p_1 d^2 C_1}{T_1} \ \text{kg/s} \tag{4.6}$$

This then gives the solids loading ratio as:

$$\phi = \frac{\dot{m}_p}{3.6\,\dot{m}_a} \tag{4.7}$$

since material flow rate is traditionally expressed in terms of tonne per hour.

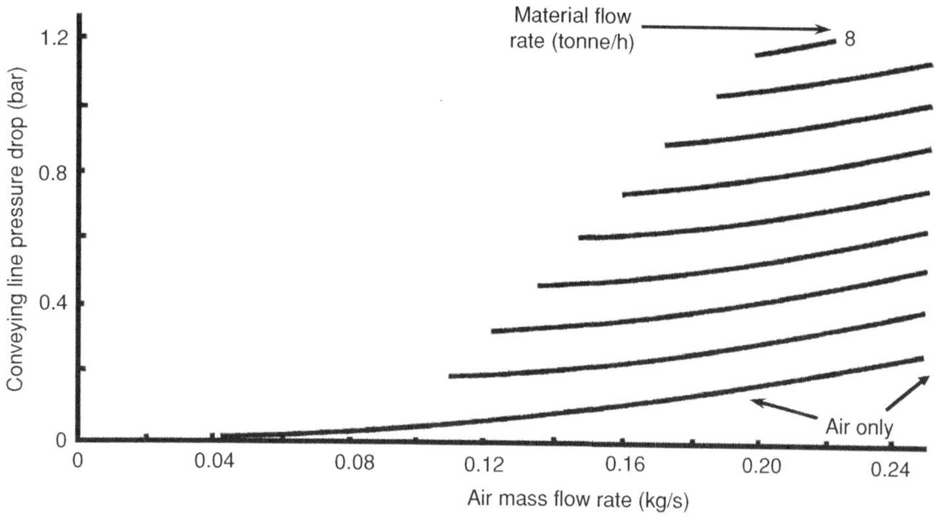

Figure 4.40 Typical pressure drop relationship for pipeline with material flow.

4.5.4 *The air only datum*

The pressure drop required to convey the air alone through the pipeline provides a datum for the conveying system. Only when the conveying line pressure drop exceeds this datum value will any material be conveyed, but then the greater the excess over this datum pressure drop the greater will be the material flow rate. This is illustrated in Figure 4.40, with the zero material flow rate line being the air only pressure drop for the pipeline.

This air only pressure drop can be calculated reliably from basic fluid mechanics. The equation used here is that derived by Darcy:

$$\Delta p = \frac{4fL}{d} \times \frac{\rho C^2}{2} \text{ N/m}^2 \tag{4.8}$$

The friction factor, f, is a function of the Reynolds number, *Re*, for the flow and the pipe wall roughness, and can be obtained from a Moody chart. Because air is compressible, both air velocity and air density will vary along the length of the pipeline and so Equation (4.8) should be integrated between limits in order to get an accurate value. This would be recommended in any situation where the air only pressure drop represents a high proportion of the available pressure drop for conveying, such as for long distance conveying.

It will be seen from Figure 4.40 that there is approximately a square law relationship between pressure drop and velocity. Conveying line inlet air velocity values, therefore, should not be too high or there will be an adverse effect on conveying performance, as will also be seen from Figure 4.40. The conveyed material in Figure 4.40 was a granular grade of potassium chloride and so was only capable of dilute phase conveying. The pipeline used was 95 m long, of 81 mm bore, with nine 90° long radius bends and was almost entirely in the horizontal plane.

4.5.5 *Acceleration pressure drop*

The material that is fed into the pipeline will essentially have zero velocity at the feed point and so will have to be accelerated to its terminal velocity at the end of the pipeline. The pressure drop for this can be approximated with:

$$\Delta p_{acc} = (1 + \phi) \times \frac{\rho_2 C_2^2}{2} \text{ N/m}^2 \qquad (4.9)$$

The density and velocity terms are those of the air at the end of the pipeline. The approximations lie in the fact that the air will have an initial velocity, and the terminal velocity of the material will be below that of the air, the actual velocity depending upon particle size, shape and density.

4.5.6 *Scaling parameters*

If conveying data for a given material is available, from a test pipeline or another plant pipeline, this data can be scaled to that for any other plant pipeline. Any differences in pipeline length, bore, orientation and number of bends between the two pipelines can be taken into account by means of scaling parameters. Any data coming from another pipeline will automatically include pressure drop elements for straight pipeline, bends, air only and acceleration, and so it will be a matter of determining the differences in values between the two pipelines. The evaluation can be carried out in three parts, with one for the air only, another for the acceleration and the third for all the pipeline elements, such as straight sections and bends, considered in terms of an equivalent pipeline length.

4.5.6.1 *Conveying mode* In scaling from one set of pipeline data to another, on no account should the conveying limits derived for the new pipeline exceed those of the original data, unless there is positive evidence that the material is capable. This means that the conveying line inlet air velocity derived should not be lower than that for the data to be scaled. If the pressure gradient available for the new pipeline is greater than that for which the original data was derived, the solids loading ratio may be much higher (see Figure 4.34), and hence a much lower conveying line inlet air velocity may appear possible (see Figure 4.33). If a material has no natural dense phase conveying potential, however, there will be no possibility of conveying the material at a lower velocity, and hence in dense phase, if a higher pressure gradient is available, unless there is a change in the type of conveying system used.

4.5.6.2 *Equivalent length* The equivalent length of a pipeline is taken in terms of the length of straight horizontal pipeline. This means that all straight horizontal sections of pipe in a pipeline can be added together and effectively have a weighting of unity.

4.5.6.3 *Vertical pipeline* For material flows vertically up the pressure drop will be approximately double that for horizontal pipeline. For vertically upward pipeline, therefore, it is recommended that the length of vertically up sections is doubled to provide an equivalent length (Mills 2004; Mills *et al.* 2004).

For material flows vertically down the pressure gradient can be positive or negative, depending upon the value of solids loading ratio at which the material is conveyed. For dilute phase flows there is generally a pressure drop for the pipeline but for dense phase flows, with air retentive materials, there is usually a pressure rise (Mills 2004; Mills *et al.* 2004). If the pipeline system to be designed has any significant length of vertically downward flow, great care will have to be exercised with the design process.

It is generally recommended that inclined sections of pipeline, particularly for vertically upward flow, should be avoided and that only horizontal and vertical sections should be employed in any pipeline routing. The minimum conveying air velocity required for inclined pipeline sections can be higher than that for horizontal and vertically up sections of pipeline, and so are more vulnerable to pipeline blockage. The pressure gradient in such inclined sections is also generally higher than that for horizontal pipeline.

4.5.6.4 *Pipeline bends* Data for the equivalent length of pipeline bends was given in Figure 4.35. Although the data relates to bends having a bend diameter, D, to pipeline bore, d, ratio of about 24:1, it is generally considered that the relationship holds for D/d ratios down to about 3:1 (Mills 2004; Mills *et al.* 2004). Below this, and certainly for blind tees (see Figure 4.28c), the equivalent length can be much greater.

Although bends provide pneumatic conveying systems considerable flexibility in routing, there is a considerable penalty to pay in terms of pressure drop, and hence conveying capability. The equivalent length in Figure 4.35 is in terms of conveying line inlet air velocity and it will be seen that for dilute phase conveying, bend losses can be very significant. The total loss due to the bends is the value from Figure 4.35 multiplied by the total number of bends in the pipeline, and so every effort should be made to keep the number of bends in a pipeline to a minimum.

Little data is available for 45° and other bends. Because the primary impact of the conveyed material in making the turn is the major cause of the material retardation, and hence its subsequent re-acceleration (see Figure 4.29), bend angle is not likely to have a major influence.

4.5.7 *Scaling model*

The scaling model for equivalent length is in terms of material flow rate:

$$\dot{m}_{p2} = \dot{m}_{p1} \times \frac{L_{e1}}{L_{e2}} \text{ tonne/h} \tag{4.10}$$

This is an inverse law relationship and so for a given air supply pressure, and hence energy value, if the length of pipeline is doubled, for example, there will be an approximate halving of the material flow rate. If scaling is to a longer pipeline there will be an additional loss to take into account, because of the increase in air only pressure drop for the pipeline, for pipeline length is on the top line of Equation (4.8).

4.5.7.1 *Pipeline bore* If a high material flow rate is required it is likely that a larger pipeline bore will be needed. This scaling can be carried out independently of equivalent length. Although a higher conveying line pressure drop will give an improvement in performance, it will generally be small in comparison to that which can be obtained by increasing pipeline bore. If a larger bore pipeline is used there is likely to be an additional bonus, for the air only pressure drop for the pipeline will be lower, since pipeline bore is on the bottom

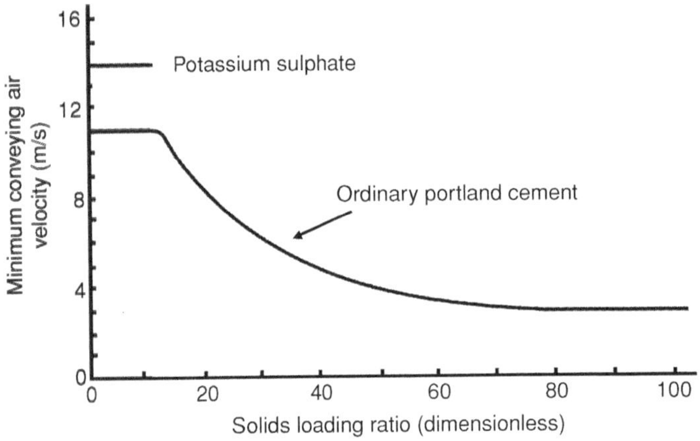

Figure 4.41 Conveying limits for materials considered.

line of Equation (4.8). The scaling model for pipeline bore is:

$$\dot{m}_{p2} = \dot{m}_{p1} \times \left(\frac{d_2}{d_1}\right)^2 \text{ tonne/h} \tag{4.11}$$

4.5.7.2 *Scaling influences* When designing a pneumatic conveying system to achieve a given material flow rate over a specified conveying route there is always a wide selection of conveying parameters that can be employed. If a large bore pipeline is selected, then a low pressure will be required and if a high pressure is used, then a smaller bore pipeline will meet the duty. There is an almost infinite combination of pipeline bore and air pressure combinations that could be used, limited only by the availability of pipeline in regular increments in size of bore, and an upper limit on air supply pressure of about 5 bar gauge, for positive pressure conveying systems delivering materials to reception points at atmospheric pressure, due to the problems of air expansion.

Where there is such a choice available in selecting conveying parameters the question arises as to the possible influence on power requirements, since this does tend to be rather high for pneumatic conveying systems. For materials that can be conveyed in dense phase there is the additional question of conveying capability if a high pressure air supply is not employed. To illustrate these points ordinary portland cement, being typical of powdered materials, and potassium sulphate, being typical of granular materials, are used. Cement has very good air retention properties and is capable of being conveyed in dense phase and hence at low velocity, whereas the potassium sulphate considered is a coarse granular material and can only be conveyed in dilute phase suspension flow. The conveying limits for these materials are presented in Figure 4.41.

Conveying data for the two materials was obtained from a pipeline 95 m long, of 81 mm bore and included nine long radius 90° bends. The influence of pipeline bore only is investigated, with a material flow rate of 40 tonne/h for the cement and 12 tonne/h for the potassium sulphate considered by way of example. The results are shown in Figure 4.42.

With a wide range of pipeline bore and air supply pressure combinations capable of achieving a given material flow rate, the obvious question is which bore or air supply

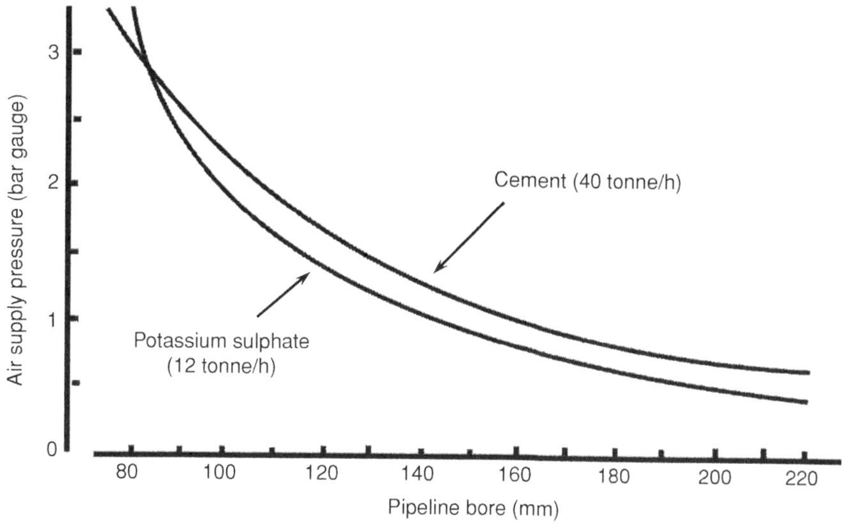

Figure 4.42 Influence of pipeline bore on air supply pressure for given parameters.

pressure results in the most economical design? Plant capital costs could vary considerably, for with different pipeline bore and air supply pressures there are differences in feeder types, filtration requirements and air mover types, apart from widely different pipeline costs, and so a major case study would need to be carried out. Power requirements, and hence operating costs, however, are largely dependent upon the air mover specification and so these can be determined quite easily by using Equation (4.1).

The approximate power requirements for the cases considered are presented in Figure 4.43. In most cases the power required for the air mover represents the major part

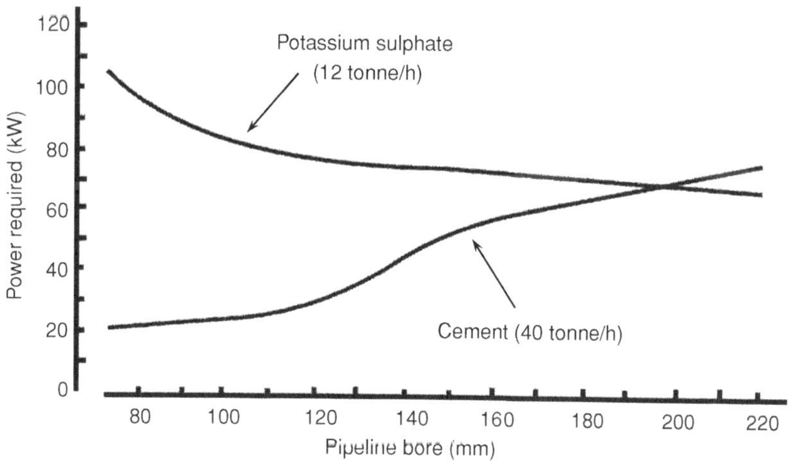

Figure 4.43 Influence of pipeline bore on power requirements for given parameters.

of the total system power required, although for screw pumps a major allowance must also be made for the screw drive. Figure 4.43 presents an interesting trend for both of the materials considered. For the cement the smallest bore pipeline is clearly the best, but for the potassium sulphate it is the largest bore pipeline.

For the potassium sulphate the decrease in power requirements with increase in pipeline bore can be explained in terms of the decrease in velocity through the pipeline. With a conveying line inlet air pressure of 3.2 bar gauge the conveying line exit air velocity will be about 68 m/s, and this reduces to 27 m/s with the much lower air supply pressure required for the 200 mm bore pipeline. Pressure drop increases significantly with increase in conveying air velocity and so the pipeline with the lowest velocity profile will generally give the lowest power requirement for a material such as potassium sulphate.

For the cement the increase in power with increase in pipeline bore can also be explained in terms of velocity profiles, but in this case it is values of conveying line inlet air velocity that are relevant. Since cement is capable of being conveyed in dense phase, the relationship between minimum velocity and solids loading ratio, as shown in Figure 4.41, dictates. In an 81 mm bore pipeline the inlet velocity is only 4.2 m/s, since the solids loading ratio is 109. In the 200 mm bore pipeline the solids loading ratio is reduced to 14 and so the inlet air velocity is 12.0 m/s.

The above relationship holds equally for negative pressure conveying systems. Because of the natural limit on conveying line pressure drop available, however, the situation is limited to short distance conveying.

4.5.8 *Scaling procedure*

The primary objective of a system design is generally to achieve the given material flow rate. The first stage in the design process, therefore, is to scale the available data approximately to the required conveying distance using Equation (4.10). For this it will be necessary to evaluate or specify a value for the conveying line inlet air velocity. With both pressure drop and pipeline bore having an influence on material flow rate a decision will have to be made both on the air supply pressure, or vacuum, and the pipeline bore. In the first instance a linear relationship can be used for conveying line pressure drop and Equation (4.11) for pipeline bore.

Since pipeline bore comes in incremental sizes, fine tuning and spare capability need to be considered in terms of reserve pressure available. With first approximation values for pressure and pipeline bore, the available conveying data can be scaled more precisely to take account of differences between pipeline geometries. Conveying air velocities and the solids loading ratio can be evaluated so that differences between air only pressure drop and acceleration pressure drop values can also be taken into account. This is an iterative process, as there are many inter-dependent variables, and so in the initial stages approximations can be made.

If data on minimum conveying air velocities is to be used in the design process, such as that shown in Figure 4.41, it would generally be recommended that a 20% margin be allowed for the value of conveying line inlet air velocity to be employed:

$$C_1 = 1.2 \times C_{\min} \text{ m/s} \tag{4.12}$$

If an air supply pressure greater than about 0.8 bar gauge, or a vacuum of more than about 0.4 bar is to be employed it would be recommended that the possibility of stepping the pipeline to a larger bore towards the end of the pipeline should be considered. This point was mentioned earlier and illustrated for a positive pressure conveying system in Figure 4.30. Stepping the pipeline will generally result in an improvement in material flow rate, as well as reducing problems of erosive wear associated with the conveying of abrasive materials, and reducing problems of material degradation associated with the conveying of friable materials.

Ideally stepping of the pipeline should be incorporated into the pipeline design process, but this does require an additional level of iteration. In the cases shown in Figure 4.42, lower air supply pressures could have been employed for the smaller bore pipelines to achieve the material flow rates quoted. This would then have resulted in an approximate 30% reduction in power required in both cases in Figure 4.43 for the 80 mm bore pipeline, gradually reducing with increase in pipeline bore, and hence air supply pressure.

4.6 Troubleshooting

Despite being very simple in concept, and having been in use for well over 100 years, the influence of material type on conveying performance is not fully understood. System design involves the use of compressible flow equations and these are not always understood. The influence of changing conveying parameters is not always obvious and it is not consistent either, being dependent upon material properties. Conveying air velocity is the major variable in system design and operation, but it is equally the major element in most operating problems. If the velocity is too low the pipeline is likely to block. If the velocity is too high; the conveying potential will be reduced, abrasive materials will cause considerable damage to the pipeline system, and the pipeline system may cause considerable damage to the conveyed material.

4.6.1 *Material flow rate problems*

Flow rate problems tend to be either pipeline blockage or systems in which it is not possible to achieve the desired material flow rate. If pipeline blockage occurs it is likely to be a design error. In the case of under performing systems it may be due to over design or poor design. In either case it would always be recommended that a check should be made on the value of the conveying line inlet air velocity for the pipeline. This can be evaluated from either the free air flow rate (Equation (4.5)) or from the air mass flow rate (Equation (4.6)) presented earlier. Re-arranging Equation (4.5) gives:

$$C_1 = 0.448 \times \frac{T_1 \dot{V}_0}{p_1 d^2} \text{ m/s} \tag{4.13}$$

4.6.2 *Pipeline blockage*

The value required depends very much upon the materials being conveyed and the mode of conveying. Typical values for C_{min} were presented in Figure 4.41 with a recommendation that C_1 should be about 20% greater than this (Equation (4.12)). For fine powders capable of being conveyed in dense phase the value of the solids loading ratio also needs to be determined. For coarse granular materials having a mean particle size of about 500 μm, a narrow size distribution and a particle density of about 1000 kg/m^3, a minimum velocity of about 16 m/s would be recommended. This will need to be increased for larger and higher density particles.

A critical parameter in Equation (4.13) is p_1. This is on the bottom of the equation and so any increase in its value will cause a corresponding lowering of C_1. A pipeline is quite likely to block, therefore, if the compressor is operated above its pressure rating, even for a short period of time. It must be stressed that absolute values of pressure must be used throughout in these equations. If p_1 is taken as a gauge pressure, or the pressure at the end of the conveying line, the design is likely to be seriously in error.

The conveying line inlet air temperature must also be expressed in absolute terms. This is on the top line of the equation and so if the value is below the design value this could be a cause for concern. This may occur in winter on start-up if insufficient time is allowed for the system to warm up. The volumetric flow rate is also on the top line. If a conveying system fails during commissioning this value should be checked. During use gradual wear of the air mover may occur and so any deterioration in performance could result in pipeline blockage after a period of time.

4.6.3 *Conveying limits*

Conveying systems are particularly vulnerable when operating in a region close to the conveying limit for the material. The problem here is generally one of air compressibility. In the majority of cases, conveying systems operate with a compressor operating at a fixed speed, which means that a constant flow rate of air will be delivered. Because there is a slight reverse flow of air through positive displacement compressors the air flow rate will reduce slightly with delivery pressure, as illustrated in Figure 4.25, but the percentage change is small and does not represent a problem. Air compressibility has an overriding effect. The position is illustrated in Figure 4.44, which is a plot of conveying data for a granular material (coke fines) and so is typical of dilute phase conveying situations.

Case 1 represents a low pressure conveying system, operating at about 0.7 bar gauge, and at that point the system would operate quite reliably. If the conveying air pressure is increased, however, to increase the material flow rate above 1.3 tonne/h, the pipeline will block at about 1.6 tonne/h, for the operating point will then cross the material conveying limit line and enter into the 'no go area'. The conveying limit for the material on Figure 4.44 represents a constant conveying line inlet air velocity of 14 m/s (see the data for granulated sugar in Figure 4.37). The slope of the line relates directly to the compressibility of air. At point 1, C_1 is about 12% in excess of C_{min}.

The operating point for case 2 is at about 0.8 bar gauge and C_1 is about 36% in excess of C_{min}. If the air supply pressure is increased the material flow rate will increase to a maximum of about 3 tonne/h at 1.35 bar gauge and then the pipeline will block. If the air is provided by a positive displacement blower having a maximum delivery pressure of 1.0 bar gauge, the capability of the blower itself may cause the system to fail due to over-loading at a much lower material flow rate.

The situation with regard to dense phase conveying in sliding bed flow is illustrated in Figure 4.45. Since the operating point is likely to be well above the minimum conveying conditions, as with case 3, increasing air supply pressure is unlikely to be a problem, in

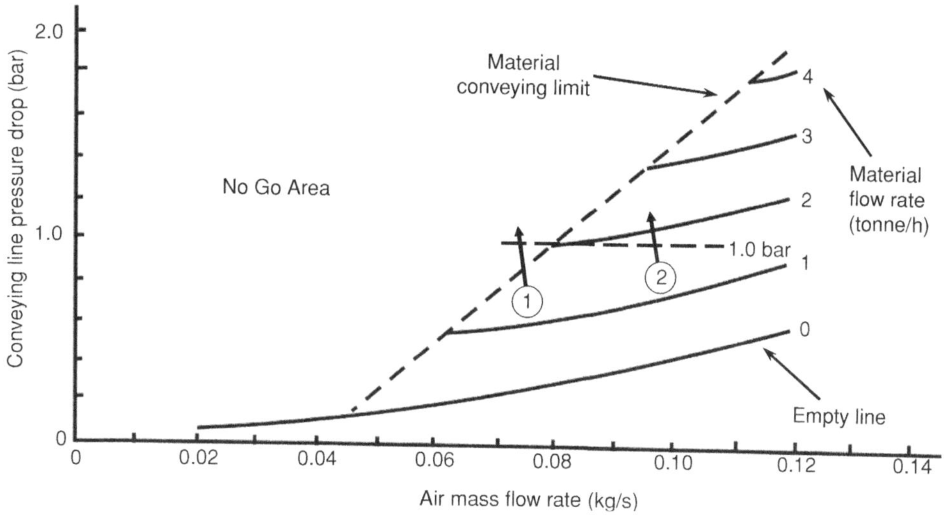

Figure 4.44 Typical performance data for material conveyed in dilute phase.

terms of drifting into the 'no go area'. A problem that can arise, however, with systems in which the material flow rate may be varied, is that of crossing into the 'no go area' at low material flow rates, and illustrated with case 4. With no change in air flow rate and a lowering of material flow rate, the solids loading ratio will reduce. If the operating point drops below the transitional line on Figure 4.33, the pipeline will be prone to blocking under these conditions.

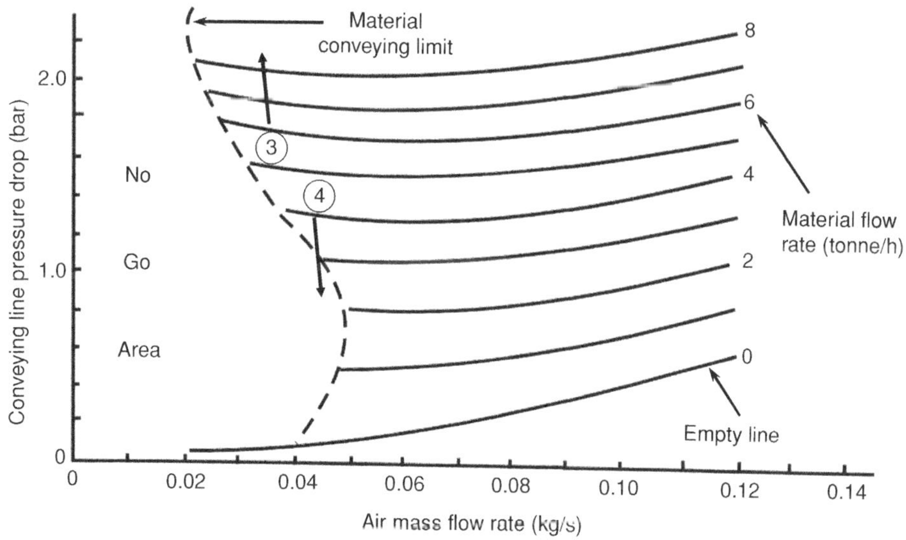

Figure 4.45 Typical performance data for material with sliding bed dense. Phase conveying capability.

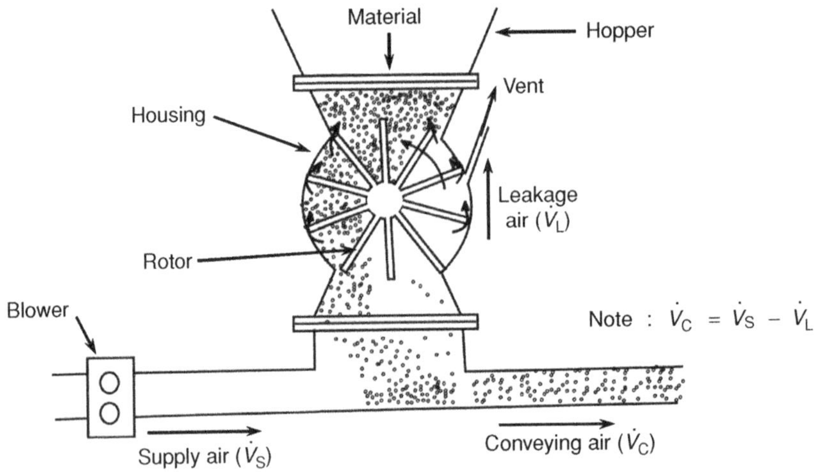

Figure 4.46 Air leakage from a positive pressure conveying system.

4.6.4 *Air leakage*

With positive pressure conveying systems the material has to be fed into the pipeline with air maintained at pressure. If the supply hopper is maintained at atmospheric pressure it is possible that air will leak across the feeding device. This is certainly the case with rotary valve type feeders, as illustrated in Figure 4.46.

This leakage air must be taken into account in the specification of the compressor since it is lost from the system and is not available for conveying. Manufacturers of rotary valves will generally provide this data. It would always be recommended that rotary valves should be vented, as shown, to prevent the leakage air from interfering with the smooth flow of material into the pipeline. The rotary valve is a positive displacement type feeder and so for a given valve and speed of rotation the material feed rate will be almost directly proportional to the bulk density of the material.

If the leakage air does interfere with the material feed it could result in a significant reduction in the bulk density of the material, and hence affect the material flow rate. It is important, therefore, to ensure that the vent is correctly designed and maintained. On start-up, after a shut down, the bulk density of the material in the hopper could increase considerably and this could result in pipeline blockage at this time due to over feeding. If this is a possibility the material in the supply hopper should be aerated prior to start-up.

A similar situation can occur with vacuum conveying systems. There is not likely to be any air leakage in the region of the feeding device, of course, but with the reception hopper maintained under vacuum there is the possibility of air leaking into the conveying system from the material off-loading valve. This air will also by-pass the conveying pipeline but will have to be dealt with by the exhauster and so this must also be taken into account.

If there is air leakage across a feeding device, such as a rotary valve, and the material being conveyed is abrasive, considerable erosive wear of the feeding device could occur. The wear will generally result in an enlargement of the operating clearances between moving parts, and hence cause a gradual increase in air loss across the feeder. If there is no increase

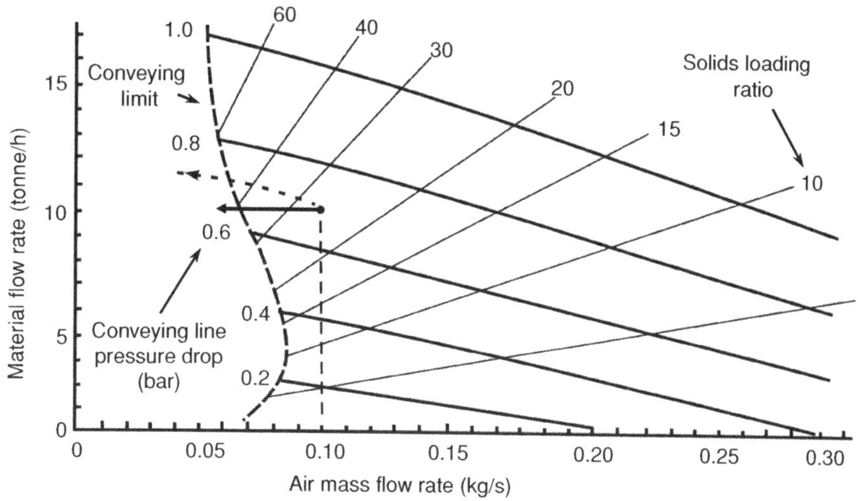

Figure 4.47 The influence of a gradual reduction in conveying air flow rate on system performance.

in the air flow rate available for conveying to compensate, pipeline blockage is likely to be the ultimate result, as illustrated in Figure 4.47. The same situation will occur if there is gradual wear and deterioration in performance of a compressor or exhauster as a result of ingesting abrasive dust laden air.

4.6.5 *Performance monitoring*

It would always be recommended that a pressure gauge be located near the start of the pipeline to monitor performance, as shown in Figure 4.48, or similarly near the end of a vacuum conveying system. Because of the high speed of conveying, particularly with

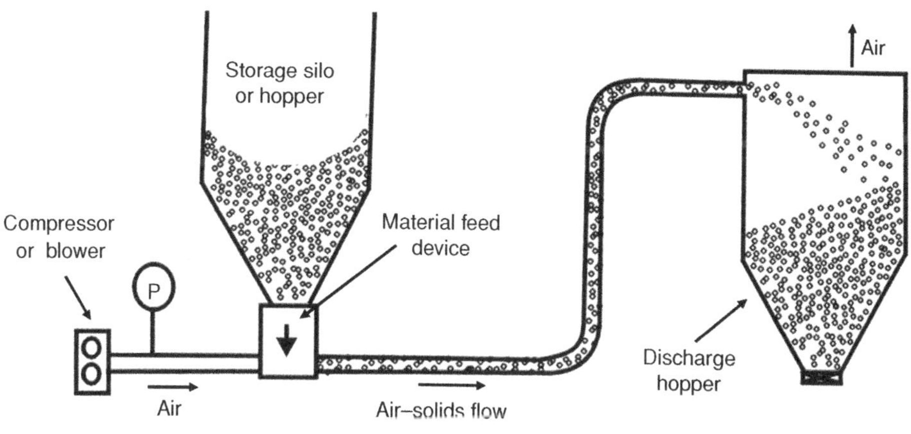

Figure 4.48 Performance monitoring of a positive pressure conveying system.

dilute phase conveying, changes in performance can happen quite quickly. If, on start-up, the pressure gauge rapidly reaches the compressor pressure, the pipeline is likely to be over fed. Similarly the pressure gauge will indicate if the pipeline is under fed. Ideally the pressure gauge should be in the conveying line, but this is too arduous a location for reliability. Just prior to the feeder in the air line is a more reliable location and the pressure drop across the feeder is not likely to be a significant percentage of the total.

A particular note of the pressure reading should be taken when air only flows through the pipeline. This value should be checked periodically because it should not change with time. If it does increase over time it might be due to a gradual build up in the pipeline. If this is allowed to continue it could ultimately result in pipeline blockage, since the increase in air only resistance could use up any safety margin available. This may be due to condensation in the pipeline. If dry air is not used for conveying it would be recommended that air be blown through the empty pipeline for a short period to dry it out, just in case, before starting to convey material.

4.6.6 *System optimising*

Many systems are deliberately over-designed, simply to ensure that they can be guaranteed to perform reliably and to minimise commissioning time. This will be achieved by employing higher conveying air velocities than necessary. Any increase in velocity, or air flow rate, however, will for most materials generally result in a decrease in material flow rate, as can be seen from Figures 4.32 and 4.37. From Figure 4.36, however, it will be seen that this only applies to the high velocity dilute phase conveying for this particular material. An increase in material flow rate with increase in air flow rate will generally only occur with low velocity dense phase conveying; of materials having very good permeability, with certain polymeric powders, and with some innovatory systems.

With an existing pneumatic conveying system the system could be checked by placing a t-piece with a control valve in the air line between the air mover and the conveying system. This is illustrated for a negative pressure conveying system in Figure 4.49. In a positive

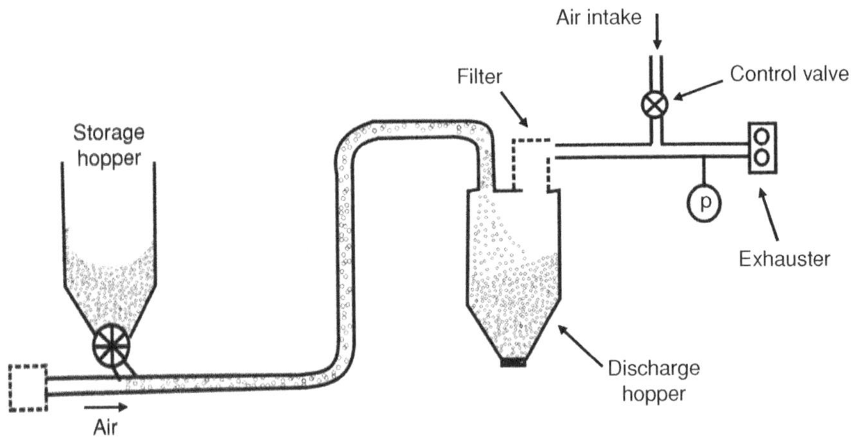

Figure 4.49 Air by-pass for checking performance of a negative pressure conveying system.

pressure conveying system the t-piece would be positioned between the compressor and feeding device. In either case it is suggested that there should initially be no change to the compressor or exhauster. By opening the control valve with the negative pressure system, air will be drawn into the system close to the exhauster inlet and so by-pass the conveying line. In a positive pressure system the air would be discharged from the system just after the compressor and so also by-pass the conveying line.

If the system continues to perform reliably with a small proportion of air by-passing the system, the valve can be opened a little more. Observation of the pressure gauge, in terms of degree of fluctuation, should give some indication of when the air flow rate is getting close to the minimum value. If the air flow rate so by-passed reaches about 10% of the rated value it would be recommended that the speed of the air mover be reduced to compensate and the process continued. If too much air is by-passed in this way the system could be susceptible to premature blockage, for if a pressure fluctuation occurs this is likely to cause considerably more air to flow through the by-pass, as a line of least resistance.

If the volume of air drawn in or discharged could be measured at each adjustment of the control valve, together with the material flow rate and conveying line pressure drop, a small part of the operating characteristics could be mapped for the system. Equation (4.13) should be used to calculate a value for the conveying line inlet air velocity for reference. A sight glass in the conveying line, just downstream of the feeding device would also be useful as this would give a good indication of the system performance.

4.6.7 *Erosive wear*

Abrasive wear is associated with sliding contact between surfaces. In bulk solids handling plant abrasive wear is a major problem at hopper walls and in chutes, where materials slide over such surfaces. Erosive wear results from the impact of particles against surfaces. Typical erosive wear situations in bulk solids handling plant are in the loading and off-loading of materials, and with free fall onto surfaces. The blowing of materials into cyclones; their loading into hoppers and onto chutes; and off-loading from hoppers, conveyor belts and bucket elevators; are common examples. These are all cases where particles can impact against surfaces and result in erosive wear, rather than slide against a retaining surface and cause abrasive wear.

In pneumatic conveying systems, bulk particulate materials are physically transported by air. Bends in pipelines, therefore, are particularly vulnerable to erosive wear, as are diverter valves and any other surface against which particles are likely to impact, including the pipeline itself to a limited extent. Where a pressure difference might exist on a plant, in the presence of abrasive particles, erosive wear will also occur, if there is a flow of air. A particular example here is with rotary air locks and screws used to feed materials into positive pressure pipelines. Even isolating valves will wear if they are not completely air-tight or fully shut.

4.6.7.1 *Impact angle and surface material* The curve shown in Figure 4.50 illustrates the variation of erosion with impact angle for two different surface materials. Both surface materials showed very significant differences in both erosive wear rate and the effect of impact angle. These materials do, in fact, exhibit characteristic types of behaviour that are now well recognised. The aluminium alloy is typical of ductile materials: they suffer

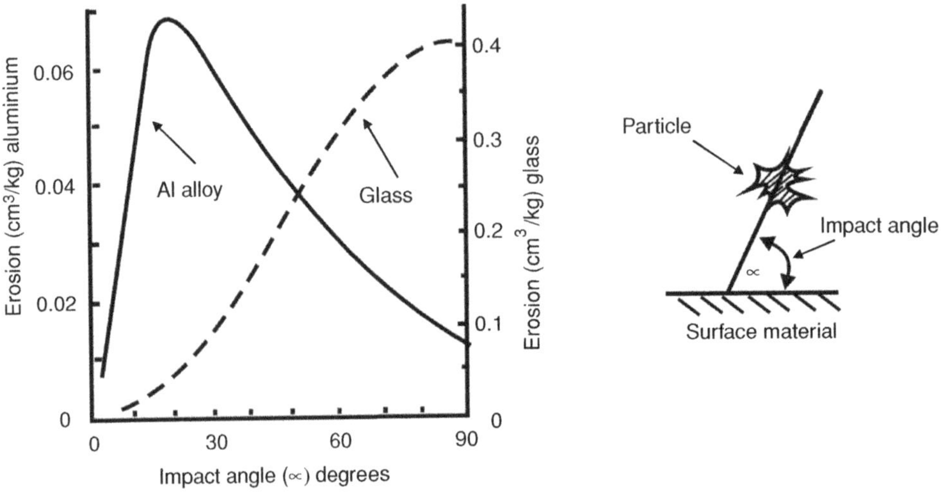

Figure 4.50 Influence of impact angle and material on erosive wear.

maximum erosion at an impact angle of about 20° and offer good erosion resistance to normal impact. The glass is typical of brittle materials: they suffer severe erosion under normal impact but offer good erosion resistance to low angle, glancing impact.

The vertical axis on Figure 4.50 is in terms of the volume of material eroded, in cm^3, per kg of particles impacted against the surface. It will be seen that both are extremely important variables and that they should be given due consideration in the design and specification of the system and its components if an abrasive material has to be conveyed. Erosive wear of straight pipeline is not generally a problem with powdered materials since particles tend to flow in almost straight lines at high velocity and impact angles are very low. With granular materials impact angles are higher, and as the slope of the line on Figure 4.50 for ductile materials is very steep at low impact angles, ordinary steel pipeline is not likely to be satisfactory for abrasive materials. This is particularly the case at the outlet from bends in the pipeline where there is considerable turbulence in the flow. With large abrasive particles, gravity will also play a part, and severe wear along the bottom of the pipeline is likely to occur.

Bends in pipelines are particularly vulnerable. Long radius bends will give a low impact angle, but not low enough to allow mild steel to be used if an abrasive material has to be conveyed. Short radius mild steel bends cannot be recommended either unless they are significantly reinforced, but as a consequence of gradual wear, deflecting flows are likely cause even more problems downstream. This point is illustrated with the wear and flow patterns for a long radius bend in Figure 4.51.

The ultimate solution to the wear problem is to use a blind T-bend, as illustrated in Figure 4.28c. Material is trapped in the blind end of the bend and primary impact is material upon material. The bend will ultimately wear out downstream due to the turbulence, but this is generally straight pipeline and so a short length of thick wall pipeline can be used to prolong life. There are two major problems, however. One is that the material trapped in

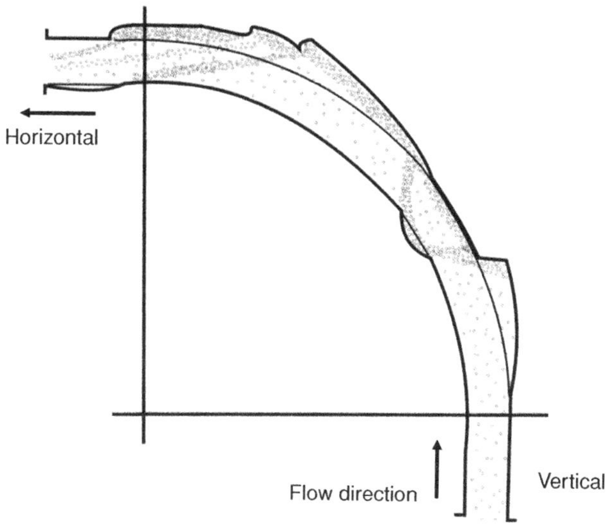

Figure 4.51 Typical pattern of deflecting flows following erosive wear of a bend.

the bend will take a long time to purge out if the pipeline has to be cleared of material, and hence the bend shown in Figure 4.28d. The other is that the pressure drop for such mitred bends is generally significantly higher than that for radius bends and so the performance of the conveying system could be seriously reduced as result of their use.

4.6.7.2 *Velocity* The model for the erosive wear of pipeline bends in terms of velocity is:

$$\text{Erosion} \propto (\text{Velocity})^{2.5} \tag{4.14}$$

This means that bends at the end of a single bore pipeline, with a 1 bar pressure drop, for example, will be more than five times that of bends at the start of the pipeline. This is another reason for using stepped pipelines to keep the velocity profile along the length of a pipeline as low as possible, consistent with always keeping above the minimum value for the given material and conveying conditions.

4.6.8 *Particle degradation*

Many materials that have to be conveyed are friable, and particles are liable to be broken when they impact against retaining surfaces, such as bends in the pipeline. The problem can be particularly severe if the material has to be conveyed in dilute phase and hence at high velocity, since particle degradation also has a power law relationship with velocity. There are, however, means by which the problem can be reduced to an acceptable level. Stepping the pipeline to a larger bore, as mentioned above, applies equally to minimising particle degradation and is an obvious starting point.

Any reduction in the normal value of the impact force on the particles will help and so long radius bends are to be recommended. Reducing the magnitude of the deceleration force on the particles as a result of impact will also help. The use of rubber hose for pipeline bends is a possibility here, if the use of such material is compatible with the material to be conveyed. Rubber can also be used to reduce erosive wear, particularly with fine particles.

Rubber is a resilient material and so is capable of absorbing the energy of impact. As a consequence both the damage to the particles and the damage to the rubber itself will be reduced. If the energy of impact, however, is above the threshold for the rubber, due to particle velocity, size or density, rubber will wear rapidly if the material is abrasive. A particular advantage with rubber is that the material is flexible and can be rotated when used for bends and so the life can be prolonged.

References

Mills, D. (2004) *Pneumatic Conveying Design Guide*, 2nd edn. Elsevier Butterworth Heinemann, Amsterdam.

Mills, D. & Agarwal, V.K. (2001) *Pneumatic Conveying Systems – Design, Selection, Operation and Troubleshooting with Particular Reference to Pulverised Fuel Ash*. Trans Tech Publications, Clausthal-Zellerfeld, 386 pp.

Mills, D., Jones, M.G. & Agarwal, V.K. (2004) *Handbook of Pneumatic Conveying Engineering*. Marcel Dekker, New York. (Note that this book is in US Engineering Units)

Nomenclature

C Conveying air velocity (m/s)
d Pipeline bore (m)
D Bend diameter (m)
f Pipeline friction coefficient (–)
L Pipeline length (m)
$\dot{m}$ Mass flow rate (kg/s) for air; (tonne/h) for material
p Conveying air pressure (kN/m^2)
R Characteristic gas constant (kJ/kg K = 0.287 kJ/kg K) for air
T Conveying air temperature (K)
$\dot{V}$ Volumetric flow rate of air (m^3/s)
ρ Conveying air density (kg/m^3)
ϕ Solids loading ratio (–)

Subscripts

$_a$ Air
$_p$ Conveyed material
$_0$ Free air conditions
$_1$ Pipeline inlet conditions or pipeline 1
$_2$ Pipeline outlet conditions or pipeline 2
$_{acc}$ Acceleration
$_{min}$ Minimum value

Prefixes

Δ Difference in value

Dimensionless

Re Reynolds number

5 Screw conveyors

LYN BATES

5.1 Introduction

The ubiquitous screw conveyor used in countless industries for moving virtually all types and condition of bulk materials is a descendent of the oldest form of a continuous mechanical conveyor in recorded history. The principle of moving a material by a helical screw is attributed to Archimedes (287–212 BC) for elevating water from the hold of a King Hiero of Syracuse ship. The simplest form of *Archimedes' screw* consists of a pipe in the shape of a helix with its lower end dipped in the water. As the device is rotated the water rises up the pipe by gravity flow in the separate helical sectors. Equipments based on similar helical principles have been used since ancient times for raising water to irrigate fields, operated manually, by animals, wind power and more recently by electric and internal combustion engines. An alternative construction of a helical blade within a circular casing forms a more compact volumetric transfer device and this form of rotating casing is now commonly employed for long kilns and rotary driers as the drive and support bearings are external to the product conveyed. This method is widely employed for irrigation and sewage handing, although generally now using a rotating screw within a static, close-fitting casing.

The basic transport mechanism is that the material resting between two adjacent screw flights on the same axis is promoted to slip down the face of the 'rising' side of the flight as the screw rotates. This action moves the product forward at the rate of one pitch per rotation of the screw, provided the material does not spill over the centre shaft to fall back into the proceeding pitch space as when the cross-sectional loading exceeds the height of the centre tube or the machine axis is excessively inclined.

It will be apparent that the angle of the blade to the horizontal must be greater than the angle of slip of the media on the flight face for this motion to occur. It is also clear that the dynamic repose condition attained by the moving media determines the transfer capacity of a given screw geometry. With a non-viscous liquid, such as water, surface slip will occur at a very shallow angle and, for screw that is rotating slowly, the surface profile will be virtually horizontal. Inclination of the screw axis therefore allows a free-flowing liquid to be elevated by inclining the screw axis but the axial transfer capacity reduces with inclination because the reduction in effective flight face inclination reduces the volume of the pocket of liquid that can be held between the flight pitches before spilling back over the centre shaft. The awkward geometry of the skewed, 'orange-segment' shape of the moved volume of an inclined screw is analysed in 'The Design of Archimedean Screw Conveyors', a small, but useful, publication of the now defunct Draughtsman and Allied Technician Association.

Many hundreds of years passed before the need arose to handle other than water in a continuous mechanical manner. The demand to process large quantities of grain from the newly exploited vast American plains led to the mechanisation of flour mills, an early type of which is thought to have been made by Evans around 1742. His grandson, Oliver Evans, built a fully mechanised mill in 1785 that included belt conveyors, bucket elevators and a form of screw conveyor that used a series of wooden blades fixed in a helical pattern around a central wood shaft. These blades were later replaced by formed metal sheet section.

The successful operation of this equipment found many replications and a supply industry was created to make such conveyors available for wider use.

Screw flights were, and still are, produced from circular discs with a hole in the centre, being split across one radial width and the helical form created by pressing the two edges apart with a twisting action, to form slightly more than a complete pitch. These sections were initially riveted together on a wooden shaft or steel tube to form a continuous helix. The proportions of the original blank may be calculated on the basis of equal stretch of the inner and outer periphery during forming, to require minimum work input to the metal. The final flight stretches to slightly more than one pitch around the shaft and can be 'snapped' onto the shaft into a stable location for the butt welding and welding to the centre shaft. A rough approximation of the size of the blank diameter and inner hole for a standard flight form of pitch equals outside diameter of the screw can be calculated on an 'equality in peripheral length' basis:

$$D_B = 1/\pi[(\pi D_F)^2 + P^2]^{1/2} \quad \text{and} \quad d_B = 1/\pi[(\pi d_F)^2 + P^2)]^{1/2} \tag{5.1}$$

where D_B is the outside diameter of blank; D_F is the outside diameter of final flight; P is the pitch of flight; d_B is the diameter of hole in blank; d_F is the diameter of screw shaft or tube.

The flight does not usually need to be continuously welded to the centre shaft in order to transmit the torque required to move the material. A few short runs of 'stitch' weld per pitch are normally quite adequate for strength purposes. Continuous welding is, however, often specified to eliminate crevices and small residue pockets in the flight corners for hygienic reasons or to oppose the onset of fatigue by eliminating stress concentrations. It should be noted that the alternating tensile and compressive stresses that are created as the shaft rotates are a major source of fatigue failure. This is because screw conveyors in regular use rapidly attain the number of revolutions, and hence stresses cycle range, where fatigue is a major factor of the metallic strength. As stainless steel is especially prone to fatigue, special weld preparation, surface finish and mechanical treatment of the surface of welds may be undertaken on screws that are considered vulnerable or of strategic importance in a plant. Many other criteria influence the surface finish of stainless steel (Bates, 1999, www.ajax.co.uk), the contact friction of the material handled against the face of the screw flight being an important design parameter.

A means of making a continuous helical flight from a strip of steel was patented in 1898 by Frank C. Caldwell (US patent 601429). This process allowed extended lengths of continuous screw flights to be produced quickly and cheaply, and was quickly adopted for use in standard screw conveying machines. The concentration and growth of other manufacturing industries during the industrial revolution drove users to employ screw conveyors for handling other materials than agricultural products, from base mineral products and chemicals to refined powders and processed food products. The main virtues and limitations of this ubiquitous tool were rapidly established, usually by practical trials, and design refinements incorporated until it became a major form of bulk material handling equipment, particularly in the process industries where materials need to be moved relatively short distances between operations.

5.2 Classes of screw equipment

Screw conveyors are but one class of screw type solids handling device, albeit a major form in industrial applications. Other types of helical screw-based solids handling machines are commonly described as 'screw feeders', 'screw elevators', 'hopper discharge screws' and 'metering screws'. Many forms of processing operations also utilise helical screws in their composition and many of the features described will equally apply to their operating circumstances. The boundaries between these classes are often blurred, as machines with a predominant characteristic of one form of machine may undertake some of the functions combined with broad performance features of others. For the convenience, it is useful to classify the machine according to the principal mode of mechanics utilised to promote the movement of the product. However, where more than one mechanism prevails, each section should be examined and appropriately designed within the composite entity.

The mechanism prevailing in a 'screw conveyor' is the 'gravity' mode of sliding down the inclined surface of the screw flight. An essential feature of this process is that the level of cross-sectional fill is less than 45%. The material progresses along the screw axis one pitch distance for each revolution of the screw (Figure 5.1).

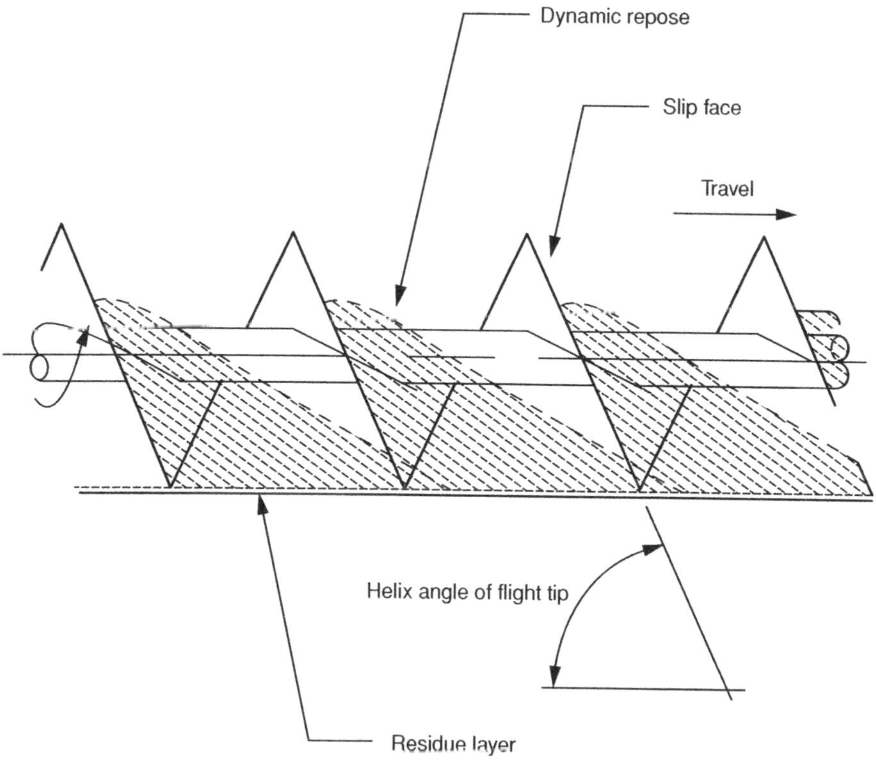

Figure 5.1 Gravity mode of material movement in screw conveyor.

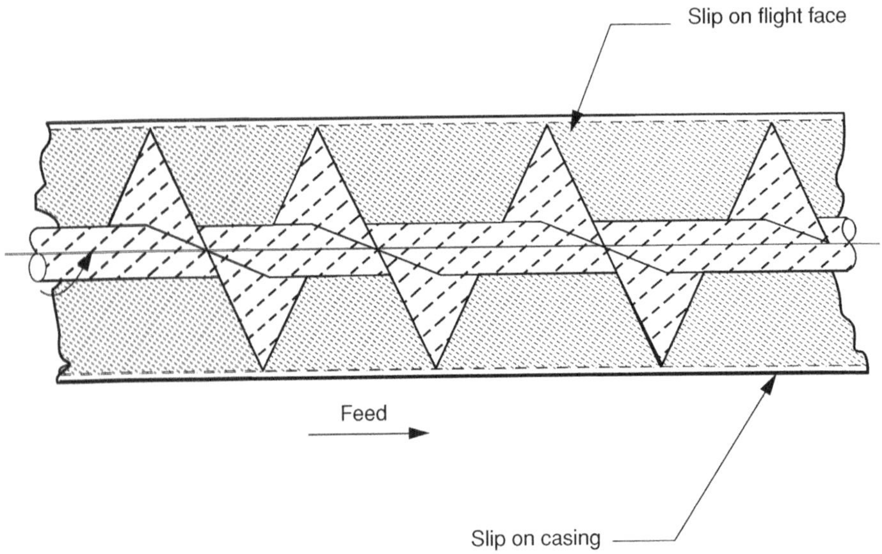

Figure 5.2 'Flooded' mode of material movement in a screw feeder.

A 'screw feeder' invariably has a flooded inlet such that the cross section of the screw has, in principle, a 100% fill, although there may be minor regions of voids behind the working surface of the flight according to the filling efficiency of the flow system. The movement of material along the screw axis follows a helical path according to the helix angle of the screw and the angle of friction between the bulk material and the flight contact surface in a confined condition (Figure 5.2).

A 'screw elevator' works in a dynamic mode, Figure 5.3, with material swirling around the screw axis and advancing along the axis in a similar pattern to a screw feeder, except that:

1 The level of fill is limited by the in-feed conditions and back-leakage, so rarely exceeds 25% of the swept volume of a pitch space.
2 The unconfined state of the bulk material allows significant dilatation of the bulk and the product moves in a vortex pattern with radial shear due to differential velocities at dissimilar radii.
3 There is boundary leakage in the flight tip clearance space that detracts from the elevating rate.

A prime determinate of the handling capacity of the screw elevator is the inlet conditions, where product has to enter tangentially with sufficient radial pressure to overcome the tendency of the screw and its prior contents to resist inwards movement. As the feed system is normally a conventional loading hopper leading into an inclined chute, only relatively free-flowing materials are able to develop sufficient transverse pressure to secure a practical degree of cross-sectional fill. Various techniques are employed to supplement the feed pressure, such as a 'flinger disc' facing the elevator inlet or a separate horizontal screw feeder,

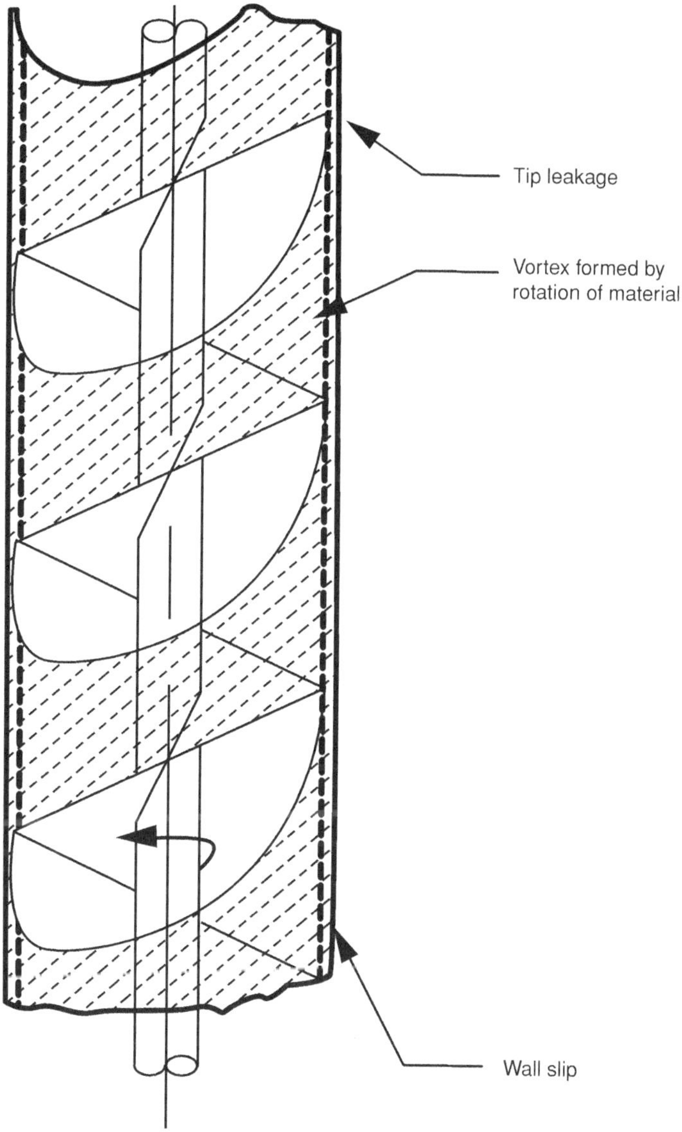

Figure 5.3 'Dynamic' mode of material movement in a screw elevator.

which then incorporates additional benefits such as enlarging the feed hopper capacity (Figure 5.4).

5.3 Standard screw conveyor features

Standard sizes of screw, casing and component details are now readily available for general use and equipment for simple applications can be specified by non-experts from guidance

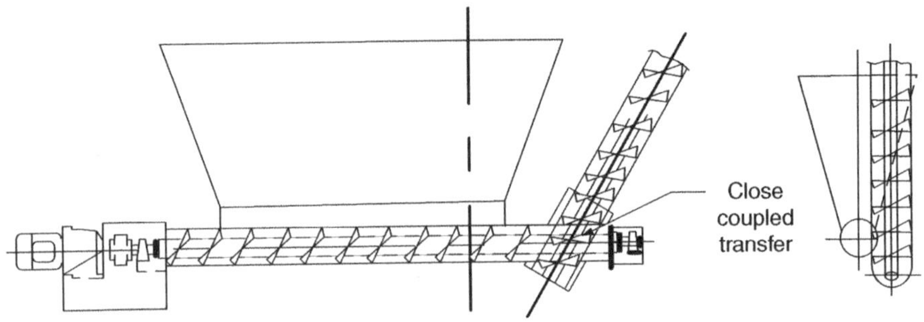

Figure 5.4 Feeding of screw elevator by screw feeder.

provided in various equipment supplier catalogues, as Forcade (1994), many of which replicate information from CEMA Book 350, produced by the US Conveyor Equipment Manufacturers Association. This publication lists dimensional particulars for 6″, 9″, 12″, 14″, 16″, 18″, 20″ and 24″ diameter screw conveyors, with standard thicknesses of troughs and covers for standard and heavy duty, together with a mine of application information, much of which is based on empirical data and specific to the handling of well-defined substances. The outstanding success of screw conveyors in various industries, such as flour milling, has lead many users into specifying them in a standard form for duties for which they have not been suitable. Their apparent simplicity is easy to interpret as a capacity to move any bulk material entered between the flight pitches.

Unfortunately, there are various limitations that counter this illusion, although special designs are often available to overcome some of the drawbacks. Screw conveyors do offer considerable operating virtues, but will not convey every class of bulk material or be suitable for every solids handling application. Misapplication can result in an unsatisfactory installation, in some cases damage to product in others a high maintenance cost or short working life. One point of major significant is whether to utilise intermediate bearings on screw conveyors that are too long to span with a single screw.

A feature of the screw construction, it is normally required, is that the screw does not deflect more than the allowance of clearance between the flights and the casing of the machine to avoid flight wear and product contamination. As this working clearance is limited, to minimise unmoved residue, the length of screw between support bearings is also limited according to the stiffness of the centre tube. There are practical restrictions on the size of tube permissible for a given screw diameter and, as deflection increases with the fourth power of span, the overall length of screw that can be employed between support bearings is severely inhibited. Long conveyors therefore require the use of intermediate bearings within the casing, yet these are a perennial source of operating and maintenance problems because they are obstructions to the passage of material and tend to wear rapidly due to contamination by the product. A wide range of bearing materials is used for centre bearing material to suit different selection criteria. Protection from product contamination is compromised by the limited axial and radial dimensions available, the screw having to push product past a gap in the screw flights in the remaining annular space. The maximum deflection of a conveyor screw should not normally exceed 6 mm.

The deflection, δ, of a screw supported at both ends in simply supported bearings is:

$$\delta = \frac{5\,WL^3}{384\,EI} \tag{5.2}$$

where W is the total weight of the screw; L is the length between bearings; E is the modulus of elasticity; I is the moment of inertia of pipe or shaft.

Excessive screw deflection may sometimes be avoided by using a larger pipe or one with a thicker wall. A larger diameter tube is generally more effective and economic, but the proportions of the screw flight should not be distorted by the use of standard pitches with very shallow flight up-stands or high compressive states may be induced in the pitch contents. For long conveying distances, a more positive way to avoid excessive deflection or the need for intermediate bearings is to use two or more single span machines in a cascade arrangement. Whilst this approach incurs a higher initial cost, it is often more economical in applications that demand reliable operation with minimum maintenance.

It is therefore necessary to understand the limitations of this class of equipment perhaps even more than the many and varied advantages that can be exploited. These mainly depend on the intrinsic nature of the bulk material that is to be handled, therefore the first task of an engineer faced with the selection of equipment for a solids handling duty is to recognise those features that may give rise to unsatisfactory operation.

The engineer's second task in the process of conveyor specification is to be scrupulously diligent in investigation-relevant physical properties of the product under the full range supply and operating conditions pertaining whilst in the equipment. For this work, it is essential to recognise that the name of the material is rarely an adequate guide as to how the material may behave. A 'representative sample' may also be misleading, as it is necessary to deal with all conditions that may be offered. A 'representative sample' may reflect a typical condition, but only a single condition and not necessarily the 'worst' that has to be handled. Likewise, all ambient, process and operating variations must be known.

It is vital for reliable operation that the physical properties of the bulk material are known and within the capability of the screw to handle. An I.Mech.E document (I.Mech.E. 1994) outlines the interesting aspects of a bulk solid that should be determined in all bulk solids handling applications.

5.4 The many operating benefits of screw conveyors

1. They are normally totally enclosed, protecting the product from contamination and the local environment from dust and spillage by containment of dust, vapours, gasses and internal or external pressure.
2. They are of compact cross section because they have only one working element that does not require a return path.
3. They are flexible in capacity up to their maximum transfer rate, which is essentially self-limiting, preferably with suitable flight variation between the feed region and the conveying length. Their potential as a feeder is widely exploited and offers a great number of benefits when properly integrated with a feed hopper design.
4. Intermediate inlets and outlets of various sizes can be fitted.

5 Different sizes, loading and materials of construction and surface finish may be used to serve applications that demand strict levels of hygiene or to deal with corrosive, hot or abrasive products.
6 A screw is reversible, to deliver in either direction from a central inlet, or right- and left-hand flights can be mounted on the same shaft to bring products to a central outlet from two inlets or deliver to two outlet at the same time from a central inlet.
7 Can deal with lumpy products and, with suitable flight construction, deal with wet, damp, cohesive or sticky products, filter cakes, centrifuged products and the like.
8 May be equipped with heating and/or cooling jackets, have heat transfer fluid through the centre shaft and through special flight constructions.
9 Can incorporate mixing elements of flight variations to undertake continuous mixing and blending duties.
10 May be designed for dust suppression or liquor addition, surface coating or agglomeration.
11 Compacting or pre-breaking functions can be undertaken.
12 May be inclined for elevation and de-watering applications.

Against these, and other uses for these versatile machines, must be recognised some general limitations to avoid misapplications.

5.5 General limitations of screw conveyors

Although some may be overcome by special design.

1 Screw conveyors are not self-cleaning as product will rest in the clearance space essentially allowed between the screw flights and casing to prevent metal contact. Cross contamination, product degradation or unwanted residue may result.
2 Shaft deflection limits the single span length of a screw. Whilst intermediate bearings are commonly fitted to secure longer conveying lengths, they introduce extra pockets of residue, routine maintenance demands and in some cases can be a perennial nuisance.
3 They are mechanical inefficient compared with belt conveyors and less cost-effective on extended lengths and for very high transport capacities.
4 High power loads may be experienced when starting under full conditions and running empty or under light load conditions is mechanically inefficient.
5 Hard pieces, particularly taper shaped or materials that laminate, can jam in the clearance space between the screw flights and the casing.
6 Products that set to a hard ‘caked’ condition can impose a high ‘rim drag’ and/or cause wear on the tip of the screw flight.
7 Sticky or stringy products can build up around the centre shaft to clog the working volume of the screw pitch.
8 Very delicate products may be damaged in transit by the repetitive dynamic repose effect of the material sliding down the front of the advancing pocket of product moved forward in a pitch space.
9 The interaction between the screw and the media handled introduces a behavioural relationship that must be satisfied for efficient operation. Whereas simple conveying

duties can generally be reliably sized and assessed for power requirements, many forms of screw equipment for feeding, elevating, compacting, mixing and other process duties require specialised knowledge or representative tests to prove their performance. Many mechanical designers have limited experience of particulate solids technology, powder testing and bulk processing; hence some forms of screw equipment are best undertaken by specialist manufacturers.

Within this balance between the advantages and limitations of screw equipment, empirical developments and advances in the technology are expanding the frontier of applications for both standard and custom-built machines.

'Custom'-designed conveyors take two forms. Standard components and their variants can be assembled in many ways to satisfy specific lengths and inlet and outlet requirements, to tailor a conventional conveyor to suit a specific installation. The other extreme is an individually designed machine that incorporates special features or techniques to be purpose-built for a particular function or duty. The cost of the latter type may be significantly more than a unit constructed from standard components but, as with many items involved in the handling of bulk solids, reliable and efficient performance is paramount. Notional savings on capital cost can involve losses that far exceed any purchase cost difference if the equipment does not work well.

5.6 Screw conveyor capacity

By definition, the supply of material to a screw conveyor is controlled by prior equipment. If the inlet is filled with material the machine will take away an amount determined by the screw geometry and its rotational speed, and it is essentially a 'screw feeder', of which there are many variations as described later. The handling capacity of a horizontal screw conveyor depends on three main factors: machine size, operating speed and the degree of cross-sectional loading. The cross-sectional loading is usually selected according to the nature of the bulk material, with margins allowed for supply fluctuations as appropriate.

Manufacture's literature abounds with qualified empirical formulae, mainly based on CEMA 350, which denotes a 'material factor' for a recommended level of cross-sectional loading of the conveyor. These factors are meant to take account of the abrasiveness of the bulk material and its 'flowability'. Wear due to abrasion is a power function of contact pressure and speed, so the objective should be to minimise these values. For a given mass transfer rate it is normally better to run 'slow and full', rather than fast at a lower cross-sectional loading. Apart from making an allowance for feed variations, the maximum loading that can be accommodated depends on the friction of the bulk material on the face of the screw flight and, to a lesser extent, how readily the material deforms in a dynamic state. A high friction product will cause material in the coarser helix angle region near the centre of the screw to be lifted during rotation and hence reduce the volumetric handling capacity of the machine.

The peripheral helix angle of a screw with pitch equal to screw diameter gas is approximately 72° from the horizontal at the point level with the centre of the screw shaft. This is steep enough to promote slip on all but very surface-adhesive forms of material. However, with a typical centre shaft being approximately 25% of the outside diameter of the screw, the inclination to the horizontal of the flight face adjacent to the shaft with the same screw is only about 38°. Materials with a high value of wall friction will tend to be lifted in this region by centre tube friction and corner effect between the flight and the tube and carried over the shaft to a previous pitch space. Fortunately, the area a region's cross section is linearly related to the diameter, so the dominant conveying area is the most effective region at the outer portion of the screw flight.

Based on a normal max screw conveyor loading of 45%, the max transfer volume for the standard flight ratio of pitch equals screw diameter, is given by:

$$\text{Effective conveyed volume per revolution} = 0.45 \times 3.142/4[D^2 - K^2]P \qquad (5.3)$$

where K denotes whichever is the largest of d or $D \times \text{Tan} \times \Phi_w$; D is the screw diameter; P is the flight pitch; d is the shaft or tube diameter; Φ_w is the wall friction of the bulk material on the flight face.

Within the normal operating range of screw conveyors the transport velocity is linearly related to the cross-sectional loading and the rotational speed, the speed generally being limited by considerations of wear with abrasive materials and the condition of the bulk for other applications.

Screws are essentially volumetric devices, so it is necessary to know the 'effective bulk density as conveyed', to convert the volume to a mass-based value. Product moved by a screw is usually in a state of mild agitation, rather than displaced as a coherent mass, therefore the effective density for conveying purposes tends to be less than when the material is in a settled condition. This particularly so with fine powders that entrain air when disturbed, so a useful guide to the effective density that a bulk material will attain when conveyed by screws to shake a small sample of known weight in a jar and check the instantaneous volume that it occupies when deposited into a container. The bulk condition attained at slow speeds of rotation equate to instantaneously settled conditions of an agitated product. Higher speeds of rotation induce more agitated bulk conditions. This may not be significant, other than its effect on bulk density when conveying horizontally, but when the machine is used at an inclination the loose state of fine powders will leak back down the conveying axis to seriously impair the transfer capacity.

5.6.1 *The effect of machine inclination*

A screw conveyor will handle material up gentle inclinations with only a small loss of transfer capacity, provided the material is not in a fluid condition. However, the ability to elevate loose solids in a conventional conveying mode falls off progressively as the inclination is increased above 15° to the horizontal, such that it will only move about 30% of its horizontal capacity when used at a shaft inclination of 30° and above this angle of slope the conveying rate reduces very rapidly to zero. At low angles, the transfer rate is only affected by the change in geometry of the pocket of material moving within the pitch spaces of the screw flight. The surface repose angle of the material remains constant to the horizontal as the flight face angle reduces. This has a twofold effect:

1. The volume of material that can be carried forward in the pitch space decreases with the change in geometry to determine the maximum level of the pocket of material that can be moved.

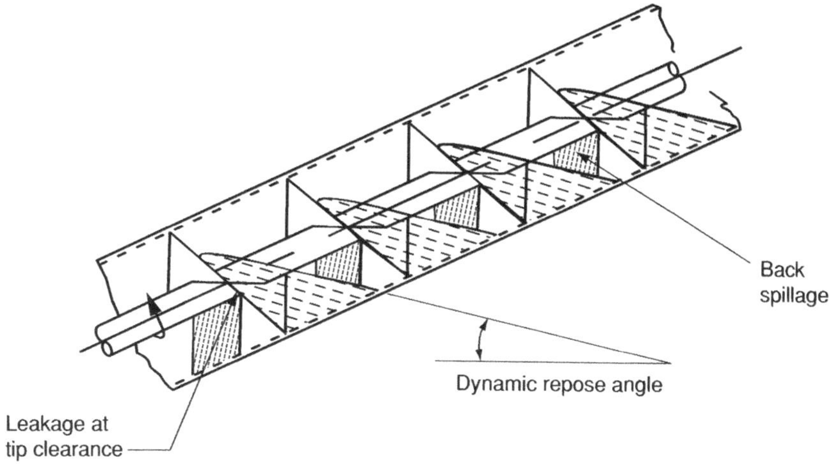

Figure 5.5 Effect of machine inclination of handling capacity.

2 Material spills over the shaft back into the previous pitch space due to the lowering of the effective angle of the central region of the flight from below the angle of slip.

It must be remembered that the normal inclination of the inner diameter of the flight adjacent to the centre shaft is only about 38° when the screw axis is horizontal, so at 20° slope of the conveyor this surface reduced to about 18° to the horizontal, a slope at which many bulk materials will not slip on metal surfaces, apart from the drag induced by the centre tube and sharp corner between the flight and the tube. This also indicates how surface friction of the material on the flight face is a crucial design parameter, particularly for inclined screw conveyors (Figure 5.5).

At operating angles steeper than about 25° to the horizontal, the situation deteriorates even more rapidly. Above this inclination material can also start to slide back in the casing clearance, particularly at the sides where the combination of slope angle with the radius of the casing forms a steeper wall surface than the screw axis (Figure 5.6). The position is aggravated by the bias of cross section fill produced by the rotation of the flight, which is intensified at steeper inclinations by the extra drag induced by the flatter flight. The potential to elevate material by the conventional mechanical mode of sliding down the face of the screw flight effectively ceases at axis inclinations above around 35°.

It is possible to run a conveyor screw at steeper inclination at low speeds, but the cross-sectional loading rapidly attains a virtually full condition and the mechanics of transfer changes radically to that of a screw feeder. In these circumstances the material is acted on by the circumference of the screw flight and restrained from rotation with the screw by the frictional drag of the outer casing. Back leakage is prevented by the lack of space for the material to pass into and the torsional loading increases significantly compared with a horizontal screw conveying action.

Apart from the heavy torque demands, the main operating drawbacks to this form of elevating are that the full screw resists feed at the inlet, so the machine will only work

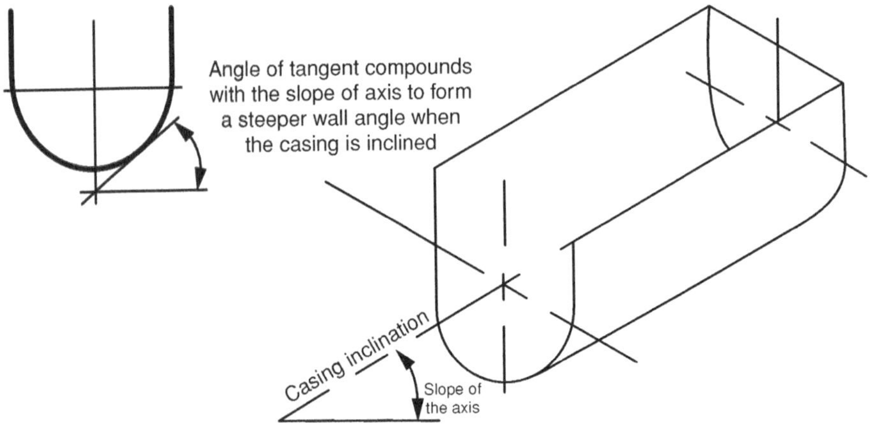

Figure 5.6 Combination of casing inclination with casing curvature.

with free-flowing products and a well-designed inlet chute, and the machine will not clear of product, remaining full when material ceases to be fed into the machine. This form of operation is not recommended, except for relatively short units that are not sensitive to permanent residue.

Material can be carried at a steep angle by a screw elevator, where the screw is surrounded by a circular casing and runs at a faster speed, but this introduces another different mode of conveying, where rotation of the mass as a dynamic vortex is retrained by friction on the casing wall and a degree of back leakage is an acceptable loss of the material transfer capacity. The selection and specification of screw elevators is generally the domain of specialists, partly because the number of variables involved in dealing with industrial applications require the machines to be custom-built. The other main reason is that reliable operation is very sensitive to the bulk material properties and equipment geometry. There are two exceptions to the generalisation. Agricultural screw augers are made in standard sizes to handle grain and other products on the basis of 'proof by trial'. Some units are built into machines, such as combined harvesters and agricultural driers. Stand-alone machines are also user on farms and like duties.

Another crude form of screw elevator is the so-called flexible screw elevators. The very limited flexibility these machines possess is permitted by the use of a plastic tubular casing and continuous screw flight that does not have a centre shaft. These units mainly have small diameter 'coreless' screws, i.e. without centre shafts, running at relatively high speeds.

A more recent development, an astonishingly long overlooked innovation in view of the maturity of the basic screw technology and its international application on a massive scale, is the 'static screw elevator', invented by Peter Olds in Maryborough, Australia, in 2004. The patented technique exploits the peripheral drag of the boundary layer held in a vertical casing by virtue of its rotation, whilst product is promoted to move up a static helical screw by the boundary contact friction. The process is somewhat similar the mechanic of a conventional screw elevator, except that there is no agitated vortex causing internal shear in the material in transit and no leakage in the boundary layer between the screw and the casing. These operating advantages are supplemented by numerous convenient installation features, probably the most valuable being the exceptional low pick-up facility and the lack of a bottom seal and bearing, which are a perennial service problem with most ordinary screw elevators. Further details of this remarkable machine are given elsewhere (Bates 2005).

5.7 Power requirements

The drive of a screw conveyor must overcome the torsional resistance of an empty screw, the work content of moving the material and the restraint offered by shearing over a bed of residue in the flight tip clearance space. An allowance must also be made to deal with starting and for exceptional operating conditions that may arise during the lifetime of the equipment. However, the mechanics of the machine, which may be of a considerable length and working under a light or no load condition, permits all the available excess power and overload capacity of the drive to be concentrated at any point, or on the final flight, should a local jam or blockage occur. Were such a situation likely to cause damage to the screw or other items, it is prudent to incorporate a detection or protection device.

5.7.1 *Empty running*

Sources of initial power loss occur within the drive train, bearings and seals. These are normally relatively low and figures may be secured from manufacturer's catalogues and basic engineering principles. The exception is packing glands, which can introduce high resistance if misaligned or over-tightened. It must be appreciated that shaft ends on screw conveyors are not like stiff machined axles or gearbox shafts but are usually fit, and sometimes bolted, into the ends of tubes that are subjected to weld distortion and deflection. Any slight deviation from axial trueness can result in a force being transferred from the gland to the adjacent end bearing, with a consequent increase in load and torsional resistance on both of them. Weld distortion is corrected in manufacture with well-made screws.

A clear sign of shaft misalignment is shown by any cyclic movement of the bearing mounting or oscillation in tension of a chain drive, if fitted. The writer did have an experience of a partially stretched chain that gave all the symptoms of a bent drive shaft, except that the frequency of drive–mount distortion did not synchronise with the screw shaft speed. This clue, provided by the ratio of the chain length to the two sprockets, allied with stable dial gauge reading on the screw conveyor shaft, solved both the mystery and a client's concern for the quality of the screw auger.

It is comparatively easy to over-tighten a packed gland, a procedure often adopted if any leakage takes place, whereas a prime cause of gland leakage, it is usually the former problem of gland misalignment, due to the eccentric motion of the shaft that does not cured the leakage problem. However, zealous gland tightening of a well-oriented gland gives rise to a firm grip of the packing on the shaft. This condition can be readily detected by the frictional heat generated but, if not corrected, can give rise to serious power losses.

A working figure for completely empty running may be taken as:

$$\text{kW loss} = \frac{L \times D^2 \times S \times B}{200 \times E} \tag{5.4}$$

where L is the length of conveyor, m; D is the diameter of flight, m; S is the speed of rotation, RPM; B is the bearing factor: 1 for ball bearing, 2 if hanger bearings fitted; E is the drive efficiency: 0.95 for geared motor and coupling, 0.9 if V belt or chain drive fitted.

5.7.2 *Power to transport material*

The material in transit slides down the inclined face of the screw flight, so one element of the energy loss equates to the loss of potential energy of the mass of material sliding down an inclined slope. The actual slope of the flight face varies from the tip to the root of the flight, hence it would be ultra-conservative to rate the energy loss according to the effective slope of the flight tip. A more realistic figure is to base the equivalent slope over the conveyed distance as being around 10° more than the angle of wall friction between the material in transit and the contact surface of the screw flight.

A second factor is the sliding of the contents in the conveyor over the static residue of material resting in the casing clearance. The screw will also experience a degree of tip drag as the flight rotates in contact with the static deposit left in the clearance space between the screw flight and the casing. This will normally impose little torsional resistance on the boundary of the flight. As normal empty running torque is considerably less than that in the working condition, this aspect can be ignored. However, when dealing with fine damp materials, products that are cohesive, matt, wedge, bind, or with crystalline material that 'cakes', this deposited layer may turn into a hard bed that is progressively packed with fine or trapped particles imposing a high binding force on the contact surface of the flight tip.

Special consideration should be given to any applications where the bulk material being conveyed can trap or form a firm, unyielding packed condition in the residue left between the flight tip and the conveyor casing. This effect may be countered by forming a flexible casing, typically of a conveyor best type material, within a framework to hold the belting to the required 'U' shape. Local yielding of the casing under high contact pressure forms an unstable base for the thin residue layer to retain its coherency and it will tend to break away. An alternative is to use a thin flight or dress the flight tip to a narrow edge such that the bearing area is minimised and the dead layer can be scraped away under high local pressure. The flights should be constructed of an abrasive resisting material having a hard weld deposit on the tips if dealing with a product likely to cause wear.

The energy content of the 'sliding' component of the conveying power is given by:

$$\text{Loss of potential energy} = C \times L \times \tan[\varphi + 10^\circ] \tag{5.5}$$

where C is the mass moved, Te/h; L is the effective conveyor length, m; Φ is the friction angle between product and flight face.

The energy content of the 'shear' component of the conveying power is given by:

$$\text{Shear energy of transport} = C \times L \times \tan\psi \tag{5.6}$$

where C is the mass moved, Te/h; L is the effective conveyor length, m; Ψ is the internal angle of friction of the product handled.

5.7.3 *Conveyors subjected to 'flood feed' conditions*

Some applications, such as the discharge of plate filter presses and dust collectors, may be required to deal with solids loading conditions that fill cross section of the screw along its full exposed length in the supply hopper, often with an overpressure due to the pile of material above the screw. Whilst the prime function of the equipment is to collect material over an extended length of screw and then discharge the contents in a controlled manner, the order of discharge of the material is usually immaterial. It is necessary to provide the power needed to drive the screw under these conditions and probably also to start the screw from a settled state, which may be a much more onerous task. A further condition is that the subsequent transfer capacity alone the flow route is adequate to cope with the surge feed of a flooded condition when the average rate at which the bulk material is delivered to the system may be much less. Such applications must be designed as screw feeders, even though the period of operating as such may be very limited.

5.8 Screw feeders

There are two essential elements to a screw feeder, a supply hopper and a dispensing screw. At one end of the spectrum screws are employed to discharge bulk storage silos and hoppers, sometimes in multiple banks. The rate and accuracy of dispensation may or may not be important in such cases, but the primary function is to ensure a reliable and controllable discharge. Usually at the smaller end of the scale, feed screws are required to provide a definitive discharge rate, with an accuracy determined by the application needs. These machines may have a large or continuous re-supply system, but more often comprise a relatively small, integrated hopper that is replenished as required.

5.8.1 *Hopper discharge screws*

There are various reasons why screws may be specified for discharging hoppers or silos:

(i) To generate extraction over the full outlet area of a storage container
(ii) To provide a large, effective outlet size for the container to ensure reliable flow
(iii) To save headroom/give extra storage capacity
(iv) To control the rate of discharge
(v) To deliver to two or more receiving points from a hopper outlet
(vi) To enhance the discharge rate of a flow inhibited product and counter 'flushing'

A Screw feeder that is incorporated to discharge a bulk storage facility is part of an integral design where the selection of width, length and extraction characteristics of the feeder inlet impinges on the overall hopper or silo specification and performance. The minimum width of opening is set by the flow characteristics of the bulk material, determined by powder testing if not free flowing. If the length of the feeder inlet is less than three times its width, the flow channel size must be considered as a compromise between a square and a slot, according to its proportions. Likewise, unless the side faces above the inlet have parallel ends, an extra allowance must be made on the slope of plane flow inclination for mass flow

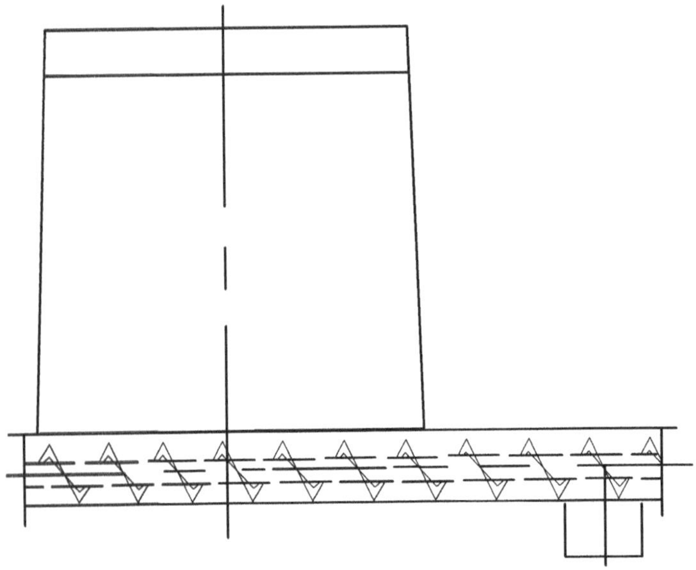

Figure 5.7 Basic form of screw feeder.

criteria towards the radial flow values, according to the degree of transverse convergence. A typical basic screw feeder is shown in Figure 5.7.

The first step in the design of any bulk storage container is to determine the optimum flow regime that is to be formed when discharge occurs. The basic choice is 'mass flow', 'expanded flow' or 'non-mass flow', often called 'core' or 'funnel flow'. These patterns are described in detail elsewhere (University of Florida 1999). 'Mass flow' is mainly selected to avoid indeterminate storage time with products that deteriorate with age or segregate. 'Expanded flow' is selected for poor flow bulk materials and 'core flow', i.e. all other flow patterns, for products that are unchanged by storage time and generally free flowing.

The benefits and drawbacks of mass flow are summarised in Table 5.1.

Having selected the appropriate form of flow regime to suit the application, the next stage is to complete a systematic design process, such as in Figure 5.8, to determine the proportions of the hopper, key dimensions and the type of discharge control that is suitable:

(i) *Generating extraction over the full outlet area of a storage container.* It is crucial for mass flow that extraction takes place over the total screw feeder inlet area.

Progressive extraction at the screw feeder inlet is not essential for non-mass flow installations and applications where the container is totally empties before being refilled but is useful to provide at least a degree of progressive extraction on long feeder screws to reduce the drive power requirements and avoid extended length of 'dead' material over the outlet slot.

Progressive extraction can be secured on feeder screws in many different ways. An early form of generating continuous extraction was by tapering the outside diameter of the screw along the inlet length of the feeder. This is effective in its objective, but has various operating and commercial drawbacks. The major practical issue is that the width of the screw varies

Table 5.1 Benefits and drawbacks of mass flow.

Benefits	Drawbacks
No 'dead' regions of flow	Tall headroom/*reduced storage capacity*
More predictable storage times	Potential wear on walls
Secures *flow through smaller outlets*	The outlet *must* be fully 'live'
Significantly reduces segregation	Powder tests are essential to determine design
Resists 'through-flow flushing'	High wall pressure generated at the hip joint
The flow pattern is predictable	The design relates only to the product tested
Flow is roughly 'first-in, first-out'	
Flow can be exploited to blend contents	*Any* property change may negate mass flow
Proven design *guarantees reliable flow*	Flow rate is less than non-mass flow

Notes: It is *essential* that flow occurs over the total area of the hopper outlet.
The flow velocity is *not* uniform across converging sections of the hopper.
Final portion of the discharge can concentrate segregation.
'Expanded flow', i.e. mass flow at the outlet region only, secures the outlet size benefits of mass flow and may suffice for bins that refill at a low level.
'Mixed flow', i.e. where the non-mass flow channel expands to the container walls below the stored surface level. This can appear to be mass flow, but is not generated 'kick' stresses on the walls at indeterminate locations.
Conical hoppers require walls approximately 10° steeper than Vee hoppers.
Additional benefits of lower wall angles and smaller outlets may be gained by relaxation of the transverse flow channel to a slot outlet, provided that the slot develops fully live flow.

over the length, so the effective outlet width for flow is the smallest diameter. The cross-sectional area of the screw increases as the square of the diameter, so the extraction is not even unless the pitch is reduced, which is usually contrary to practice and would lead to problems of deep pocket spaces between the flights. Whilst this progressively increase in extraction may be acceptable in general use, it has the effect of taking more from the outlet end of the inlet. It is this location that determines the discharge rate so, as the container will preferentially empty at this end, the discharge will progressively be controlled by the reducing size of the screws that are covered with product.

A further drawback of this construction is that it is not convenient to have a taper outlet slot on a container so the practice is to slope the feeder casing from the screw diameter to a parallel hopper outlet size that matches the largest diameter of the screw. All most invariably, this results in a casing wall inclination at the smaller end of the feeder that is inadequate for product slip. In combination with the gap at the side of the screw to the casing wall that fills with static product and opposes wall, the effect is to create a narrow flow channel with non-mass flow characteristics. The economics of manufacture also detract from widespread adoption of this technique as both the screw and the casing have taper components that demand extra fabrication time.

The simplest of providing progressive extraction is by way of pitch variation. This has very strict limits because very short pitches form deep and narrow pockets between the screw flights that are ineffective in conveying and tend to 'log', that is completely fill the pitch space with product. On the other hand, the mechanics of the screw flight limits the pitch length when the helix angle of the flight causes excess product rotation with the screw. It is rarely effective to increase the screw pitch to a dimension greater than the screw diameter. Incremental transfer of product along the screw axis can cease prior to attainment of this pitch length and is certainly much reduced at longer pitches. It should also be noted that

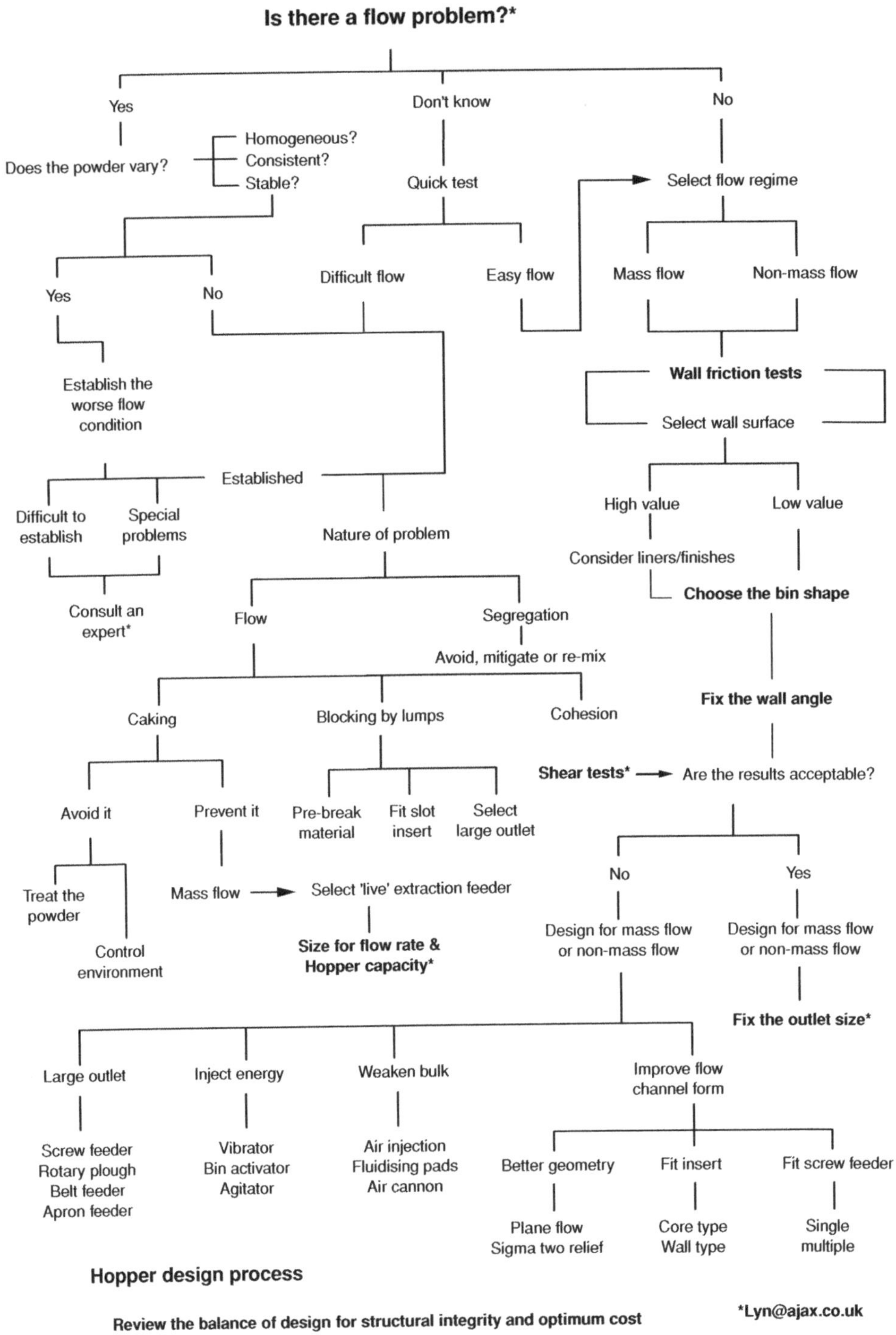

Figure 5.8 Hopper design.

the initial section of screw flight exposed to the product as the start of the feeder inlet will virtually fill with material, so subsequent extraction is dependent on the marginal gain in axial transfer capacity.

These features tend to limit the effective length of feeder inlet for a mass flow application to about four screw diameters when the pitch only is varied. This construction is, however, effective on long screws, where intermittent changes of pitch will relieve the shear under a 'dead' region of flow and require less torque to shear through a length of flowing media.

A more effective progressive extraction construction is to make the centre shaft taper from a large diameter at the start of the feeder to a small diameter at the discharge end. Denial of the central region is relatively ineffective at small diameters, because the area concerned is proportional to the square of the diameter. Best results are secured with changing the pitch as a uniform pitch would require a shallow flight face to promote the motion of a long annulus of product. This method can utilise a uniform inlet slot and parallel casing, but is relatively expensive to manufacture. It is well justified on long feeder screws, where the combination of pitch increase and tube size reduction can be changed to a pitch variation only at a point along its length. A more economical design is to employ a 'stepped shaft' and variable pitch construction that can serve length to width ratios up to around 8 or 10.

(ii) *Providing a large, effective outlet size*. The use of a screw feeder inevitably offers a slot shape of outlet, which is equivalent in flow generating terms to a circular outlet twice the width of a slot. It also provides a flat side for the hopper connection, and the plane flow channel of a V-shaped hopper can be about 10° less inclined for mass flow than a conical hopper. Twin or multiple screw feeders are often used to service wide hopper outlets that may be required to countering a tendency to arch, either with mass flow or non-mass flow hoppers. These work best with progressive discharge, but will also extract material from under an arching condition until the arch size is unstable and collapses, provided the critical arching size is less than the area of the screw feeder inlet. If this situation repeats over a period, the static regions will be subjected to time consolidation and the design must allow for any such gain in strength when the remaining material has to be emptied from the hopper.

Multiple small screws offer many operating benefits over single or fewer large discharge screws. They take considerably less power, less headroom and give a much smoother discharge. They are, however, more expensive because of the multiple bearings and seals, extra drive train and profiled casing but, in general, the choice is usually very clear as the penalties of using very large screws can be quite high. Multiple drives can also be useful to facilitate discharge in the event of a screw or drive problem or offer feed rate change for finer discretion on batch fill duties.

(iii) *Headroom saving/extra storage capacity*. Considerable headroom saving or additional storage capacity can be secured by employing screw feeders. The volume of a V section is 50% more than the equivalent height of a cone and widening the inlet of the screw feeder, by fitting twin or more screws of diameter D adds:

$$D \times L \times (H_1 + H_2) \qquad (5.7)$$

extra to the volume of the storage capacity for each extra screw, where L is the length of screw feeder inlet; H_1 is the depth of converging V section; H_2 is the depth of hopper body section.

(iv) *Feed control.* A screw feeder not only provides the facility to control on/off and the volumetric rate of discharge, but also allows some discretion on where the material is delivered in relation to the centreline of the storage hopper. It is often convenient, for example, to discharge into a process vessel, vehicle or subsequent conveyor that would not readily fit under the storage container. Variable speed drive permits a range of discharge rates to be easily adjusted to suit the requirements, be integrated with process control or provide a fast initial discharge and slow trimming feed to accurately dispense batch weights.

(v) *Multiple discharge locations.* Good hopper shapes are commonly spoilt by fitting an extra outlet to the side of a cone section of a hopper. This introduces eccentric flow and regions of 'dead' storage when the original outlet is not in use. Feed screws can be made reversible, to deliver to either of two locations, which is a useful facility provided only to one discharge point is in use at any one time. Another inhibiting feature of this technique is that the screw must be made with uniform pitch construction to avoid compacting the contents when reversed. This limits the 'live' extraction length of a feeder to about two exposed pitches with a standard form of screw construction. This does not allow the full slot benefit to be secured as the slot length should beat at least three times the width to overcome end effects for mass flow. A proprietary method has been devised to overcome this limitation. A further refinement by the same company developed the technique of utilising twin cantilever-mounted screws, facing end to end in the same casing. By arranging the individual drives to run at half speed of the partner when operating in the reverse direction, two standard forms of screw will serve a slot length of four screw diameters with progressive extraction from the full length of the feeder inlet when operated to discharge to either end outlet independently or to both end outlets at the same time. Incorporating the above 'Lynflow', slot enhancing technique, allows the effective 'live' inlet of a screw to be up to six screw diameters long and able to serve two outlets independently or together, securing both the full flow benefits of a slot construction with the extra capacity and flow benefits of a V form of hopper construction in the crucial outlet region of a storage hopper.

(vi) *Enhanced discharge rate and counter 'flushing'.* A problem encountered with the handling of fine powders is that the change in voidage from a dilated to fully settled condition, and vice versa, involves the passage of the ambient media, usually air, through a mass of tortuous interstitial passages to bring the void pressure to ambient conditions.

This time for this process to attain equilibrium may be extensive, particularly in warm or hot conditions as the viscosity of air decreases with temperature. The phenomenon of 'flushing' is directly related to the presence of excess air in the voids of a mass of fine particles that sustains a pressure that prevents a more intimate particle-to-particle contact for the development of shear strength. The contrary condition arises when the bulk is well settled. The bulk must initially expand to commence to flow though a hopper outlet by gravity. All the expansion must be given by enlargement of the void space that has very limited initial air content. Expansion is, therefore, inhibited by the differential pressure of ambient and that of the void space in the expanding bulk and satisfying this void demand of expansion is resisted by the narrow passages between the particles through which this air has to pass. The tendency is for a slow, but progressive arch to collapse above the outlet until the arch size attains a dimension larger than the 'critical arching span' for the strength of the bulk solid. At this point there is a dramatic collapse of the arch and the falling mass can either entrain air to flush uncontrollably through the outlet or fill the void of the arch for the cycle to repeat. Either event is not conducive to controlled discharge.

Enlarging the effective area of flow by means of fitting a screw feeder with an elongated slot reduces the rate of expansion and gives more time for the bulk to attain a gravity flow condition in a controlled manner. Similarly, the larger flow channel generated in the supply hopper gives more time for air to settle from an initial fluidised state to a stable feed condition. In this context it is important to note that a non-mass flow hopper will enable a short-circuit route to develop for fresh material in a fluidised condition to emerge at the outlet through a deep bed of stored bulk material. Such a hopper has a relatively small cross section of flow channel through a bed of static material and the residence time for fresh material is relatively short. This not only gives little opportunity for the material to de-aerate, but also the high product flow velocity is counter-flow to the rising gas of de-aeration and tends to take the trapped air down with the bulk to maintain the loose condition.

Even a mass flow hopper has a significant velocity gradient across the converging section of the container and can preferentially draw-down fluidised material through the bed if this velocity differential reaches near the surface. The reason is that flow pressures of a fluidised mass are hydrostatic, whereas the lateral pressures of gravity flow of a normal media are much less than the vertical pressures. A depression in the flow channel that admits a bulk material in a fluid condition will therefore exert radial pressures mush in excess of the non-fluid product and progressively penetrate the flow channel. To counter this tendency, the cross section of the live flow channel should be as large as practical and have minimum velocity differences. Alternatively, an accelerated de-aeration device should be fitted in the centre of the flow channel.

5.9 Power needs of screw feeders

Empty and running conditions may be generally ignored when sizing the power requirements of a screw feeder because the main load to be faced is starting conditions, where the screw has to commence rotating in a settled bed of confined bulk material.

It must be emphasised that this duty is completely different to the sizing of a screw conveyor drive as the bulk material is subjected to a compacting overpressure at the feeder interface and completely surrounds the screw. Determination of the effective pressure acting on a screw feeder inlet is beyond the scope of this book as it is influenced by the hopper geometry and filling conditions, but a crude assessment may be given by considering the effective compacting load, C_L, to be of the order of:

$$C_L = 2W \times L \times \gamma \qquad (5.8)$$

where W is the width of slot; L is the length of slot; γ is the compacted density of product.

The shear strength, f_s, then equals to $C_L \times \tan\delta$, where δ is the internal angle of friction of the bulk.

The effort required to rotate the screw is that needed for the inclined face of the flight face to shear the bulk at the peripheral radii of the flight and also move the weight of the screw contents along the screw axis. It will be seen that the effort will differ along the axis for a variable geometry screw flight, but a conservative rating can be attained by considering the longest pitch construction to apply along the full length and the shear stress to apply around the whole circumference of the screw. A separate assessment may necessary for the section of screw projecting beyond the inlet region if, as is usually the case, the screw geometry is changed after the inlet section to dilate the material to avoid wasting energy. This portion can be considered as a highly loaded screw conveyor, only requiring to move the contents of the screw.

The torque, T, required to turn a pitch section of a feeder screw is therefore:

$$T = f_s D/2[\sin(\tan^{-1} P/\pi D + \tan^{-1}\varphi)\pi DP + \pi/4(D - d^2)P\gamma] \tag{5.9}$$

where f_s is the shear strength of product; D is the flight diameter; P is the flight pitch; ϕ is the contact friction angle of material on flight face; γ is the bulk density of the product.

The axial thrust, A, generated by this pitch length is:

$$A = f_s[\cos(\tan^{-1} P/\pi D + \tan^{-1}\varphi)\pi DP + \pi/4(D - d^2)P\gamma] \tag{5.10}$$

5.10 Dispensing screws

A major use of screw feeders is for controlling the rate of feed on powders and bulk materials to batch or process operations. Feed rates from many tonnes per hour down to less than a kilogramme per hour are quite common. Lower feed rates are generally more demanding of design in order to secure reliable flow and an even rate of dispensation. This is partly because low feed rates demand small diameter screws, and hence narrow widths of flow channel that may experience flow difficulties. Also, low feed rates usually mean low screw speeds and the evenness of discharge may be affected by cyclic variations because of the screw geometry, or erratic fluctuations due to the avalanching of the slope of free-flowing products or 'cohesive breakaway' of materials that are not free flowing.

Screw feeders are essentially volumetric discharging devices and used solely as volumetric dispensers are widely used for the controlled metering of bulk solids. They have to be calibrated to secure a specific mass flow rate and, for stable accuracy, depend on the product being handled flowing reliably and in a consistent state of density. Screw feeders can sustain uniform accuracy, typically within +2% of a set figure, when dealing with a firm, granular material, assuming that the screw speed is steady and no changes take place in the product. It is not unusual for finer limits to be held with hard-grained, free-flowing materials. The main virtue of volumetric feeding is that it is considerably more economical than when the additional cost of gravimetric control is added and is perfectly satisfactory for many applications where fine accuracy is not important. Screw speed change can be exercised on volumetric feeders for example, manual control of a batch dispensation can change from a high to low discharge rate to provide fine discretion on fill quantity or simple feedback provided by some process control to adjust the supply to an optimum rate.

The main drawbacks of volumetric feed control are:

(i) Feed rate deviations cannot be detected or automatically rectified should they occur
(ii) Density variations do occur in many bulk products, particularly fine powders
(iii) Accuracy cannot be guaranteed or held to fine limits
(iv) It is not possible to record the actual rate of feed

When dealing with relatively high value products, sensitive formulations or a traceable record is necessary, a gravimetric system is virtually essential.

A popular technique is the loss-in-weight system, whereby a batch load of product is filled into a hopper equipped with a weighing device and the reduction in weight, measured at very frequent intervals, is compared with the change that should have occurred according to a pre-set value. Corrective action is applied by changing the screw speed to swiftly compensate for any deviation to ensure that the feed rate does not fall outside close limits. When the contents of the supply hopper reduces to a low level the feeder control alters from gravimetric to a fixed speed for a short period, the fixed speed being based on the immediate past controller experience, during which period the hopper is refilled. There is sufficient product in the feeder flow channel to meet the discharge quantity during the re-fill period, after which the feeder resumes gravimetric control that can compensate for any density difference between the refill material and that previously held.

In some cases it is more convenient and/or accurate to weigh the receiving vessel, as this may have a lighter tare value and permit finer load cell discretion. Feedback control may also be derived from process conditions, such as heating of reactors or ability of the process machine to accept product. Advances in load cell design and control technology has considerably improved the capability of gravimetric feeders to deliver an accurate feed rate. It remains that no electronic wizardry can compensate for unpredictable variations in bulk density, as may occur in non-mass flow hoppers. An efficient system demands that the material arrives at the screw, not only reliably, but in a consistent bulk density condition. Much can be done in the way of advanced hopper design to promote reliable flow through narrow flow channels and condition the product to a consistent bulk density condition.

5.11 Special screw feeders

A variety of special features can be incorporated in screw feeder design to secure operating convenience, easy-clean facilities, compacting, force feeding against a resisting pressure, providing a 'seal plug' against vapours, heat, contamination and the like. A novel, proprietary technique to generate progressive axial extraction with a uniform pitch construction screw has been developed by Ajax Equipment Ltd. Within their Lynflow trademark of innovative techniques that exploit fundamental principles within state-of-art technology. Users and designers of screw conveyors and feeders continue to find new, novel, innovative and efficient applications, so the subject is far from a complete technology.

References

Ajax Equipment Ltd. (1998) *'Lynflow' Slot Enhancing Technique for Uniform Pitch Screw Feeders*. www.ajax.co.uk.

Bates, L. (1999) *The Surface Finish of Stainless Steel*. Ajax Equipment Ltd. www.ajax.co.uk.

Bates, L. (October 2005) The static screw elevator. *Proceedings of the 8th International Conference on Bulk Materials Storage, Handling and Transportation*, University of Wollangong.

Forcade, M.P. (1994) *Screw Conveyor 101, Basic Training Manual for Screw Conveyors*. Goodman Conveyor Company.

I.Mech.E. (1994) *Guide to the Specification of Bulk Solids for Storage and Handling Applications*. Bulk Materials Handling Committee.

University of Florida (January 1999) Flow regimes in bulk storage hoppers. In: *Educational Resources for Particulate Technology*, Vol. 1, No. 1, Art 3.

6 Trough conveying

DON McGLINCHEY

6.1 Introduction

A very common transport method is to confine or restrict a granular or powdered material by cross-sectional boundary and utilise an external force such as that produced by gravity or an 'internal' force, usually mechanical, to effect movement of the material in the axial direction. Typical geometries that provide these boundaries are a 'U-section' trough or a 'tube' or 'pipe' (see Figure 6.1). Examples which will be covered in this chapter are:

- Chutes
- Air slides
- Vibratory conveyors
- Chain and flight conveyors

The other principal technologies that fall into this category are screw conveyors which are covered in Chapter 5 and aeromechanical conveyors which are discussed in Chapter 7.

The application of these technologies is typically restricted to relatively short distances of about 1–20 m; however, the range of product mass flow rates that can be accommodated may be from a few tonnes per hour to many thousands of tonnes per hour. Although the most obvious examples will have a linear geometry, changes in both velocity and direction can be achieved to match particular handling or process requirements. The industries to which these technologies may be applied include: power generation, plastics, pharmaceuticals, bulk and fine chemicals, bulk ores and aggregates and agriculture.

Operational difficulties arise, as with most solids handling problems, from a mismatch between bulk material properties and the plant item. Segregation and wear issues are common concerns; however, these problems can be mitigated by the correct choice of design and operational parameters.

Most operations can be completely enclosed providing an environmental advantage and many can combine transport with a process objective such as heating, cooling or screening.

6.2 The chute

On a superficial level the concept of the bulk solids chute resembles a playground slide. Its apparent simplicity, however, does not guarantee trouble-free operation or sympathy to careless design. The basic chute design may be viewed as a surface, set at a sufficient angle to the horizontal to overcome friction by the effect of gravity alone (see Figure 6.2). There is an added complication in that bulk solids also have an internal friction angle to be considered.

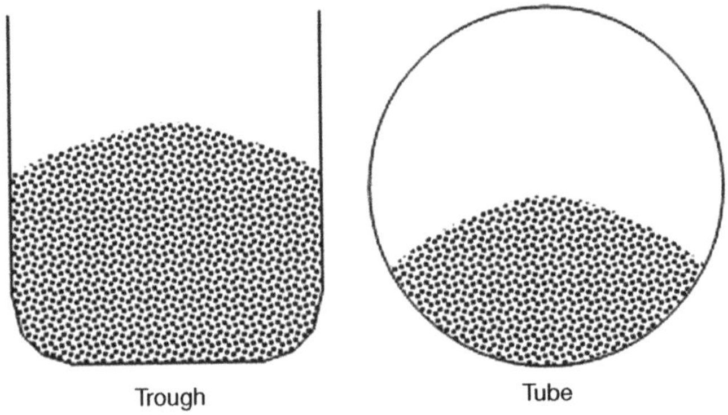

Figure 6.1 U-section and tube trough geometries.

The pressures on the walls of the chute, see Figure 6.3, can be approximated as the hydrostatic case if the depth of the bed is less than the width of the chute.

$$\sigma = \rho_b g H \tag{6.1}$$

A more complete description of stresses in confined bulk solids is given in Chapter 3 and is used in the analysis described later in this section.

Chutes may be used to transfer material from one process vessel to another, from storage into a process vessel or as an intermediate between one transport system to another or from storage to a transport system. They can be used 'on-axis' or be used to redirect the flow through 90 or 180°; they can be used to change the velocity (usually accelerate) of a flowing material. The change of direction can be achieved by a series of linear chutes or a single-curved chute (see Figure 6.4).

The flow of a bulk solids down a chute can be categorised as either 'slow' or 'fast (rapid)'. The two conditions are illustrated in Figure 6.5 for a simple flat inclined chute. In slow flow the bulk solid moves at a constant velocity and bed height. This condition occurs for a free-flowing material where the slope of the chute lies somewhere between the angle of repose and the angle of internal friction. In fast flow the bulk solids accelerates down the chute with a decreasing bed depth. If the chute is long enough, a maximum or terminal velocity may be reached. The chute must be at an angle greater than the angle of internal friction.

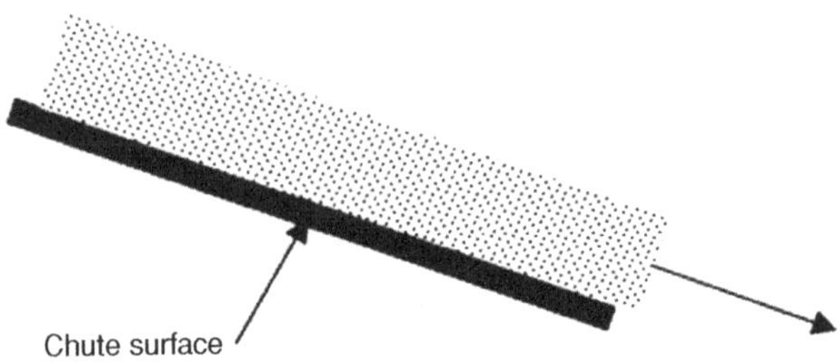

Figure 6.2 Schematic of 'block' of bulk solid sliding overt a declined chute surface.

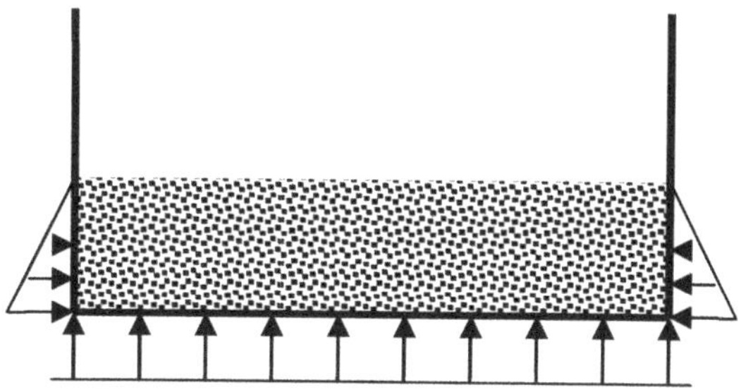

Figure 6.3 Simplified view of pressure on chute walls.

(a)

Figure 6.4 Two linear chutes and single-curved chute effecting change in direction.

(b)

Figure 6.4 (*continued*).

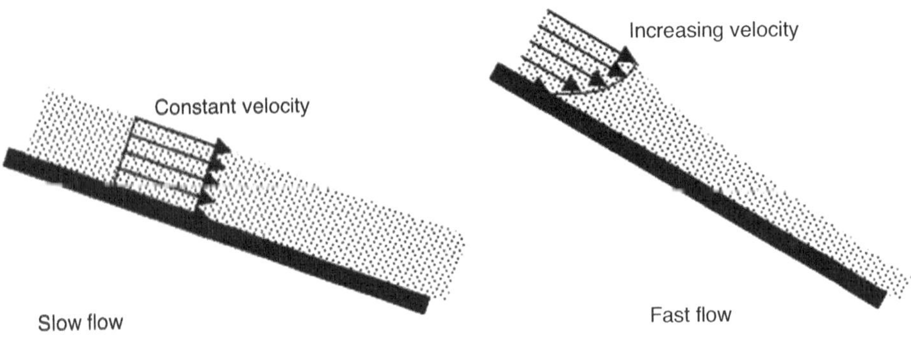

Figure 6.5 Slow and fast flow down an inclined flat chute.

Most industrial chutes, linear or curved, are designed for rapid flow. Although a general mathematical description of all rapid shear flows would be a difficult or 'perhaps an unattainable goal' (Pudasaini & Hutter 2003) a more accessible analysis for a typical industrial geometry has been performed by Roberts (2003). Here the motion of the material is described as a 'thin stream flow' where the depth of the flowing bed is always less than the width of the chute. (A common geometry would have the bed depth less than half the bed width.) In this case, Roberts uses a lumped parameter model to take account of the frictional drag at the chute boundaries (wall friction) and internal friction. Roberts has shown that under these flow conditions, approximately 82% of the energy losses may be attributed to the bulk solids sliding along the bottom of the chute, 9% due to sliding along the sidewalls and 9% due to internal friction (Roberts 1969).

The derivation of the model is not reproduced here; however, it is useful to review several of the equations.

The velocity distribution along the length of a straight inclined chute is given as:

$$u = \left[u_0^2 + 2gs(\cos\theta - \mu_e \sin\theta)\right]^{1/2} \tag{6.2}$$

and for a chute of constant curvature:

$$u = \left\{\frac{2gR}{4\mu_e^2+1}\left[\left(1-2\mu_e^2\right)\sin\theta + 3\mu_e\cos\theta\right] + e^{-2\mu_e\theta}\left[u_0^2 - \frac{6\mu_e Rg}{4\mu_e^2+1}\right]\right\}^{1/2} \tag{6.3}$$

where μ_e is an equivalent friction factor and is given by the expression

$$\mu_e = \mu\left[1 + k_v\frac{H}{B}\right] \tag{6.4}$$

From continuity of mass at steady flow conditions:

$$\dot{m}_s = \rho_b A u \tag{6.5}$$

is constant and if the variation in bulk density can be taken as negligible:

$$A = A_0\frac{u_0}{u} \tag{6.6}$$

Therefore, the variation in stream thickness, H, can be predicted given that the cross-sectional area of material is expressed as:

$$A = BH_1 + \frac{B^2\tan\lambda}{\lambda} \tag{6.7}$$

for a rectangular cross-section and

$$A = \frac{B^2}{4}\left[\tan\lambda\left(\frac{1-\cos\delta}{3}\right) + \left(\frac{\delta - \sin\delta}{2}\right)\right] \tag{6.8}$$

for a circular cross-section (Woodcock & Mason 1987).

Equations (6.5)–(6.8) can be used either to determine a mass flow rate of solids in an existing channel or aid the design of a chute for a required solids flow rate.

Material flowing in a curved chute will initially experience acceleration as gravity forces are larger than the frictional resistance; however, a position at θ is reached where the material flow may become unstable. It is, therefore, normal to see chutes 'cut off' at angles of 45°.

Not all chutes have to be flat or have a constant radius of curvature and many instances, for example in belt loading, have a parabolic profile.

6.2.1 *Operational problems*

6.2.1.1 *Wear* Wear on a chute may come from two principal sources: one due to impact as material enters the chute and two from sliding wear as the material flows along the chute.

The level of wear due to impact is governed by the magnitude and direction of the forces involved, and the chute and bulk material properties. Important properties are chute and bulk solids' relative hardness, particle size and particle shape. Process conditions such as material flow rate will obviously have a large influence on wear rates. In general, if the chute is manufactured from a brittle material then normal impacts will result in greatest wear, whereas if the chute material is ductile impact angles of around 30° will give the largest wear rates.

In sliding, rubbing or abrasive wear is a function of the normal pressure and velocity at the interface of the chute and the continuous stream of bulk solids assuming rapid flow conditions. Roberts gives an abrasive wear factor, W_c, for the chute bottom surface as:

$$W_c = \frac{\dot{m}_S K_c \tan\varphi}{B} N_{WR} \tag{6.9}$$

where

$$N_{WR} = \frac{v^2}{R} + g\sin\theta \tag{6.10}$$

is a non-dimensional abrasive wear number.

The wear on the sidewalls of a chute will be less than that of the chute bottom. Roberts gives the following equation as a means of estimating the average wear on the sidewalls assuming that the sidewall pressure increases linearly from zero at the stream surface to a maximum value at the bottom:

$$W_{csw} = \frac{W_c K_v}{2K_c} \tag{6.11}$$

Using this equation with values of K_c and K_v of 0.8 and 0.4, respectively, results in a prediction of sidewall wear at 25% of the bottom surface wear.

6.2.1.2 *Segregation* Segregation from the use of chutes can be generally found from one of two mechanisms. One mechanism is known as 'percolation' or 'sifting' where when the bed of material is in motion with a velocity profile through its depth, smaller particles can pass through the interstices made by the larger particles (see Figure 6.6). The smaller particles near the chute surface may alter the friction coefficient and can slow the material close to the surface resulting in an exaggerated velocity profile (Bates 1997). The bulk then leaves the chute with a velocity dependent on position and particle size.

The other mechanism, known as 'trajectory segregation', occurs if the discharge from the chute has a horizontal component and is a result of the influence of air drag, which is a function of particle size, on the particle's trajectory. This phenomenon can be seen if transfer chutes are positioned at an angle and the segregation is seen along the cross section (see Figure 6.7).

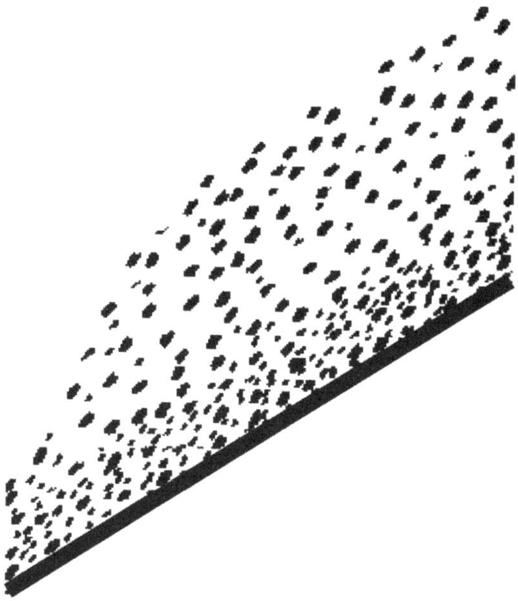

Figure 6.6 Sifting mechanism of segregation.

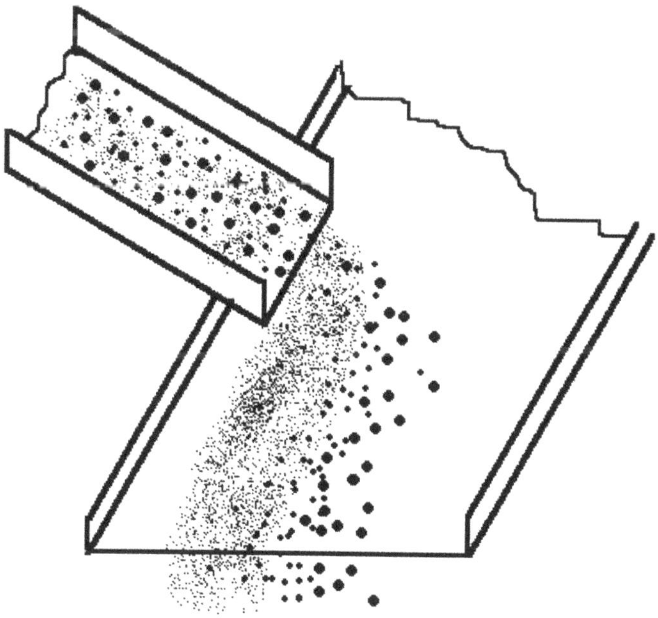

Figure 6.7 The effect of trajectory segregation.

6.2.1.3 *Caking* The term *cake* is used here to describe the residue of material which builds up on chute surfaces, that is, does not self-clear with the passage of more bulk material. Materials that form the cake are generally the fine particle sizes. Typical mechanisms are for fines which are initially dry coming into contact with a moist chute surface and adhering, then subsequent drying results in a hardened cake. Alternatively, wet fines which migrate to a chute surface, e.g. a sidewall where shear stresses are low and again adhere and subsequently dry to from a cake. These processes can be cumulative and the cake builds-up over time until it is either dislodged by the bulk or disrupts the flow. Areas where build-ups are likely are at any seam or join, at fixing locations and at low shear stress regions. Therefore, good manufacturing practice must be specified, for example welds must be ground smooth, no weld spatter etc. Good housekeeping applies, any build-up should be monitored and the chute cleaned at suitable intervals; the chute should be inspected for damage.

The application of a vibrating force on the chute (usually bottom) can occasionally be used to encourage the flow of material and prevent build-up and caking and aid self-cleaning for transient flows. This is often tried as a retrofit when problems develop. However, care must be taken as vibration may exacerbate segregation and may cause problems for structures which were not designed to withstand vibrating loads. The use of flexible materials has been demonstrated to prevent excessive build up of caked solids. The idea here being that as the material builds up and gains mass, eventually the flexible section will extend to an extent that the cake breaks off and is carried away with the normal flow.

Material flow and self-cleaning can also be aided by the application of air to the trough base to affect the strength of the material. Again this is often a retrofit as a reaction to a troublesome flow problem.

6.2.1.4 *Belt loading* Chutes are commonly used in the transfer of material from storage via a feeder to a belt conveyor. The function of the chute is principally twofold: to direct the flow of the bulk solid on to the belt without spillage and to accelerate the flow such that the horizontal velocity matches the belt speed and direction. The design of these chutes must also minimise the wear of the chute and the belt at the feed point and is the province of the expert (Wensrich 2003).

6.3 Vibratory conveyors

Eliciting motion of a bulk material in a trough is not restricted to utilising the effect of gravity alone and the application of vibration to the trough can provide the necessary motive force. Rather than using vibration as a flow aid on a gravity chute as discussed previously, the trough, in its entirety is set in motion. This section will cover the basic theory and the key features without an overly mathematic treatment.

Unlike the chute which must be set at a relatively large declined angle to ensure that the bulk will slide against the chute surface, most vibratory conveyors are used for horizontal or near horizontal conveying with the exception of vibratory elevators which will be discussed later. They can convey on moderate (about 15°) upward and downward slopes. It is possible to design in multiple inlets and outlets to the conveyor. The outlets can either be a series of gates in the trough, or otherwise the stream can be split in some way to provide a number of

Figure 6.8 Typical vibratory conveying system with electromagnetic drive.

outlets. Multiple inlets are possible simply by feeding at different points along the trough. It is also a simple matter to cover the trough should the material be particularly dusty, need protection from the atmosphere, and plastic or stainless steel troughs are available for more sanitary applications, such as food, pharmaceuticals etc. A typical conveyor geometry is illustrated in Figure 6.8, showing a flat bed trough mounted on leaf springs with an electromechanical drive. Alternative trough geometries may be enclosed or curved or tubular; drives of direct mechanical linkage, out of balance motors and coil springs rather than leaf. Regardless of particularities between devices, common principles and theory can be usefully applied.

Consider the simplified arrangement shown in Figure 6.9, which illustrates the action of the vibratory conveyor.

The displacement of the trough S_T can be described by the equation:

$$S_T = \lambda(1 - \cos 2\pi ft) \quad (6.12)$$

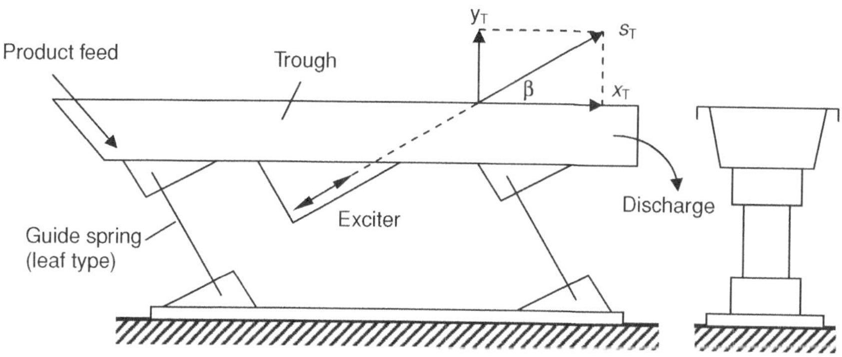

Figure 6.9 Simple model of vibratory conveyor motion.

which can be split into horizontal and vertical components:

$$x_T = \lambda(1 - \cos 2\pi ft)\cos\beta \tag{6.13}$$

$$y_T = \lambda(1 - \cos 2\pi ft)\sin\beta \tag{6.14}$$

and the accelerations found by taking the double derivatives with respect to time:

$$\ddot{x}_T = \lambda(2\pi f)^2 \cos 2\pi ft \cos\beta \tag{6.15}$$

$$\ddot{y}_T = \lambda(2\pi f)^2 \cos 2\pi ft \sin\beta \tag{6.16}$$

In order that any bulk material resting on the trough does not simply follow the same motion with no net transport, the material has to be 'thrown' forward, that is it has to have a sufficient vertical acceleration to overcome gravity and be free of the trough surface. This condition can be expressed as:

$$\ddot{y}_T > g \tag{6.17}$$

The ratio of vertical acceleration to g is known as the 'dynamic material coefficient' or the 'throw factor' Γ.

$$\Gamma = \frac{\ddot{y}_{T_{max}}}{g} = \frac{\lambda(2\pi f)^2 \sin\beta}{g} \tag{6.18}$$

Another commonly quoted value is the ratio of the maximum trough acceleration to g (as opposed to the previously defined vertical acceleration ratio), K, the 'dynamic machine coefficient':

$$K = \frac{\ddot{s}_{T_{max}}}{g} = \frac{\Gamma}{\sin\beta} \tag{6.19}$$

K typically has values of around 1–4 for conveyors and up to 12 for feeders.

For simplicity consider the motion of a single particle initially in contact with the trough surface which will follow the motion of the trough until the acceleration is larger than gravitational acceleration at which point it leaves the surface and follows a parabolic trajectory until it makes contact with the trough again and is subject to another cycle of acceleration. In this simple picture the particle moves forward in a series of hops or bounces. An important feature of the set-up of the vibratory conveyor is that the time the material spends in contact with the trough is much shorter than the time spent in flight. This is illustrated in Figure 6.10, which shows trough displacement and acceleration with time with typical points of contact and flight marked.

This brings about efficient transport and low rates of particle damage and/or trough wear; in fact, vibratory conveying can be a very gentle mode of transport. There are limitations due to the practicalities of large accelerations and typical values for accelerations and amplitudes are given in Table 6.1.

6.3.1 *Estimation of solids flow rate*

The mass flow rate or throughput of the bulk material can be estimated if the net transport velocity, trough and material geometry and bulk density are known. Assuming the displacement equation (6.12) holds, then the horizontal velocity can be written as:

$$\dot{x}_T = \lambda 2\pi f \sin 2\pi ft \cos\beta \tag{6.20}$$

Table 6.1 Normal operating ranges for vibratory equipment.

Type of machine	Frequency (Hz)	Amplitude (mm)
Vibratory feeder	13–60	12–1.0
Vibratory conveyor	3–17	50–5.0

which takes a maximum value:

$$\dot{x}_{T_{max}} = \lambda 2\pi f \cos\beta \qquad (6.21)$$

The average transport velocity of the bulk solid can then be estimated by a general expression:

$$u_S = \eta_u \dot{x}_{T_{max}} = \eta_u \lambda 2\pi f \cos\beta \qquad (6.22)$$

where η_u is the efficiency of transport.

This 'efficiency of transport' term is a function of the conveying conditions through two equipment terms, the dynamic material coefficient (Γ) and the vibration angle (β), as well as material-dependent terms. The material dependence will be based on wall and internal friction, material shear strength or cohesion, particle size and density and almost

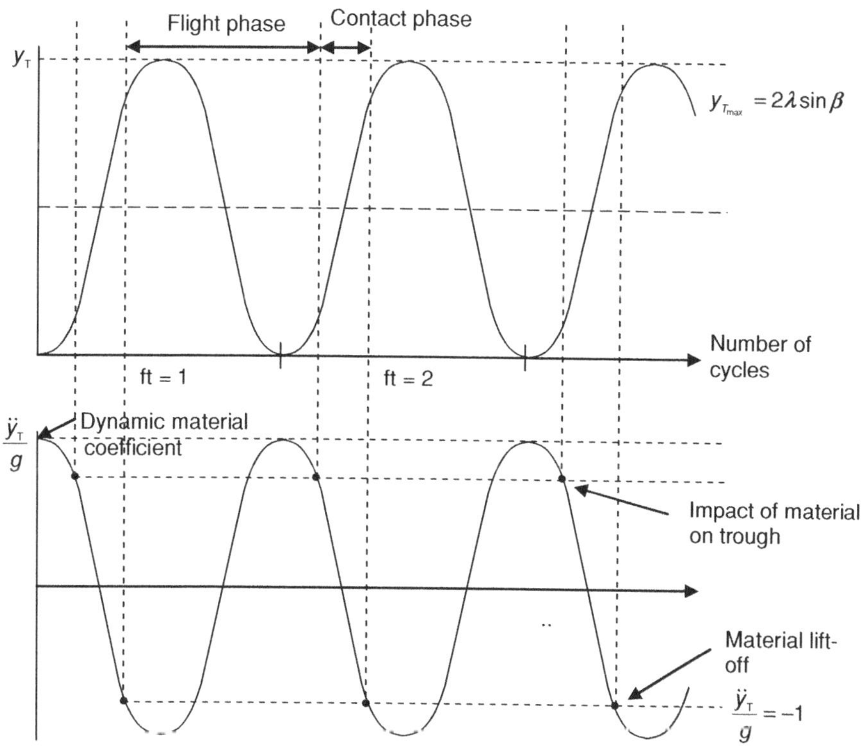

Figure 6.10 Variation of trough displacement and acceleration with time.

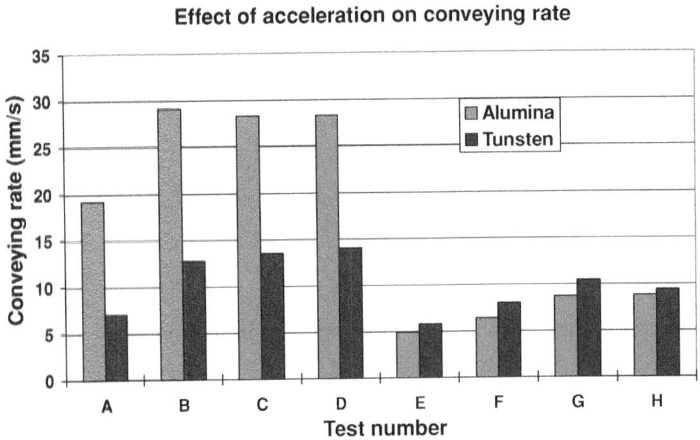

Figure 6.11 The effect of changing acceleration on conveying velocity for two materials.

any number of effects due to temperature, ambient gas etc. However, the effect of material-related properties on transport efficiency can be expressed as a single term F_m, which must be determined experimentally (very roughly 0.9 for dry granular materials, but is very dependent on particle size). Other empirical factors allow for various depths of material bed (F_h with values ranging from 1 at small bed depth to around 0.75 at bed depth of 300 mm) on the trough and the slope of the trough (F_j with horizontal conveying having a value of 1, incline less than 1 and decline slightly more than 1), resulting in the formula:

$$u_S = \eta_u F_m F_h F_j \lambda 2\pi f \cos\beta \tag{6.23}$$

The mass flow rate or throughput of the conveyor can be estimated by:

$$\dot{m}_s = \rho_b A u_S \tag{6.24}$$

where A is the cross-sectional area of the bulk material bed (assumed constant) and ρ_b the bulk density.

The effect on flow rate of different materials being conveyed under the same vibrational conditions is shown in Figure 6.11 and the ability of the methodology to predict flow rates under different vibration conditions from an empirical knowledge of the relevant factors for a particular material, alumina in this case, is shown in Figure 6.12.

6.3.2 *System choices*

The principal choices to be made in selecting a conveyor are trough dimensions which can be estimated using the approach given above based on a target mass flow rate, the drive mechanism, the method of mounting the trough to allow movement, whether the system is resonant or not, and whether it is intended or desirable to process the material in some way in addition to conveying.

The drive for a vibratory conveyor can be direct mechanical, eccentric mass mechanical, electromagnetic or hydraulic.

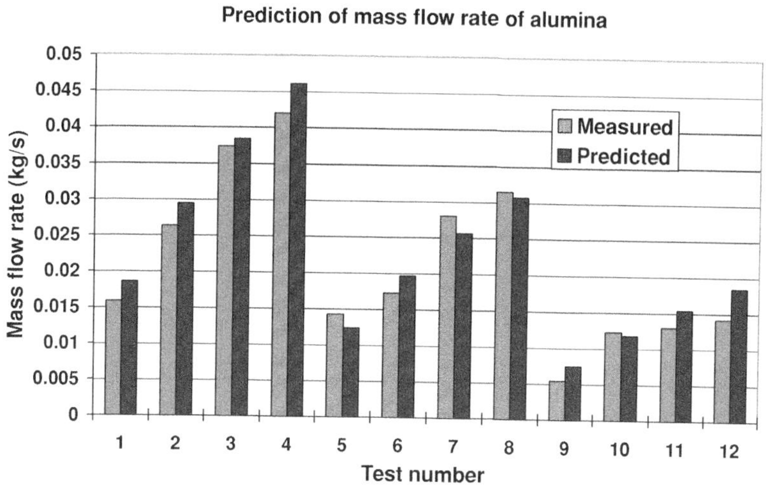

Figure 6.12 Predicted versus measured solids mass flow rate.

The direct or positive mechanical drive has a direct mechanical linkage from a drive motor to the conveyor trough. This type of drive is normally used for longer heavy duty conveyors. The general arrangement is illustrated in Figure 6.13 with an oscillating motor connected by a mechanical linkage to the trough, which is mounted on some type of spring system.

A major problem with the direct drive type arrangement is the transmission of vibration to the support frame and surroundings. There are two methods of overcoming this problem: the first is to use a counter-weight that balances the vibration within the trough structure and isolates it from the support, and the second method is to use a contra-vibrating double trough where each trough vibrates exactly out of phase with the other.

Typical operating characteristics of this type of drive are for an operating frequency in the range of 5–15 Hz, a conveying distance of anything between 5 and 30 m, a vibration amplitude between 3 and 15 mm and an average conveying speed of between 0.2 and 0.8 m/s.

The second type of mechanical drive is the eccentric mass type, as shown in Figure 6.14. This drive configuration is applied to conveyors and large feeders. The most common arrangement is to use two contra-rotating masses of equal size. This two-mass system is able to produce an oscillating linear motion perpendicular to the axes of the motors. This

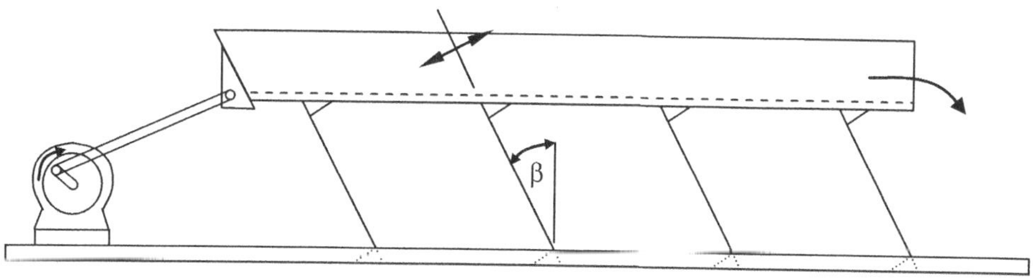

Figure 6.13 Positive mechanical drive – simple direct drive.

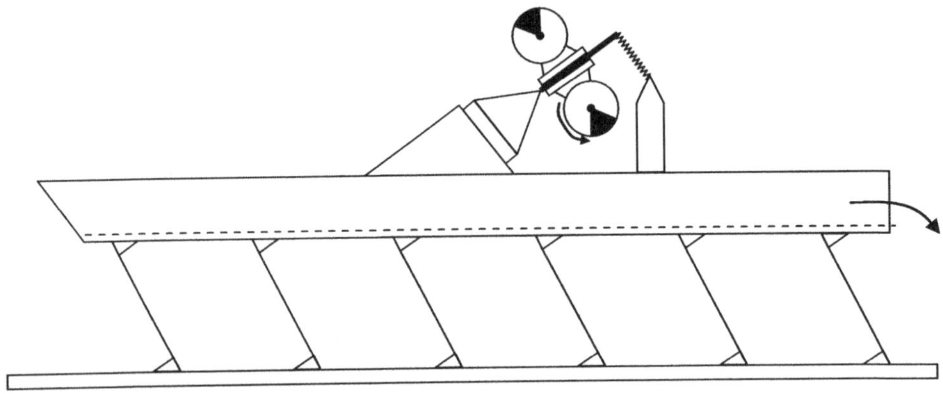

Figure 6.14 Vibratory conveyors with rotating eccentric mass mechanical drive.

gives some flexibility of control and allows the vibration angle of the system to be altered to suit the particular product being conveyed. This is particularly useful for feeders, as the rate can be changed to account for material variations.

The leaf spring design restricts motion to the longitudinal direction and the trough will move back and forward in this plane under the action of the eccentric masses. A coil spring mount will result in the motion being simply governed by the direction of oscillation as defined by the eccentric mass movement.

The typical operating parameters are: operating frequency of around 15 Hz for conveyors and up to 30 Hz for feeders; distances approximately the same as the direct mode for conveyors, but slightly less for feeders. Vibration amplitude is variable but is usually in the range of 1–10 mm, and a conveying speed that is slightly less than the positive drive type.

A final note about this type of drive involves the use of variable frequency controls with this type of conveyor. These relatively new devices allow control of the conveyor while it is in operation, and thus can be used to modify the feed rate while material is being conveyed.

6.3.2.1 *Electromagnetic drive* The third type of drive and the first non-mechanical type is the electromagnetic drive which is illustrated in Figure 6.15. The basic principle of operation for this type of drive involves the cyclic energisation of one or more electromagnets, with

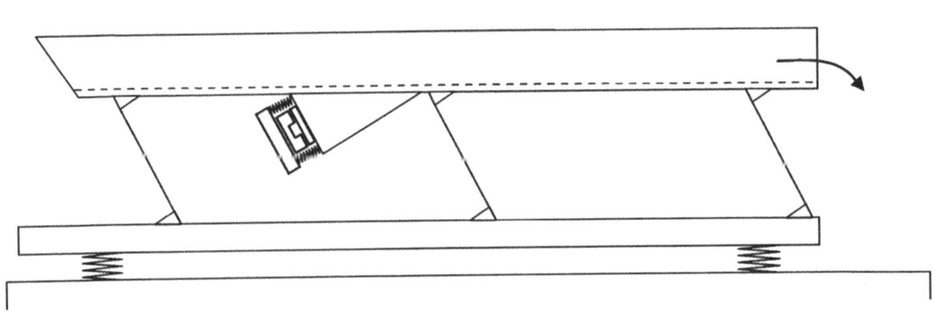

Figure 6.15 Typical arrangement of electromagnetic drive.

at least one attached to the conveying trough. The electromagnets are not in contact, which has advantages in terms of there being no mechanical fatigue in the design. The operating frequency does tend to be restricted to twice the electrical supply frequency, which is either 100 or 120 Hz depending on the region of installation. However, it is possible to control this value down to about 50% by using either a rectifier or thyristor control. The most common application of this drive type is for short vibratory feeders. The vibration amplitude is correspondingly low at about 0.1–3 mm and thus so is the speed, which is rarely higher than 0.3 m/s.

6.3.2.2 *Hydraulic drive* The final type of drive, and perhaps the least common, is the hydraulic drive which is typically associated with short heavy duty feeders and is used mainly in explosion-sensitive environments. The hydraulic drive is ideal for this situation as a remote pump unit can be used, well away from the hazardous environment and the hydraulic fluid is supplied to the vibrator at pressure from this remote location. Speed control of the pump unit allows a variation of the oscillation frequency, and the capacity or speed can be controlled by the use of pressure control valves on the supply lines.

6.3.3 *Mount*

The type of mounting system selected depends on three key considerations. The first of these is the drive mechanism as different drive mechanisms will put different loads on the trough and these must be considered when selecting a mounting system. The second consideration is the loading on the trough. This will affect the size of the mounting springs, links, etc. as well as the type of mounting system chosen. Finally, the transmission of vibration must be considered.

6.3.3.1 *Directionally constrained systems* The directionally constrained mounting system consists of a number of leaf springs or hinged links connected to the trough. The stiffness of these springs/links results in a restriction of the oscillation in a perpendicular direction to the supports. In general, these systems are ‘tuned’ to avoid natural frequencies, which have the advantage of producing a system that is relatively insensitive to trough loading. This system is principally used for conveying.

6.3.3.2 *Non-directionally constrained systems* The second type of mounting system to be discussed is the non-directional type. These systems are freely mounted on isolator springs, and thus have a freedom of movement in both longitudinal and lateral directions. These systems are more easily tuned than other systems due to this advantage, but in all other ways are similar to the directionally constrained type.

6.3.3.3 *Natural frequency systems* Finally are the natural frequency systems. These systems are designed to resonate at the natural frequency and so a low power input is required to maintain it in this oscillation. However, a disadvantage of this system is that it is highly load sensitive due to the fact that the natural frequency of the system is dependent on the total mass. Therefore, it is only really suitable for applications where either the material mass is insignificant relative to the trough mass, or the mass in the trough is carefully controlled by closely monitoring the feed rate. However, it is possible to produce a relatively consistent transfer rate from a vibratory feeder with varying load, for example from the feed hopper head load, by making use of the combined effect of a change in natural frequency and damping produced by the increased mass of the vibrating system (trough and bulk solid). When interfacing a hopper to a vibratory feeder care must be taken to allow the bulk solid to flow freely from the hopper outlet and in such a way as to keep the entire outlet live. This is done by matching the two geometries as discussed in Chapter 2. A problem seen on plant is when the skirts which are used to prevent spillage or dust migration are too stiff and have a severe effect on the trough's motion and therefore on the transport of solids. It has also been known for the vibratory conveyor to be directly bolted on to the hopper outlet with predictably unsatisfactory outcomes.

The vibratory conveyor can also offer possibilities for a simultaneous processing operation. Typically, wide flat conveyors are used with a mesh or meshes to simultaneously screen/sort materials and convey to different locations. The large surface area in contact with the bulk solids also make effective thermal transfer for heating or cooling.

A rather clever variation in a technology developed for near horizontal transport by forming the trough into a spiral allows the material to be raised vertically within a reasonably small footprint. The vibratory motion is generally applied via the core support structure and a total elevation of around 10 m is possible.

6.4 Air slides

Air slides or air-assisted gravity conveyors are mainly used to convey fine low cohesive materials, such as alumina, cement or pulverised fuel ash, at low angles of declination (slight inclines are possible with special designs). The principle of operation is to reduce friction or material strength by the use of interstitial air introduced along the length of the conveying trough via a porous membrane. The material can then flow at much lower angles than that of a chute which relies on gravity alone. They are employed for transfer operations of a few metres and for much longer conveying runs with throughputs anywhere from 10 to 1500 tonnes/h.

The operation of the air slide is highly dependent on the properties of the material being conveyed and in particular the fluidisation characteristics. To determine the viability of using an air slide with a material it is useful to look at the Geldart classification methodology (see Chapter 1). Knowledge of the material's particle density and particle size enables the material to be assigned to a particular group based on the location of these parameters on the Geldart chart shown in Figure 6.16.

If the material falls into Group B then it is likely to be a good candidate for conveying on an air slide with a low angle of slope and with some materials and appropriate design the material can be transported at a slight incline. These materials fluidise easily for conveying and de-aerate quickly when the air supply is halted so giving a good degree of control. Group D materials will also fluidise but require excessive amounts of air and are therefore not preferred materials for an air slide. Group A materials will fluidise well with little air; however, some of these materials may have long de-aeration times and will continue to have fluid-like behaviour for some period after the air supply has been stopped which may lead to difficulty in controlling the flow. Group C-like materials have cohesive properties which make fluidisation difficult and so are generally not suitable for an air slide. However,

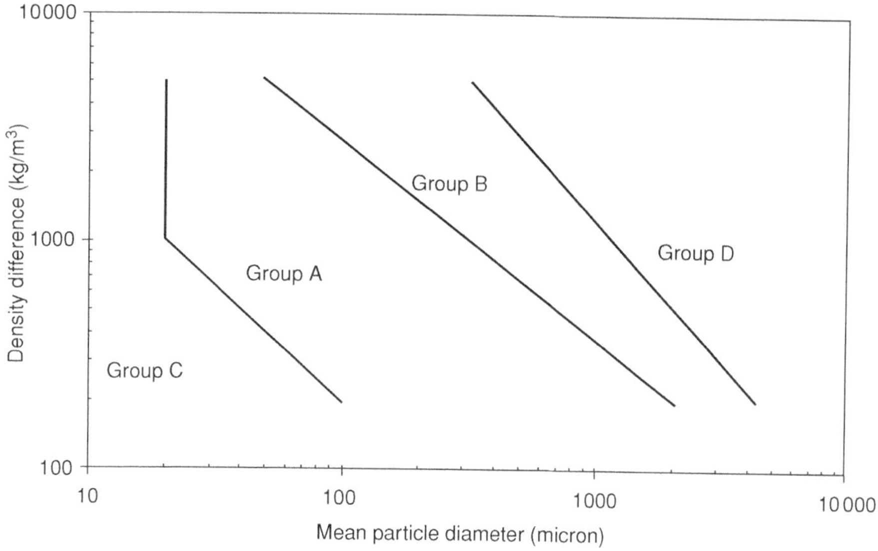

Figure 6.16 Geldart classification chart.

some of these materials may still be capable of transport in an air slide as the air passing through the plenum, rather than fluidising the relatively impermeable bulk, will form an 'air cushion' which reduces the local friction between the wall and the bulk solid and so aids flow. The slope of the slide may be higher at around 6–8° for this type of material.

The conceptual design of an air slide is fairly simple with a porous membrane sandwiched between two 'U' troughs, as illustrated in Figure 6.17.

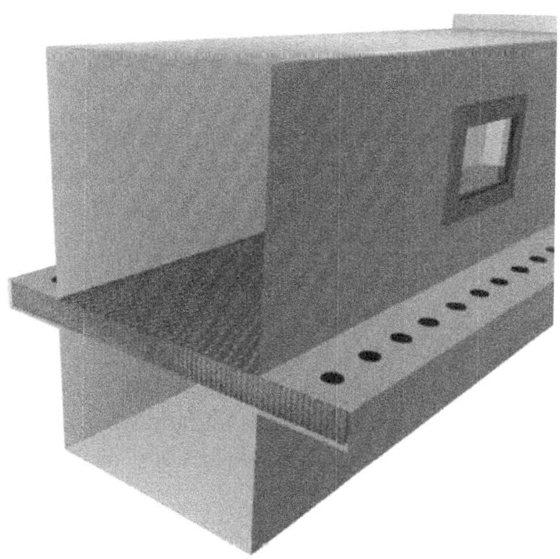

Figure 6.17 Schematic of typical air slide construction.

Table 6.2 Porous membrane media.

Membrane media	Advantage	Disadvantage
Cotton fabric	Low cost; light and relatively strong	Little rigidity and may require support; performance adversely affected by the presence of moisture
Polyester fabric	Low cost; light and relatively strong; less affected by moisture; higher temperature range than cotton	Little rigidity and may require support
Woven steel fabric		Relatively expensive
Sintered metal	Good for hygienic or high temperature applications; generally gives good fluidisation; manufactured with a high degree of uniformity	Very expensive
Sintered plastic	Smooth surface and more rigid than woven fabric	More expensive than equivalent woven fabric; not for very high temperature applications
Ceramic tiles	Good fluidisation possible; good for high temperatures	Tiles must be fitted and sealed properly; relatively easily damaged by impacts

The area below the porous membrane forms a plenum chamber and the area above the membrane, the conveying channel. The membrane can be made from a variety of materials; a selection of common media is given in Table 6.2 along with advantages and disadvantages.

A major function of the plenum chamber is to ensure an even distribution of fluidising air across the entire area of the conveying channel base. This is achieved by having a suitable pressure drop across the porous media such that the pressure drop across the bulk solids is not significant in determining the superficial air velocity and that the pores of the membrane are of a size and distribution to ensure reasonable performance and are not easily blocked. The pressure drop across the membrane should be available from manufacturer's literature or a simple test and the expected pressure drop and fluidising air required for the bulk solids is obtainable from fluidisation tests (see Chapter 1). Limiting the area of the membrane (and plenum chamber) to mitigate the effects of pressure variation has encouraged the development of modular units.

The air-assisted gravity conveyor or air slide is generally constructed from modular components giving a reasonably wide flexibility of design. For example, it is relatively easy to construct a system with multiple inlets and outlets, to divert oversize material and curve the chute around corners to deliver to all parts of the plant. Of course, the declination must be maintained over the length or else the flow will halt.

The principal elements of an air-assisted gravity conveyor are:

- Solids feed – generally a special bolt on unit included by the supplier to interface between the delivery hopper or feeder and the main body of the conveyor.
- Inspection cover – for the location of blockages and/or monitoring the condition of the porous membrane.
- Supply of filtered air – the motive air supply at the appropriate pressure and volumetric flow rate.
- Plenum chamber – a powder-free area for the containment of the motive air.

- Porous membrane – discussed above.
- Solids discharge – generally a special bolt on unit included by the supplier to interface between the main body of the conveyor and a receiving vessel.
- Vent to suitable filter unit – this can be optional on sealed units where the volume of air is low and the discharge is directly to a hopper by a sealed route. Otherwise the unit filters any entrained product from the air. The filtered material can be cleaned from the filter unit and discharged directly back into the process.

There are a number of ways to feed an air-assisted gravity conveyor. The most common feed method is simply a flood feed from a mass flow hopper. A non-mass flow hopper may cause problems with inconsistent feed which may result in blockages. That is, if too much material is fed onto the conveyor, then there will be insufficient air to fluidise the material, and if too little material, or none at all, then conveying may halt due to a maldistribution of air from the plenum chamber. While an air-assisted gravity conveyor does not work independently as a feeder, if the bulk solid is metred onto it by the use of a screw feeder or rotary valve, it will then convey this material at a constant rate. Another method of controlling the flow of material is to use a gate or baffle at either end of the conveying duct.

6.4.1 *Selection considerations*

When selecting an air-assisted gravity conveyor, the following four areas are key considerations:

- The slope of the channel
- The distance that the material is required to be conveyed
- The width of the conveying channel, which is the ultimate constraint on the capacity of the conveyor
- The air requirement to keep the material in a fluidised condition

6.4.1.1 *Slope* The optimum value for slope of the channel is dependent on the particular bulk solid, with free-flowing materials requiring a smaller angle than more cohesive materials; it is therefore recommended that bench scale tests be completed with the chosen material. If the angle is less than a critical value there is a possibility that the material will back up along the channel and eventually cause a blockage. If the angle is larger than the optimum value there is no gain in performance, and the increase in height may be detrimental to the plant design (Woodcock & Mason 1987). It is possible to convey up a slight incline if the channel is fully choked and the fluidising air also acts as motive air.

6.4.1.2 *Conveying distance* Conveying distance in most cases should not theoretically be a limitation on the use of air-assisted gravity conveyors which can be built from a series of modular units. Headroom may however be a limiting factor. For example, if a conveyor is 100 m long with a slope of 1°, the start of the conveyor would need to be 1.75 m above the base, and this would increase to 17.4 m for a 10° slope required for a cohesive material. Other operational considerations for long conveyors are, maintaining a uniform air pressure over the length and venting the fluidising gas to reduce the superficial air velocity in the space above the conveyed material.

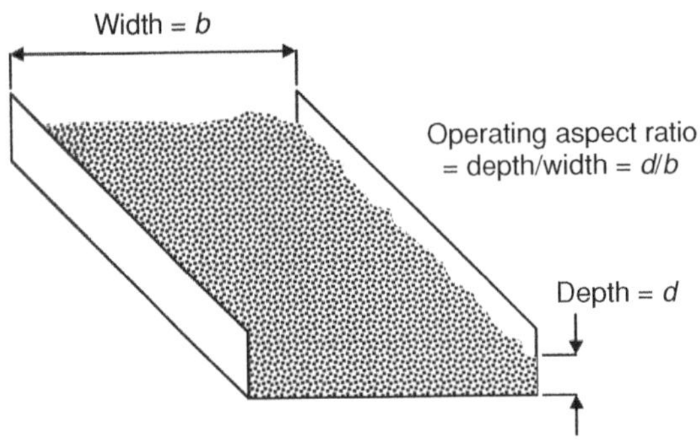

Figure 6.18 Conveying channel parameters.

The width of the conveyor is usually the main factor in determining its capacity. In the technical literature for these devices, 'typical capacity' (in volume per unit time) is given as a function of the width of the conveyor (Dynamic Air 2006).

In order to estimate the width required to achieve a desired capacity, knowledge of the average velocity and the average bulk density of the material is required. The bulk density in this case is not the aerated bulk density, as there is a further de-rating factor to account for this.

An estimate of channel width can be made from

$$b = \left(\frac{r_e \dot{m}_s}{r_a \rho_b u_s} \right)^{1/2} \tag{6.25}$$

where operating aspect ratio (r_a) – up to a maximum of about 0.5, this is the ratio of the depth of material to the width of the conveyor; expansion ratio of conveyed material (r_e) – a de-rating factor for the bulk density of the material that takes into account the aeration required for fluidised flow; average solids velocity along channel (u_s); mass flow rate (m_s) – this is usually the target throughput; bulk density (ρ_b) – the bulk density of the unfluidised material (see Figure 6.18).

6.4.2 *Air requirement*

In terms of the air requirement for an air-assisted gravity conveyor, the two parameters to be specified are the volumetric flow rate and the plenum pressure. Experience has shown that the total plenum pressure is typically in the range of 250–500 mm H_2O or 24.5–49 mbar and fluidisation velocities must be determined by bench scale tests of the proposed material together with conveyor dimensions (length and width). The slope of the channel and the solids mass flow rate need to be also considered. For free-flowing powders 2–3 times the minimum fluidisation velocity is usually sufficient, but for finer more cohesive particles, the optimum value can be up to 10 times the minimum fluidisation velocity (Woodcock & Mason 1987).

6.4.3 *Maintenance and troubleshooting*

While the air-assisted gravity conveyor has a relatively straightforward construction, and a limited number of moving parts to cause problems, there are issues with its maintenance and troubleshooting that need to be considered. Arguably the most important component in any air slide is the porous membrane. This must be kept in good order and free from damage to ensure a reliable performance from the conveyor. The most common problem with the top side of the membrane is the blinding or blocking of the pores. As pores become blocked, the distribution of air becomes uneven resulting in flow problems and eventual stoppages. Correct selection of the porous media to match the conveyed material will mitigate these problems; however, inspection and monitoring of the pressure in the plenum can give indications of problems developing. Any problems with the bottom, or clean air, side tend to be related to unclean air being used. Filtering this air for oil, moisture and particulates can go along way ensure that a minimum of damage is done to this side of the membrane. As with all bulk solid plant, blockages are practically inevitable and the most common method of clearing blockages is to use a rod or other sharp implement which must be avoided. This is particularly damaging to the delicate construction of the porous membrane; the woven types can be cut, the ceramic types cracked and even the metal ones dented out of correct working order.

6.5 Chain and flight conveyors

Bulk materials can be transported in a trough by the action of a moving mechanical component acting directly on the bulk solid. An example of this has already been covered in Chapter 5 on screw conveying where the motive force is provided by a helical flight attached to a rotating shaft along the axis of the trough. In this chapter, the use of chains with bars or flights to affect a transport of a bulk material is described.

The first example of this is known as a drag conveyor where an endless chain is dragged through a trough containing the bulk solid. Variations come with a simple chain with short flights spaced at a number of links running at relatively high speed to a double chain with large flat flights attached across the chains at each link which run more slowly. Illustrations of different chain and flights are shown in Figure 6.19 and the simple assembly in Figure 6.20.

Typical applications for this type of conveyor are in the heavy duty areas of mining or minerals processing or power generation. Their simple geometry and robust nature make them ideal for handling materials such as hot cement, clinker and ash or wet mineral ores, coal and grain. Construction materials may be malleable iron or hardened steel for heavier duties.

Typical chain speeds are in the region of 0.1–0.6 m/s although there are reported velocities of up to 1 m/s. These low speeds keep wear rates at acceptable levels even with very abrasive products. However, this low speed also means that throughputs of solids are limited to around 900 m^3/h. The estimation of mass flow rate of solids for these devices is not simple as it relies not only on the trough and flight geometry and flight speed (the swept volume) but also on material properties such as cohesion, wall friction and internal friction, which influence the slip velocity, the level of fill and the material's bulk density. The volumetric flow rate of material may be greater or less than the volume swept out by the moving flight depending on whether the material slips over the flight or is dragged along with material in contact with the flight.

The shear zone can be chosen to lie within the bulk which will leave a layer of 'dead' material on the trough bottom. This may protect the surface of the trough, however, may

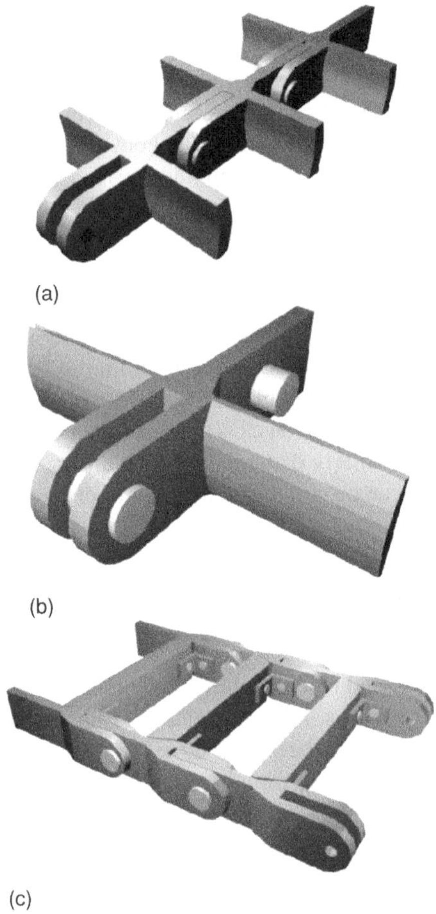

Figure 6.19 Examples of chain and flight patterns.

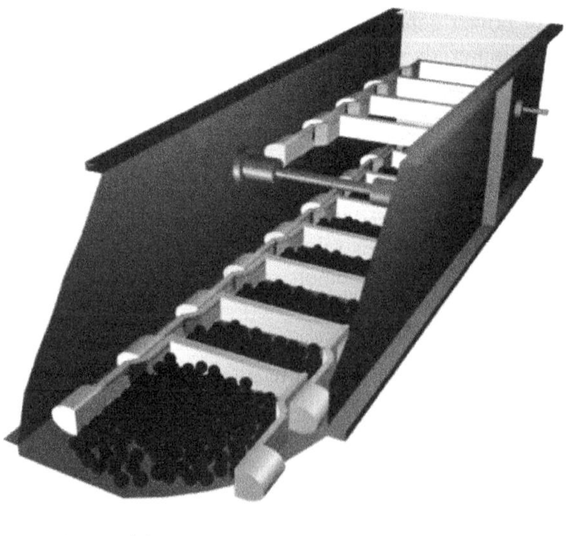

Figure 6.20 Standard single strand drag conveyor.

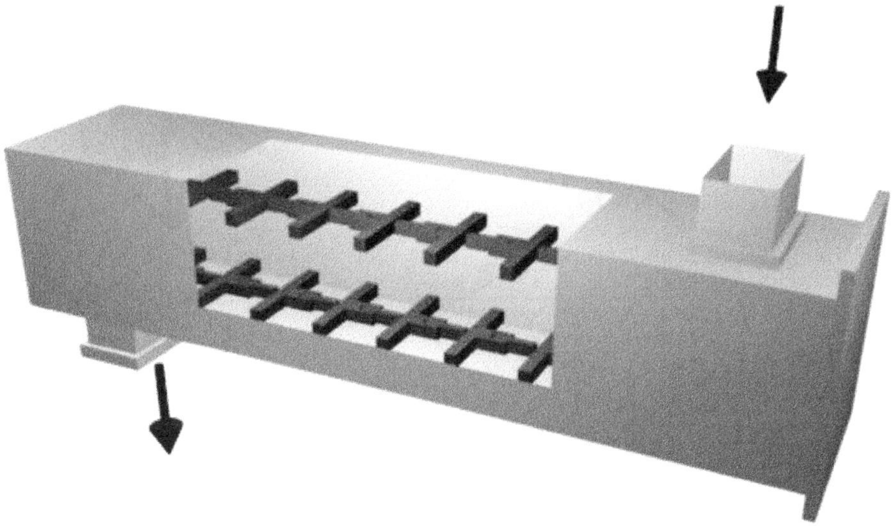

Figure 6.21 Basic en masse conveyor/feeder.

result in higher forces on the chain and reduced throughput. The simple geometry of the trough lends itself to relatively straightforward fabrication of wear-resistant linings needed.

Conveying distances for this type of conveyor are usually less than 100 m and they can be very noisy in operation. Problems can occur if the material entering the conveyor retains air and remains fluidised for some time; in this case the throughput may become very low as the effective slip velocity becomes high. A common problem occurs when large particles obstruct the movement of the chain or flight, for example by becoming jammed between the flight and the trough. The inherent flexibility of the chain system can go some way to mitigate the likelihood of a lump becoming trapped; however, if a jam does occur this puts high loads on the chain leading to high power demands on the motor or local failure of the chain. There may be enough capacity in the motor to generate local forces which will shear a relatively soft lump or fracture a brittle particle. The possibility and likelihood of this occurring should be considered at the selection stage.

6.5.1 *En masse conveyor*

The 'en masse' conveyor or the 'continuous flow conveyor' receives its name from the fact that the bulk material tends to move as a single mass along the conveyor with little particle–particle relative movement. It can carry material horizontally, vertically and on an inclined loop but each stage is usually restricted in orientation to a single plane. The en masse was developed in England in the 1920s principally by Arnold Redler and is commonly referred to as a 'Redler conveyor'; a simple unit is illustrated in Figure 6.21. Like the screw conveyor in some ways (see Chapter 5), the en masse conveyor is a 'frictional conveyor' in that it relies on friction to impart motion to the powder through its flights. Internal friction within the powder also assists the conveying process, although unless the wall friction is low enough,

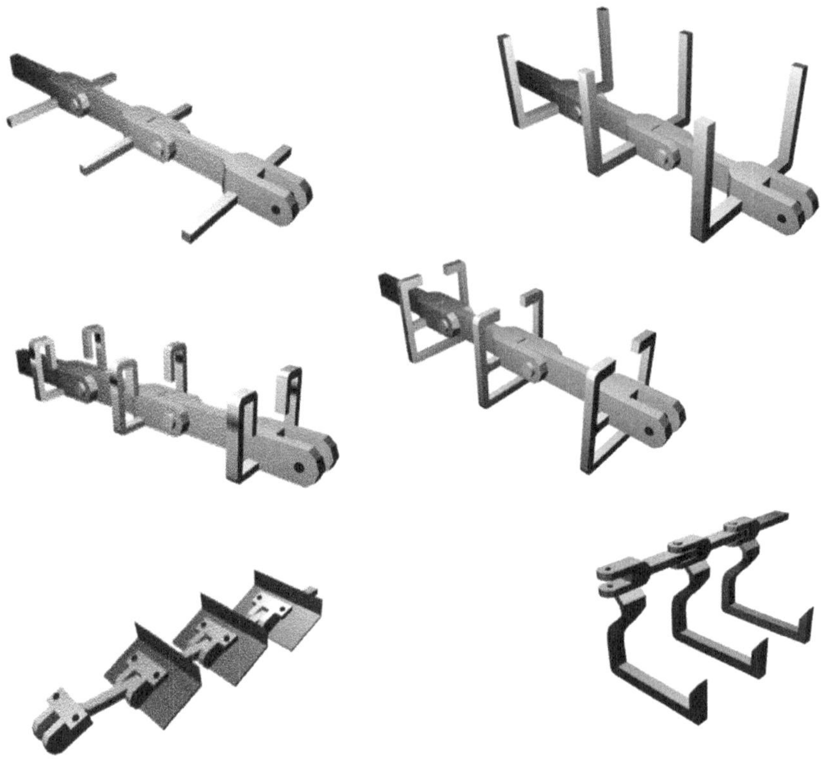

Figure 6.22 En masse conveyor flight profiles.

the material will not slide along the trough. In the ideal case the material moves effectively as a single mass along the trough at a speed approaching that of the chain.

En masse conveyors were originally designed to convey materials in the horizontal plane only, but the conveying mode has proven to be effective in vertical conveying also, provided the inlet is kept fed with powder and the trough is enclosed (similar to the action in a flexible screw conveyor, see Chapter 7).

6.5.2 *En masse conveyor design considerations*

There are a number of different design features that can be selected to achieve conveying of different materials in an en masse conveyor. There are a number of flight profiles, as can be seen from Figure 6.22.

- Flat flights normally only used in horizontal conveying
- Skeleton flights that allow a greater contact area and are thus more suitable for vertical conveying
- Suspended flights that minimise the contact of the chain with the material, and again these are used for horizontal conveying
- Solid peaked flight that gives greater surface area and prevents the material from flowing over the back of the flight

The flight spacing can also be modified. The usual pattern is to have one set of flights per chain link; however, if the flights are spaced more widely it has been shown that materials that are subject to bridging are able to discharge more easily. There are also special flight designs, such as 'wiper' flights (made from, e.g. neoprene) to improve clean-out and oversize or 'scavenger' flights to enable better handling of sticky materials. The flight material is also key consideration, such as stainless steel for a food-type application, as well as the type and size of chain that needs to be designed to withstand the conveying load.

Some key considerations in the selection of the overall design of the trough or casing, see Figure 6.23, are:

- The location and size of the feed and discharge points.
- The requirement for inspection ports or hatches for clearing possible blockages or tramp material and/or cleaning of the casing.
- The provision of explosion protection for appropriate materials.
- Whether or not a curved section is required, or whether an inclined section would suffice. When conveying abrasive materials care must be taken to ensure that high wear areas have been suitably reinforced.
- A special consideration occurs with high temperature products. The high temperature in the conveyor can lead to creep in the chain material. The effect of this lengthening is usually overcome by having a casing with an adjustable length, and by having a controlled amount of sag in the chain designed in.

6.5.3 *En masse conveyor performance calculations*

When selecting an en masse conveyor, a key consideration is the volumetric throughput of the conveyor, which can be converted to a mass throughput by the use of the appropriate bulk density.

The volumetric throughput is related to a number of factors, these being the cross-sectional area of the bed, the chain velocity and a factor r_v that is a slip factor between the chain and the bulk material. The value of r_v can generally be taken to as 1.0 for a horizontal conveyor and varies between 0.6 and 0.85 for inclined and vertical sections.

$$\dot{V}_s = A_b v r_v \tag{6.26}$$

The mass throughput is a function of this volumetric throughout and the bulk density of the material in the bed.

$$\dot{m}_s = \rho_b A_b v r_v \tag{6.27}$$

Experience has shown that the optimum velocity, the chain and flights is dependent on the type of material and can be categorised into three groups:

1. The free-flowing materials, which have an optimum velocity of greater than 0.5 m/s
2. Abrasive and/or aeratable materials are generally conveyed at less than 0.25 m/s, otherwise excessive wear and/or dust generation are possible
3. Fibrous or flaky materials are best conveyed at around 0.4 m/s

Excessive velocity has the potential to cause degradation of friable materials, particularly the fibrous or flaky materials, and the slip velocity may increase leading to inefficient conveying.

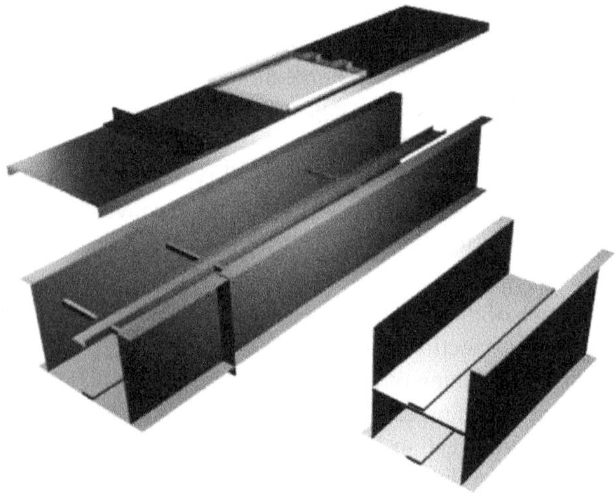

Figure 6.23 En masse conveyor box-section casing.

Because of the wide range of types of conveyor and flight, as well as the different materials conveyed, previous experience and/or laboratory testing are usually required to determine the power requirements.

6.5.4 *En masse conveyor applications*

Typical conveying distances for en masse conveyors in the horizontal plane are 1–100 m with throughputs of 1–1000 tonnes/h, conveying in combination with horizontal and vertical modes reduces these figures. The devices are effectively employed in combined processing and transport applications such as cooling, heating, drying and blended products.

6.6 Summary

The selection of a conveyor should be based on a match of the duty and material properties to the technology; Table 6.3 gives some broad guidance. Usually, the duty is given; it is therefore very important to have a good understanding of the bulk material through a determination of its properties and characteristics. The characterisation methods and accepted industry standards and practices are discussed in Chapter 1. The key properties of relevance here are particle size and size distribution, particle density and hardness, bulk density, wall friction (intended construction materials), internal friction, shear strength and fluidisation behaviour. Equipment trial should be carried out with the range of materials that will be conveyed. If the material that is to be conveyed is not yet available, then some simulation or modelling maybe appropriate. Two simulation approaches growing in popularity and reliability are those based on either a computational fluid dynamics formulation of the continuum behaviour of the bulk solid or a discrete element method where individual particles are modelled (Ma *et al.* 1993; Theurerkauf *et al.* 2003).

Table 6.3 Summary of technology.

Technology	Key function	Materials handled	Conveying length	Capacity
Chutes	Transfer	Large lump to fine powder	1–10 m; declination required	Up to few 1000 tonnes/h
Air slide (air-assisted gravity conveyor)	Transfer/convey	Powders (Geldart A/B)	Few m to 100 m; near horizontal declination; special design – few degrees incline	Up to few 1000 tonnes/h
Vibratory conveyors	Transfer/convey/feed/metre; combined process step – screen, heat, cool, dry, wash	Larger lump (unit mass) to powders – special considerations for fine/light products; abrasive/thermal	1–20 m; near horizontal; spiral vertical up (10 m)	Few kg/h to few 100 tonnes/h
Chain and flight	Transfer/convey/feed/metre; combined process step – heat, cool	Large lump to powder; special consideration for aeratable products	Few m to 50 m; horizontal to vertical up (20 m)	Few tonnes/h to few 100 tonnes/h

References

Bates, L. (1997) *User Guide to Segregation*. British Materials Handling Board, UK.

Ma, D., Eraslan, A.H. & Ahmadi, G. (1993) A computer code for analysing transient three-dimensional rapid granular flows in complex geometries. *Comput. Fluids*, **22**(1), 25–50.

Pudasaini, S.P. & Hutter, K. (2003) Rapid shear flows of dry granular masses down curved and twisted channels. *J. Fluid Mech.*, **495**, 193–208.

Roberts, A.W. (1969) *Trans. ASME J. Eng. Industry, Series B*, **91**(2), 373.

Roberts, A.W. (2003) Chute performance and design for rapid flow conditions. *Chem. Eng. Technol.*, **26**(2), 163–170.

Theurerkauf, J., Dhodapkar, S., Manjunath, K., Jacob, K. & Steinmetz, T. (2003) Applying the discrete element method in process engineering. *Chem. Eng. Technol.*, **26**(2), 157–162.

Wensrich, C.M. (2003) Evolutionary optimisation in chute design. *Powder Technol.*, **138**, 118–123.

Woodcock, C.R. & Mason, J.S. (1987) *Bulk Solids Handling*. Blackie Academic and Professional, Glasgow, UK.

Nomenclature

A	Cross-sectional area of material on a chute at a particular location
B	Width of chute channel
H	Depth of flowing stream at a particular location
K	Dynamic material coefficient
N_{WR}	Abrasive wear number
R	Radius of curvature of chute channel
W_C	Abrasive wear factor
W_{CSW}	Abrasive wear factor on sidewall
f	frequency
g	Gravitational acceleration
k_v	Ratio of lateral to radial pressure in bulk solids, typical values 0.4–0.6
$\dot{m}_s$	Mass flow rate of solids
s	Displacement
t	Time
u, v	Velocity
β	Angle of oscillation
δ	Angle of arc of contact
ρ_b	Bulk density
λ	Amplitude/surcharge angle
θ	Angle of inclination
ϕ	Wall friction angle
μ	Actual friction coefficient for bulk solid in contact with chute surface
μ_e	Equivalent friction factor which takes into account the friction coefficient between the bulk solids and the chute wall, the stream cross section and the internal shear of the bulk solid.

All other symbols as noted in the text.

Further reading

Colijn, H. (1985) *Mechanical Conveyors for Bulk Solids*. Elsevier, Oxford.

Haugland, L. (September 1996) Air assisted gravity conveyors. *The POSTEC Newsletter* No. 15.

http://www.dynamicair.com/pdf/9806.pdf, for document entitled. Dyna-Slide Air Activated Gravity Conveyor Series 126.

http://www.martinsprocket.com/material.htm, for information regarding capacity and length of drag conveyors. Accessed 28 July 2006.

http://www.redler.com/about/redlerhistory.htm, for the history of en-masse conveyors. Accessed 31 July 2006.

Inculet, I.I. & Strathdee, G.G. (1988) Electrostatic beneficiation of potash ores. In: *Industry Applications Society Annual Meeting*, Conference Record of the 1988 IEEE.

Latkovic, D. & Levy, E.K. (August 1991) Flow characteristics of fluidized magnetite powder in an inclined open channel. *Powder Technol.*, **67**(2), 207–216.

Marjanovic, P., Stanojevic, M., Todorovic, B. & Vlajcic, A. (1993) The influence of variable operating conditions on the design and exploitation of air-slide systems in thermo-power stations. In: *Conference Proceedings of Reliable Flow of Particulate Solids II*, Oslo, Norway, 23–25 August.

Schneider, H. & Stocker, D. (July, September 1998) Design of fluidized bed channels fitted with high perm-porosity sintered material. *Powder Handling Process.*, **10**(3).

7 Small-scale bulk handling operations

ANDREW COWELL

7.1 Introduction

The area of 'small operations' in *Bulk Solids Handling* is not one to be underestimated. While a large amount of equipment sales in monetary terms is to the multinational companies and their subsidiaries, there are a large number of companies that are working on a much smaller scale. Before we discuss these companies any further it is worth defining the term 'small operations'. A small operation is one where the majority of the bulk solids handling is done manually, although there will be some labour-saving devices employed. The split of manual handling and labour-saving devices will depend on the industry and the individual company, but in most cases a further bias towards the labour-saving devices would increase productivity. However, as always, this must be balanced against the economic costs of this purchase. Some examples of these 'small operations' industries include bakers, glass makers, small quarries, fish feed manufacturers, speciality food and small tobacco manufacturers. These companies can be further characterised by the size of equipment they use and the throughput of their plant. For example, a typical flexible screw conveyor would have a capacity in the range of less than 1 and up to 10 tonnes/h (for an 80 or 114 mm bore conveyor) (Bourton & Clague 2000), whereas a screw conveyor of 150 mm diameter (the smallest offered by some suppliers (Fairport 2006)) is capable of up to 50 tonnes/h. Taking a further example from pneumatic conveying systems, a power station may require in excess of 100 tonnes/h of product and for this product to be conveyed over 1 km (Wypych 1999), a smaller vacuum system typical of these smaller plants may have a maximum of between 2 and 10 tonnes/h and may convey as short a distance as 4.5 m (PIAB 2006).

The common characteristic of all these industries, and a definition of bulk solids handling, is that they all handle powdered or granular materials. The restrictions on the handling of these materials can sometimes be quite stringent, especially in terms of dust generation and contamination of the working environment. The handling of heavy loads is also becoming more tightly regulated, and thus this must also be considered when optimising the handling of powders in a small operation. This area is covered in some of the relevant UK legislation: Control of Substances Hazardous to Health Regulations 2002 (COSHH), Manual Handling Operations Regulations 2002, Health and Safety at Work Act 1974 (HASAWA) and Management of Health and Safety at Work Act 1992 (Risk Assessments) (HSE 2006). The responsible body in the European Union is the European Agency for Health and Safety at Work (EU-OSHA 2006), which was launched in 1996. While this agency does not have its own set of regulations, it does act as an advisory body and a statistical source for the member states. In the US, the relevant legislation is the Occupational Safety and Health Act of 1970 (OSH Act), and the regulations come under the Standards 29 CFR. Part 1910 of this document is entitled 'Occupational Safety and Health Standards', where subpart N covers 'Materials Handling and Storage' and subpart Z covers 'Toxic and Hazardous Substances'

(OSHA 2006). In Australia, the Australian Safety and Compensation Council (ASCC) is responsible for the development and oversight of OSH policy. The National Code of Practice for the Control of Workplace Hazardous Substances (NOHSC:2007 (1994)) and the Adopted National Exposure Standards for Atmospheric Contaminants in the Occupational Environment (NOHSC:1003 (1995)) is similar to COSHH, and Manual Handling is covered under National Standard for Manual Handling (NOHSC:1001 (1990)) and National Code of Practice for Manual Handling (NOHSC:2005 (1990)).

7.2 Equipment

The equipment that could be used to improve the efficiency of a small operation can be summarised by the following list, although a number of these items are called different names by different people:

- Aeromechanical conveyors
- Bag dump stations
- Bulk bag dischargers
- Bulk bag fillers
- Drum dumpers
- Flexible screw conveyors
- Sack fillers
- Small volume batch conveyors

The key advantage of most of these is their portability. A small operator is unlikely to want to have a number of devices performing the same function and is unlikely to be able to afford this expense either. Another advantage of all of these is that they reduce the amount of dust generated by the handling of the powders. They will also generally reduce the physical load on the plant operator, whether this is by mechanical means, as in the case of the screw conveyor and the drum dumper, or by pneumatic means for the pneumatic conveying systems.

7.2.1 *Aeromechanical conveyors*

Also called cable and disc conveyors, these devices are not unique to the small operations application, but as they are available in a portable configuration it is worthwhile discussing their operation and advantages in this section. Angus (1999) defines an aeromechanical conveyor as 'a powder transporting device, consisting of a constant-diameter tube containing a series of discs mounted at fixed intervals on a cable – with a drive mechanism that propels the discs at a velocity sufficient to cause turbulence within the tube'. Practically, an aeromechanical conveyor consists of a series of discs, mounted within a wear resistant tube, connected in a loop by a durable cord. The material is fed into the system through a small hopper that is mounted adjacent to the conveying tube. The material flows into the tube and is 'picked up' and fluidised by the conveying air that travels at a similar velocity to the discs. The durable cord is mounted on sprockets that are situated at regular intervals along the conveying track and at the 'corners' of this track. The modular nature of the system and the ability to convey at any angle provides great flexibility in conveying route (see Figure 7.1).

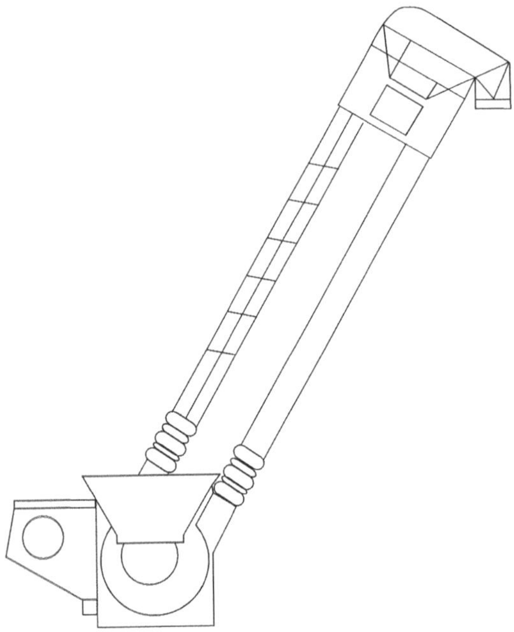

Figure 7.1 Typical aeromechanical conveyor.

Thus for the practical operation of an aeromechanical conveyor, there are a number of issues that require consideration. The first of these is the size of the discs relative to the pipe bore. To achieve the turbulence mentioned in Angus' definition above, the gaps need to be large enough to create this turbulence, but not so large that excessive amounts of material can escape through this gap. For the common tube diameters of 75 and 100 mm, a radial clearance of between 3 and 6 mm is typical. Another key parameter, as can be seen from Angus' definition, is the velocity at which the discs are moving. While there is no definitive method of calculation for this, the range of 2–4 m/s can be used as a rough guide. Another important practical consideration is the tensioning of the conveyor cable. During operation there is an inevitable stretching of the cable, so a tensioning device, usually at the discharge end of the conveyor, is an essential component (Angus 1999).

The mode of operation that makes this conveyor work is not completely understood. In principle, as the discs are pulled through the tube the powder is pulled along with these discs. Furthermore, the leakage past the edges of the discs through the clearance gap is not a significant issue as the next disc will gather up this material and convey it along the pipe. Indeed, Angus (1999) contends that 'without slippage, there would be less of the turbulence and aeration of powder on which the conveyor to a large extent depends'. This slippage leads the uninformed observer to overestimate the aerodynamic proportion of the conveying. Despite the advantages caused by the aeration of the material in this type of conveyor, the mechanical work on the powder is the prime source of conveying motivation.

Another important area for consideration is the interface areas at each end of the conveyor, that is the loading and discharge points. Starting with the inlet it can be seen that the powder entering the conveyor must displace the air between the conveying discs in order that it can

Table 7.1 Typical volumetric throughput of aeromechanical conveyors.

Drive motor speed (rpm)	Pipe bore (mm) 76	102	168
	Volumetric throughput (L/min)		
90	330	730	2500
150	540	1190	4200
190	680	1500	5300
240	870	1900	N/A

be conveyed along the conveyor. For a simple gravity feed, highly aeratable powders may cause a problem as they will not settle into the conveyor and this will reduce the capacity. One possible solution is the use of baffles over the inlet in an attempt to alleviate this problem. The discharge point has a similar problem. The volume of air in between the discs is necessarily increased once the powder has been discharged from the conveyor. This additional air must come through the discharge pipe, which creates a twofold problem. The first is the same as the inlet where the material must overcome the air resistance of the incoming air, possibly excessively aerating the material leaving the conveyor; and the second is that for closed systems, where an inert gas is used, for example, the additional gas must be supplied.

Typical conveying rate for this type of conveyor is generally given in a volumetric flow rate, quoted as 15–30 m^3/h in some literature (Entecon UK Ltd 2006; Spiroflow Ltd 2006). Thus the mass throughput, the convention for most bulk material processes, is dependent on the bulk density of the material in its conveyed condition. As the material is conveyed in a fluidised state, care should be taken to obtain a suitable value for this density, especially as the most common measurements of bulk density are either poured or tapped, which will give much higher throughputs than the aerated value. Table 7.1 gives indicative volumetric flow rates for a range of conveyors (Unitrak 2006). Included in the notes for this table are some de-rating factors for the values that can be used to estimate the mass throughput. The first of these is a recommendation that the conveyor be run at 50–60% loading by volume. This is incorporated with the second, a reduction of the bulk density to account for the aeration, as an overall de-rating factor of ¼ on the mass throughput, to give the formula below for a conservative estimate:

$$\dot{m}_s = \frac{V_s \times \rho_b}{4} \tag{7.1}$$

where $\dot{m}_s$ is the mass throughput (kg/min); V_s is the swept volume (L/min); ρ_b is the poured bulk density (kg/L).

Although the aeromechanical conveying rates are given as volumetric flow rates, an aeromechanical conveyor can run at varying volumetric loading percentages, it can indeed run empty. Therefore, it will generally need a volumetric feeder to ensure a consistent throughput.

Elsewhere (Entecon UK Ltd 2006), it is recommended that the material be conveyed at a 20/80 material/air ratio.

Equipment manufacturers (Entecon UK Ltd 2006; Unitrak 2006) quote a wide range of materials that have been conveyed by aeromechanical means, including food products such as breakfast cereal, coffee beans and curry powder; metal products such as lead shot and metal powders; chemical compounds such as aluminium oxide, calcium carbonate and titanium dioxide; materials for the construction industry such as cement, DIY plaster type products and tile dust; household goods such as soap granules, talcum powder and washing powder; plastic pellets and powders; and colouring agents such as carbon black and pigments.

This method of conveying has a number of other advantages, but is limited to materials with suitable aeration characteristics. Firstly, the system is an enclosed one, hence is suitable for dust-free transport. Furthermore, as all the material is travelling at the same speed, it is reported that there is very little segregation of the material in transport. As mentioned above, these systems are also available in a portable form. The loading hopper and receiving sprocket are mounted on a frame with castors mounted underneath (see Figure 7.2).

Although this may be unsuitable for some surfaces, it may be possible to mount more suitable wheels for uneven and/or soft surfaces. The relatively low energy use of the system means that the drive motor for the conveying system can be mounted on this frame also, while the weight of the system is still suitable for manual handling. According to one equipment supplier (Unitrak 2006) the power consumption to throughput ratio of these systems is 'much lower than pneumatic conveying' and 'few other systems come close'. Finally, as the system works with a centrifugal discharge, the dust generation at discharge is minimised and no filtration system is required for the aeromechanical conveying system – there may, however, be a requirement for a filtration system on the receiving hopper. The centrifugal force acts to hold the material in a confined stream until it reaches the discharge vessel. It is also protected by a discharge tube.

There are additional limitations to this type of system other than being limited to materials with suitable aeration characteristics. Although this is a relatively gentle method of conveying material the common material handling problems of degradation and abrasion are prevalent here. The natural slippage of material between one disc cavity and the next can result in particles being caught between the discs and the conveyor wall, a possible source of damage to fragile particles. This natural slippage can also cause abrasion on both the discs and the wall if the wall and/or disc material is softer than the material being conveyed. It has been reported that this wear effect is more significant when the conveyor is operating in a horizontal alignment as compared to conveying vertically (Angus 1999). Moreover, while the power consumption is relatively low compared to something like a pneumatic conveying system, this needs to be offset against the increased number of wear parts in an aeromechanical system; that is, the discs and cable may need replacing over and above the conveying tube required for both systems.

The materials of construction of the discs, cables and tubes can be varied to suit the application. However, the discs are usually made of a plastic material, such as polyurethane, in order to minimise the friction on what are generally steel tubes. Nylon, estane and polycarbonate are also used for the disc material. The cables are constructed of a steel wire, usually of a galvanised mild steel or stainless steel dependent on the application. The conveying tube can also be made of aluminium, which is especially useful for a portable application due to its lower weight.

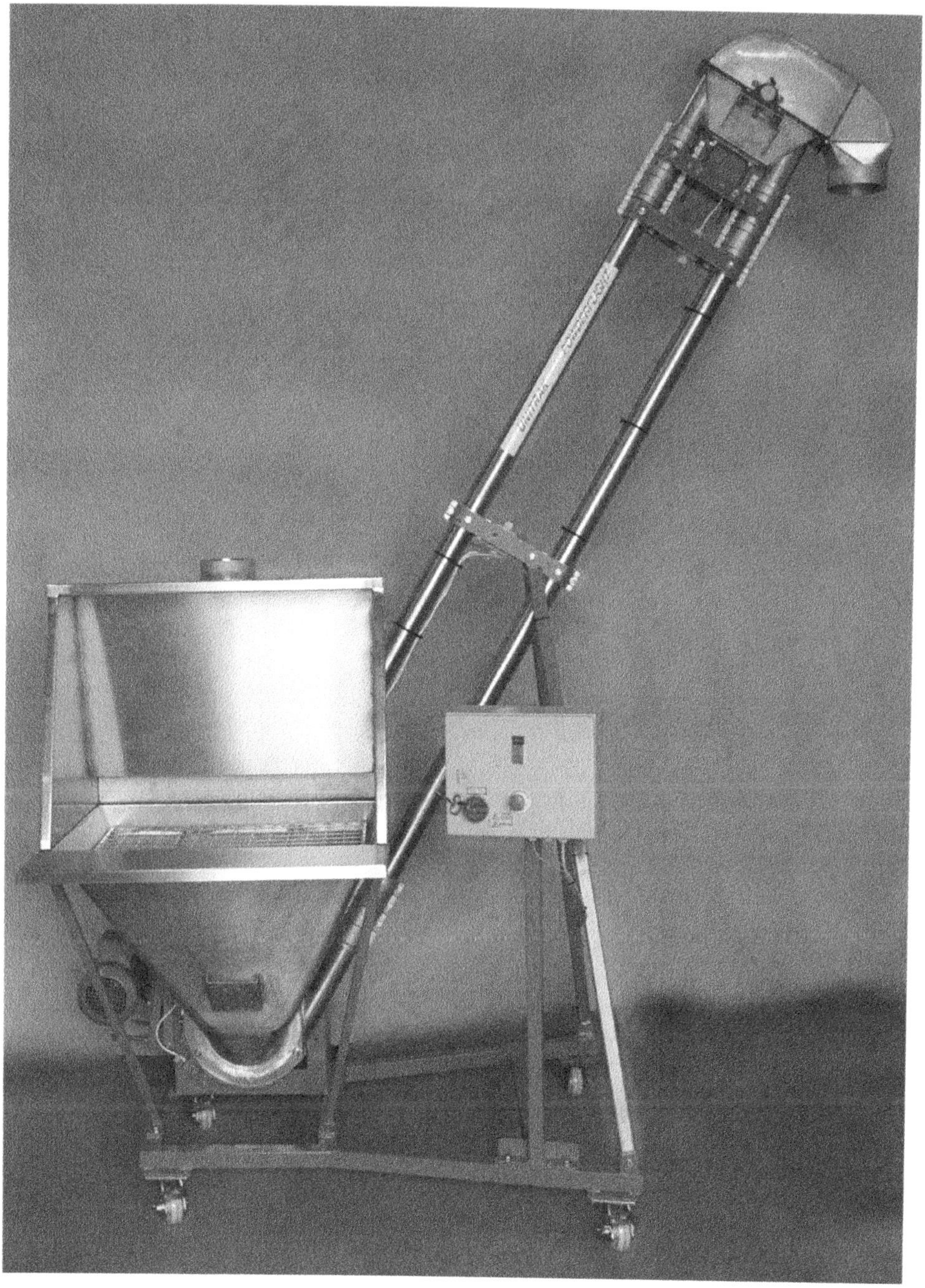

Figure 7.2 Portable aeromechanical conveyor (Unitrak Corporation Ltd).

Figure 7.3 Typical bag dump station (Flexicon Corporation).

7.2.2 *Bag dump stations*

Also known as manual dumping stations, sack tippers or bag breakers, these generally consist of a receiving hopper with an integrated vacuum driven cartridge filter dust extraction system. Some systems also have a reverse-pulse filter cleaning action that drops the material back into the receiving hopper. It is also possible to set up the hopper outlet for feed screw or pneumatic conveying discharge, as well as rotary valves and various other process equipment types. This system can be mobile and can be connected to the start of a number of conveying systems. Of course, these stations are not just limited to bags or sacks; they can also be used for bins or drums (see Figure 7.3). Some designs include a mechanical assistance device for the emptying of drums, and these are discussed further in the section on drum dumpers.

The chief advantage of a bag dump station is the reduction of dust generation in the process. This control is initiated by the extraction system mounted above the receiving hopper. There is also generally a grid situated on top of the receiving hopper that goes some way to separating tramp materials that may be in the bag, and also helps to 'screen' and de-lump or aerate the material that drops into the hopper. Once the material reaches the

Figure 7.4 Automatic bag slitter (WJ Morray Engineering Ltd).

hopper, an enclosed system is used to transfer the material to the next step in the process – another good way of minimising dust generation.

The advantage of mobility means that the station can be moved to wherever the bags are, rather than the other way around. It is defeating of their purpose to require the bags to be brought to them, as this then retains a manual handling operation that was trying to be avoided. However, this may require that the downstream conveying needs to be changed and this needs be considered when specifying a system. One alternative would be to unload the bags and transport the material to the required location in the receiving hopper. Care should be taken in this case to ensure that the material does not consolidate in the hopper. Indeed, this consolidation may be exacerbated by the vibration caused by moving the stored material.

A variation on the bag dump station that even further minimises the dust generation and manual handling requirement is the bag slitter (see Figure 7.4). Bags are placed on a conveyer belt, thus not requiring support while the bags are opened/emptied, and conveyed into a dust tight chamber where the slitting blades are situated. Depending on the configuration the bag slitter can have two or four blades. The blades work with a rotary action, spinning in the horizontal plane, slitting down the long side of the bag in the two-blade configuration, having started in the middle of the front edge (see Figure 7.5). The four-blade configuration is generally used for larger bags and all four sides of the bag are slit simultaneously. Once the bag is open, it is removed from the system and stored in a waste compaction unit. The

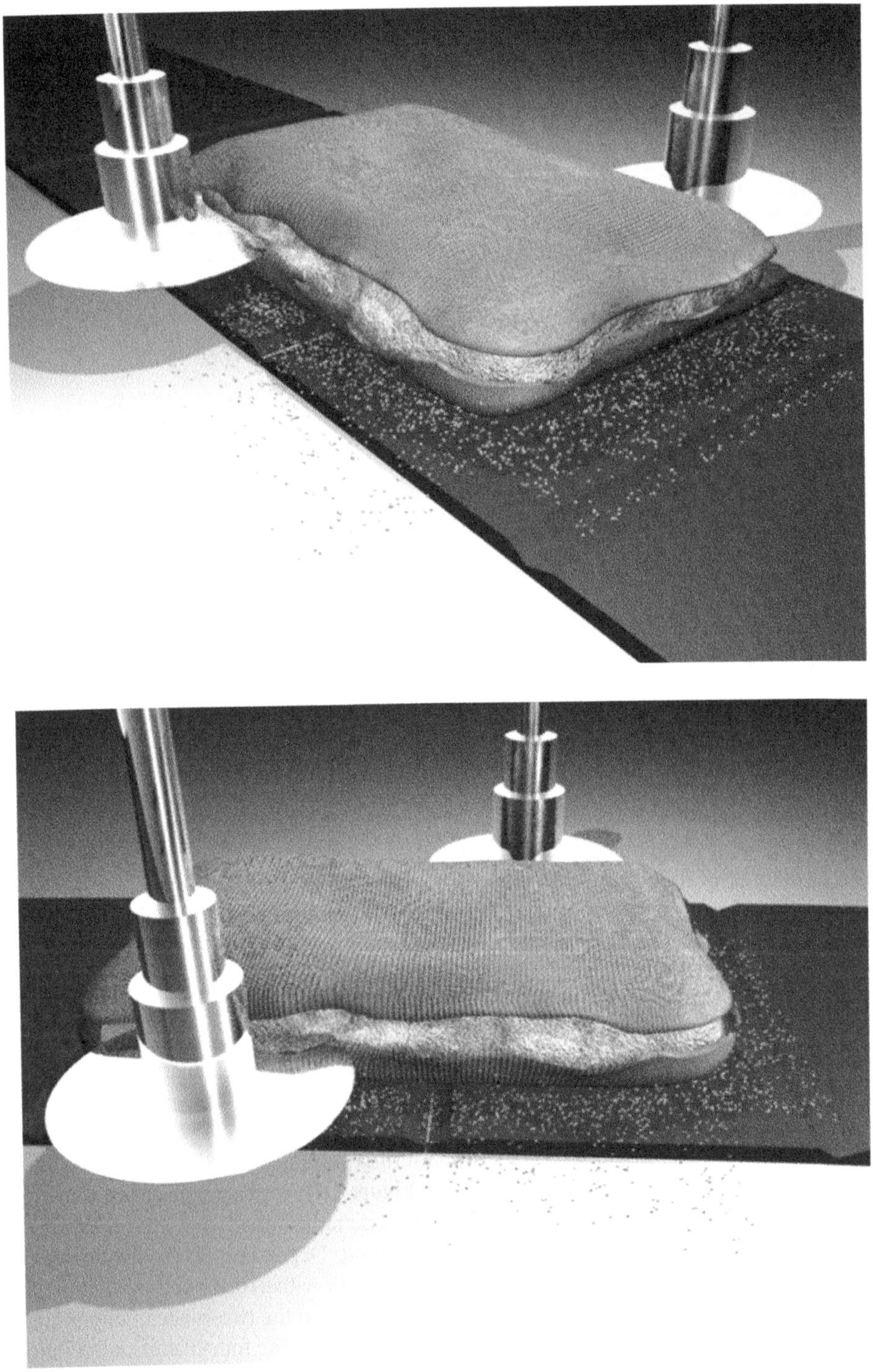

Figure 7.5 Detail of blade operation.

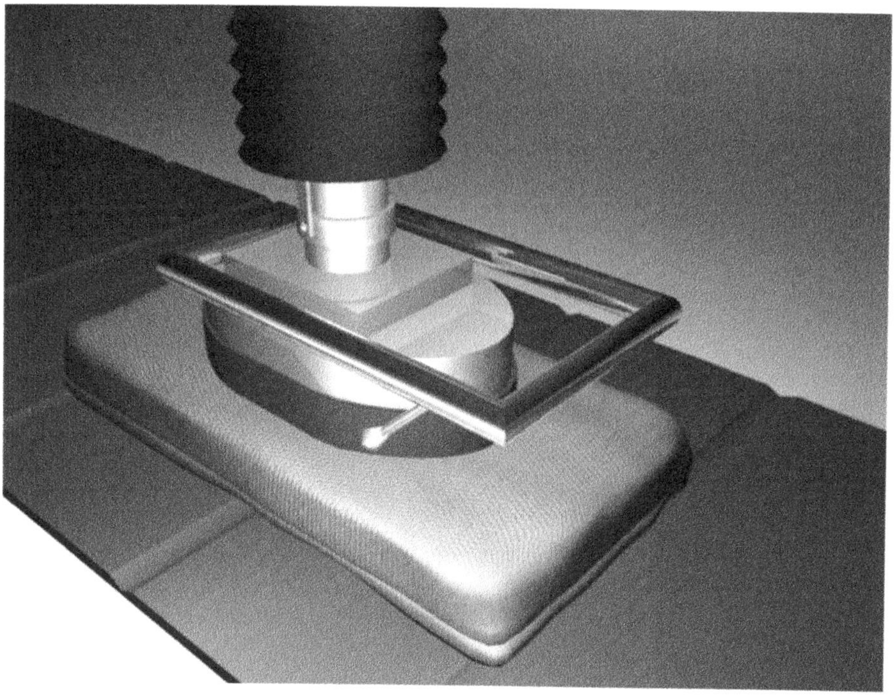

Figure 7.6 Vacuum bag lifter.

method of removal varies from a grabbing mechanism to a simple gravity drop. As with the drum dumper it is possible to fit a dust extraction system that discharges into the receiving hopper, thus minimising waste.

A labour-saving device that is sometimes used with the bag slitter, and other small operations equipment, is the vacuum bag lifter (see Figure 7.6). A suction nozzle of a similar diameter to the width of the bag is connected to a vacuum pump, and when the pump is activated the force on the bag is sufficient to lift the bag from pallet to feed conveyor, but not sufficient to break the bag. This greatly assists with manual lifting of awkward loads, but the necessity of an air supply may be restrictive in some applications.

For lower volume operations, there are also self-contained bag slitters that are of a similar configuration to the bag dumper (see Figure 7.7). In a way, they are a hybrid of the bag dumper and the bag slitter. The bags are loaded onto a grill, similar to that found in the bag dumper, and then a safety hatch is closed over the top prior to the engagement of the cutting devices. This is generally protected by a safety switch to ensure that there is no danger of injury to the person loading the device. Once the safety hatch is closed the bag is slit open, the product discharged into the receiving hopper and the bag removed for storage at the rear of the device.

An additional option for some types of bag splitter is clean in place (CIP) operation. This is particularly useful for hazardous powders, and/or processes that use the same bag slitter for a number of different powders. CIP operation means that it is possible to completely clean the piece of process equipment without dismantling.

Figure 7.7 Self-contained bag slitter (WJ Morray Engineering Ltd).

For the bag dumpers the maximum size of the sack that can be emptied is sometimes dependent on the size of the hopper that receives the material. Other manufacturers specify a maximum bag size or weight. This varies from manufacturer to manufacturer, so it is best to confirm these details as required; however, typical maximum hopper volumes are 0.14–0.20 m^3, and maximum bag weights are about 50 kg. For bag slitters the size of the bag is the more important factor, and this is generally between 1 and 1.2 m. Bag slitters are capable of handling a range of sack materials, such as paper and plastic, through to Hessian, jute and/or polywoven (Morray Beltyne 2006; Palamatic Materials Handling 2006).

Before using a particular material in either a bag slitter or a bag dumper it is recommended to check with the manufacturer as some hoppers may not be suitable. However, some examples of materials used in these devices are alumina, aluminium sulphate, carbon black, cement, cocoa beans, cocoa powder, coffee beans, flour, hydrated lime, lime, milk powder, potassium hydroxide, silica sand, sugars, talc, titanium dioxide and whole nuts (Aptech (Powder Systems) Ltd 2006; Flexicon Corporation 2006; Palamatic Materials Handling 2006; Spiroflow Ltd 2006; Vibrair 2006).

7.2.3 *Bulk bag dischargers*

As this type of process equipment becomes more and more popular, there are an increasing number of suppliers and variations of configuration. They are also known by a number of different names: Flexible Industrial Bulk Container (FIBC) Discharger/Emptier/ Discharging Station; Big Bag Discharger/Emptier; Bulk Bag Emptier/Unloader/Unloading Station. However, despite the different names and configurations, the basic purpose remains the same: to completely discharge an FIBC in a dust-free manner. It is possible, with some configurations, to discharge a proportion of the contents, stop flow and return to operation at a later date. Of course, with non-free-flowing materials, consolidation can be a problem. However, there are a number of methods of promoting and re-commencing flow that will be discussed.

The simplest configuration (see Figure 7.8a) consists of a static frame onto which the bulk bag is placed by use of a forklift, or other suitable lifting apparatus. Some designs have forklift slots incorporated into the construction. This type itself comes in a number of differing configurations depending on the type of bag used, that is, one, two or four loops. Another type of static frame design is the one with an inbuilt lifting hoist (see Figure 7.8b). This design removes the need for a mobile lifting device, as the bag discharger is mounted on a trolley for mobile operation. A dust collection device, usually with a reverse jet cleaning system, is generally fitted to these systems.

Because of the wide range of configurations and suppliers, there is no one definitive configuration description. A list of the features which are available on at least one configuration are listed below:

- Bag extension system
- Bag cutter
- Bag folder
- Bag knife
- CIP/glove boxes
- Cone valve
- Flow activation paddles/bag massaging system
- Inert gas purge
- Integral conveyor
- Integral rotary valve feeder
- Integrated sack dumping station
- Liner suction
- Low headroom operation
- Materials of construction
- Pneumatic discharge assistance
- Product screening
- Sealed system
- Spout clasping clip/iris valve
- Vibration discharge assistance
- Weigh batching

The bag extension system has a twofold purpose. Firstly, it allows different size bags to be handled by the same discharge station. This is particularly useful where bag sizes are inconsistent, depending on the supplier and/or material being discharged. The second purpose of the bag extension system is to raise the bag walls and floor into a cone shape once the majority of the material has been discharged from the bag. While some systems have flow activation paddles or a bag massaging system to promote flow, this may not be of

Figure 7.8 (a) Bulk bag discharger with forklift loading (Unitrak Corporation Ltd); (b) bulk bag discharger with integral hoist (Spiroflow Ltd).

Figure 7.9 Bulk bag discharger with glove box (Spiroflow Ltd).

assistance as the material level reaches the bottom of the bag. To aid in this discharge, the bag can be raised and the walls at the outlet will become much steeper, closer to or larger than the mass flow angle for that material.

The bag cutter is used to cut the discharge neck from single use bags. The bag folder does simply what the name suggests, and is more about keeping the plant safe and tidy than assisting with the safe and clean discharge of material from the bag. The bag knife is used to pierce the bottom of single use bags. This is either a static knife for free-flowing materials, or a pneumatically driven knife that allows greater penetration into materials that may need assistance to flow from the bag.

A CIP system allows maintenance and cleaning of equipment without significant disruption to the operation of the process. This type of system is useful for pharmaceutical and food applications that require frequent cleaning. Some of these systems have a glove box fitted as standard in order that the discharge spout can be handled without any danger of contact with the stored material (see Figure 7.9).

A cone valve is used at the discharge of some big bag systems. This has the advantage of dust-free operation, while not affecting the flow of material. Indeed, they can sometimes act to promote flow without the need for a mechanical or pneumatic stimulation of the material. As mentioned above, flow activation paddles and bag massaging devices are used to promote material discharge. This is achieved by effectively changing the local wall angle by pushing against the bag at a level just above the discharge spout (see Figure 7.10).

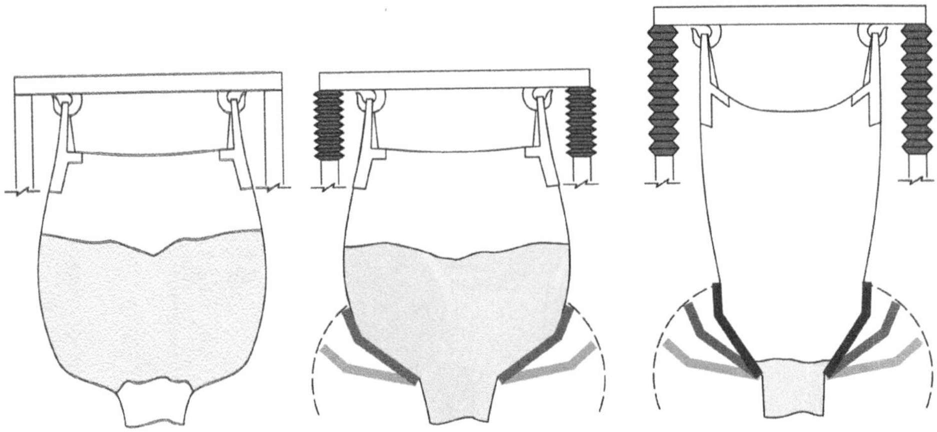

Figure 7.10 Effect of a bag massaging device (Flexicon Corporation).

An inert gas purge can be used for powders that are susceptible to explosion. A discussion of the explosibility of powders is outwith the scope of this Chapter; however, it is worth noting that the prevention of dust cloud explosions is possible using an inert gas, as the gas removes the oxygen from the bag and thus combustion is not likely. An integral conveyor is generally used to transport the discharged material to the next step in the process. The material discharges from the big bag into a reception hopper, and from here the material is generally either conveyed using a flexible screw conveyor or a pneumatic conveying system. To assist with the feeding to a pneumatic conveying system a rotary valve feeder is occasionally employed. This allows the metering of material into the system, whether it is positive or negative pressure.

Sack or bag dumping stations have been discussed previously. It is a logical extension of the bulk bag discharge station to include a hatch for the dumping of sacks. The system then is a type of hybrid combining the features of the two pieces of equipment. The liner is exposed to suction, either by opening the valve to the filter or to a vacuum pneumatic conveying system, in order to completely empty the bag and/or remove any excess dust. As space is often at a premium in processing plants most of the bulk dischargers come in a low headroom configuration. There is a minimum height that is physically required due to the geometry of the situation, but the problem is overcome in a number of different ways. The first of these is to use a half-height frame that simply consists of the discharge cone and the receiving hopper mounted on a frame. The bulk bag is then positioned over the top of this frame using a forklift or similar. Another design is the scissor lift. This is a mobile frame that allows low headroom operation by compressing down to a small height when not in operation and opening up for lifting the bag to discharge height. This option is available in a pneumatically driven type for hazardous environment operation.

When referring to optional materials, this generally refers to the material of construction of the discharge cone and the receiving hopper. These are available in a number of materials from food quality stainless steel to mild steel. When selecting for a food application, it is also important to ensure that the rubber seal materials are also compliant with regulations. Apart from the mechanical flow aids, some systems are also available with pneumatic discharge assistance. The material is aerated by injecting air at the bottom of the bag, and this assists with discharge of the material. Tramp and oversize material can also be a hazard when discharging bulk bags. To avoid these objects moving further through the process a screen between the bag and the receiving hopper is occasionally employed. For hazardous applications such as for pharmaceuticals it is usual to use a sealed system. In addition to being a completely enclosed system, these systems employ a liner on the discharging cone that is disposed of after use. In this way the FIBC is left uncontaminated and is available for re-use.

Occasionally, it is necessary to stop emptying a big bag at some point during the process. The two main methods of achieving this are the spout clasping clip and the iris valve. The spout clasping clip appears to be a relatively new development that works with a scissoring action to clamp the discharge spout and stop material flow. It would appear to allow quicker and easier clasping of the bag than the conventional iris valve. The reason for this may be the fact that the iris valve requires the rotation of one flange, whereas the spout clasping clip is a separate device that operates externally to the flow pipe. Vibration discharge aids are also attached to bulk bag dischargers. It has been shown that for some powders the application of a small vibratory load can promote flow. However, care must be taken to specify this system correctly, as an incorrect application can result in consolidation of the material, exactly the reverse of what is trying to be achieved. The final attachment that is occasionally attached to a big bag discharger is a load cell arrangement. Generally the big bag is hung from a cantilevered beam and a loss in weight system is used to achieve weigh-batching from the subsequent conveying system.

The application of bulk bag dischargers is wide and varied. A large number of different products have been successfully discharged using these systems, but as with all bulk solids plant, a knowledge of the flowability of the material is essential before specifying a system including this equipment. Some typical materials handled are breakfast cereals, carbon powder, chemical dyes, ceramic powders, cut potatoes, dried fruits, flour, limestone, pharmaceutical powders, plastic pellets, plastic powders, rubber crumb, sand, soap flakes, soda ash, sugar and yeast powder. Some typical applications of a bulk bag discharge system are to feed into a flexible screw conveyor or aeromechanical conveyor for transport further down the process: straight into a mixer, with the possibility of using a loss-in-weight system to assist with batch blending; and a particular application of three dischargers in series used to supply the three components of layered detergent tablets (Aptech (Powder Systems) Ltd 2006; Dynamic Air Conveying Systems 2006; Entecon UK Ltd 2006; Flexicon Corporation 2006; Flomat Bagfilla International 2006; Hosokawa Containment 2006; Spiroflow Ltd 2006; Vibrair 2006).

7.2.4 *Bulk bag fillers*

These devices, also called big bag fillers and FIBC fillers, come in a range of different configurations allowing varying degrees of automation and therefore process speeds. The first of these configurations, designed for minimum throughput, generally consists of either a two post or four post design with the bag supported in between (see Figure 7.11).

Another design configuration consists of a single post with a lifting frame attached, where the bag hooks are slipped over two lifting poles (see Figure 7.12). These devices are able to handle a variety of bag sizes by a simple adjustment of the bag hooks up and down the

Figure 7.11 Static bulk bag filler: (a) two post type (Flexicon Corporation); (b) four post type (Spiroflow Ltd).

supporting poles. The loading chute is mounted above the bag and material is loaded into the bag by means of either a flexible screw conveyor or a pneumatic conveying system. Some configurations make use of an intermediate hopper for storage of the material. This can be useful for pneumatically fed systems where a large amount of air needs to be removed before filling the bag.

The bag can be attached to a frame that has forklift blade slots included for easy manipulation of the bag once full. To aid with bag filling it is also possible to have load cells

(b)

Figure 7.11 (*continued*).

attached to this frame that allows accurate filling based on weight. This is particularly useful for easily aeratable powders that may require settling to ensure that the target weight capacity of the bag is achieved. Accelerated de-aeration can be achieved with some materials by mounting a vibratory system to this frame. As always, care should be taken to specify and commission this system correctly, as incorrect application of the vibration force may only serve to exacerbate the problem. Some systems include an automated controller to achieve

Figure 7.12 Single post bag filler.

this outcome. These systems can also be mounted on casters to allow portability of the system around the site. Furthermore, for sites with a height restriction, a low profile design is available that also comes with pallet truck access – removing the need for a forklift.

The more sophisticated configurations discussed next allow a degree of process automation, and allow a faster throughput of bags while still maintaining a dust-free loading process. The first of these is the static frame type that is mounted above a belt roller upon which pallets are placed to support the bags prior to and after filling. This design allows an almost continuous filling process to be achieved while maintaining all of the features mentioned for the static loading device mentioned above. Unlike the static filling arrangement, however, the weighing of the bag is achieved by load cells mounted underneath the pallet supporting the bag. There are a number of different design configurations within this subset. The cantilevered design generally allows for a variety of bag sizes through an adjustable height feature (see Figure 7.13); however, for larger bags the attachment of the bag can be at an uncomfortable height. The adjustable height feature can be either powered or manual. The problem of attachment of bags at an awkward height is overcome to a certain extent, although at the cost of slightly reduced throughput, by the use of a lowering and pivoting fill head. By lowering the fill head and pivoting it at a 90° angle, it is much easier to attach the top of the bag to the filling chute. The fill head is returned to its vertical position for filling (see Figure 7.14).

Another type consists of a two post arrangement where the bags pass under the filling device supported on top (see Figure 7.15). For applications that require an extremely high

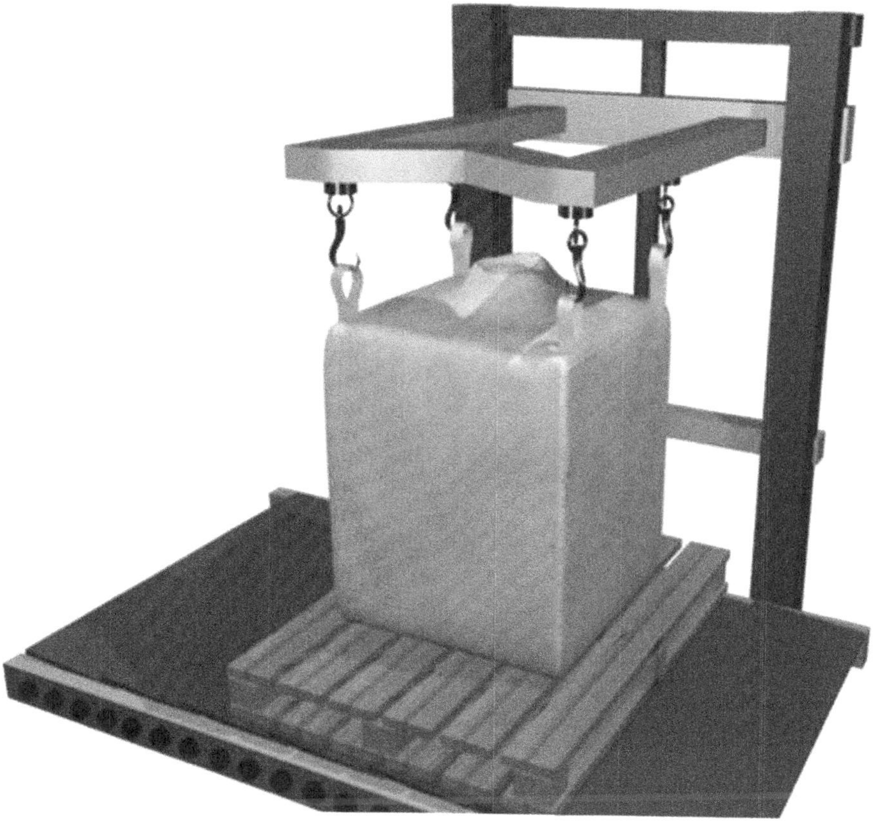

Figure 7.13 Cantilever type bulk bag filler.

throughput, up to 80 bags per hour, a carousel design could be applied (see Figure 7.16). The bags are attached to a rotating device that allows a bag to be filled while subsequent bags are attached to the carousel. This system uses hang weighing to determine degree of bag filling and features automatic bag removal. However, bags still need to be manually fitted to the carousel.

As with all of the equipment we have discussed so far, these are all available in a wide range of materials. While the frames are generally constructed from a mild steel, as they do not come into contact with the process material, a wider range of materials are available for the loading chute. These can range from high quality stainless steels for food and pharmaceutical applications to lower grade mild steels for less arduous duties. The static frame noted above can be used in applications where frequent wash downs are required for sanitary purposes.

For operations that use a variety of storage containers there are designs that are adaptable to this application. This is achieved by fitting an adapter to the filling chute used to fill the IBCs. Another option that is available for all of the configurations mentioned above is the pre-inflation of the bag prior to filling. This allows the material to flow freely into the bag without any physical obstructions. However, it is important that the material be de-aerated

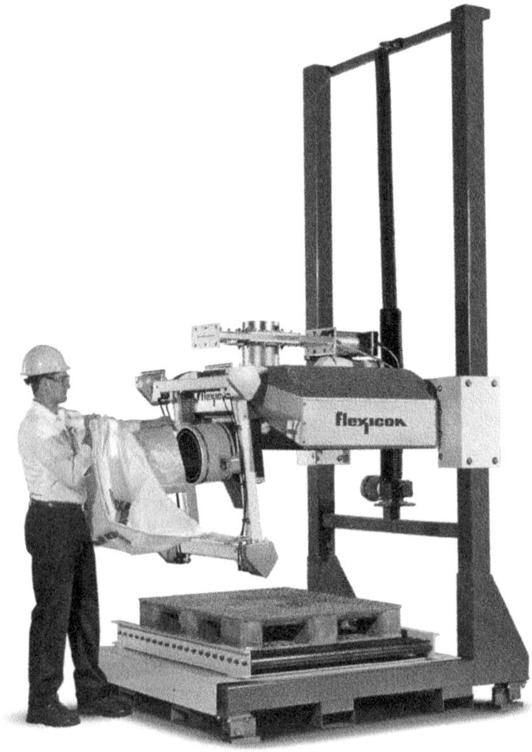

Figure 7.14 Pivoting fill head bulk bag filler (Flexicon Corporation).

upon completion of filling (see above). For materials that require an inert atmosphere, some systems can be fitted with an inert gas purge. A slightly different method of loading IBCs comes in the form of a cone valve (see Figure 7.17). There are a number of different arrangements available that offer increasing degrees of dust containment (ISL Cone Valve Technology 2006; Matcon 2006). The simplest design consists of a single valve, while other designs include an integral lower cone and others consist of a fully automated system where the valves and the containment lids do not require any user intervention.

While the integrated systems above have the advantage of being supplied in a complete package, it is not impossible to fabricate a system from scratch, even to the point of attaching a weighing system. While not such an advantage for some of the automated systems, an application such as the static system described initially would be suitable for this option. There are a number of instrumentation suppliers that are capable of providing the load cells required for the weighing system, and they may even be able to supply the associated control system required for batch filling.

As with the bulk bag dischargers, the materials used in, and the application of, bulk bag fillers is wide and varied. However, the same proviso holds; care must be taken to ensure that the system is adequately designed to take into account the flowability characteristics of the material. Some example materials and applications are the loading of polymer resins into a variety of container sizes; the batch weighing of PET; the filling of bags in a carousel

Figure 7.15 Four post type automated bag filling system.

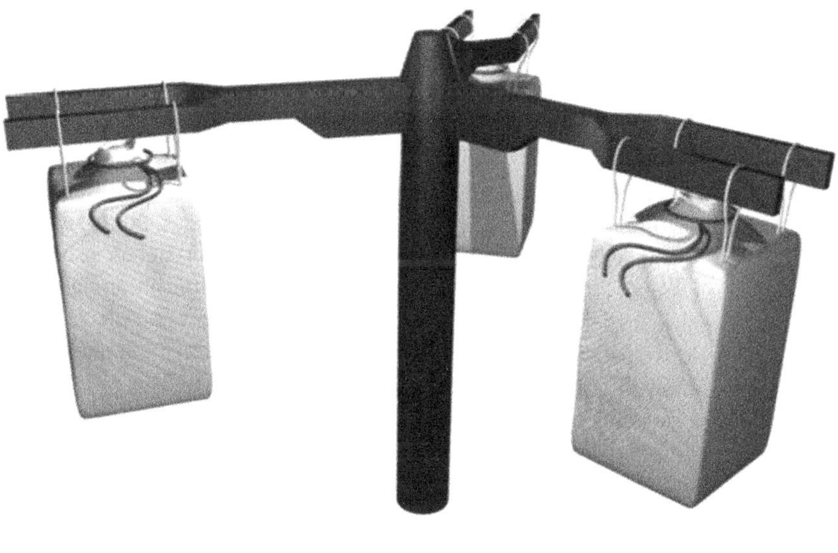

Figure 7.16 Carousel bag filler.

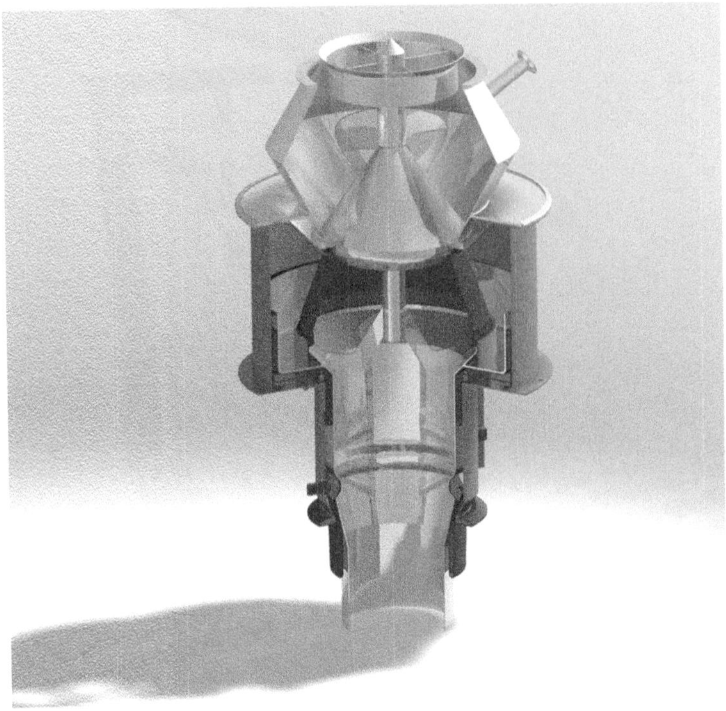

Figure 7.17 Cone valve (ISL Cone Valve Technology).

system with polypropylene and polyethylene pellets. The following materials have also been successfully used in these systems: aggregates, detergents, fertiliser, flour, grains, plastic pellets, salt, skimmed milk powder, sugar, urea and yeast (Arodo Total Bag Handling 2006; Flexicon Corporation 2006; Flomat Bagfilla International 2006; Hosokawa Containment 2006; Matcon 2006; Spiroflow Ltd 2006; Webster Griffin Ltd 2006).

7.2.5 *Drum dumpers*

A large number of raw materials are supplied in drum containers, and as the dust generation potential is quite high, not to mention the manual handling implications, this type of equipment can be applied widely. For applications where dust generation is not an issue an open chute design is acceptable. This chute must be designed at a sufficient angle to ensure material flow, but other than this, the design is relatively straightforward. Alternatively, the chute section is relatively short and only provides a guide path for the material to flow directly from the drum into a receiving hopper (see Figure 7.18a). This hopper will be attached to either a pneumatic conveying system or a flexible screw conveyor to further convey the material through the plant. For materials with a higher dust generation potential it is important that the system be completely sealed. One design consists of a cone clamped and sealed to the top of the drum before lifting begins (see Figure 7.18b). As can be seen from the figure there is a pipe fitted to the bottom of the cone section that is shaped to fit into the receiving hopper through an additional rubber seal. This ensures that all dust is

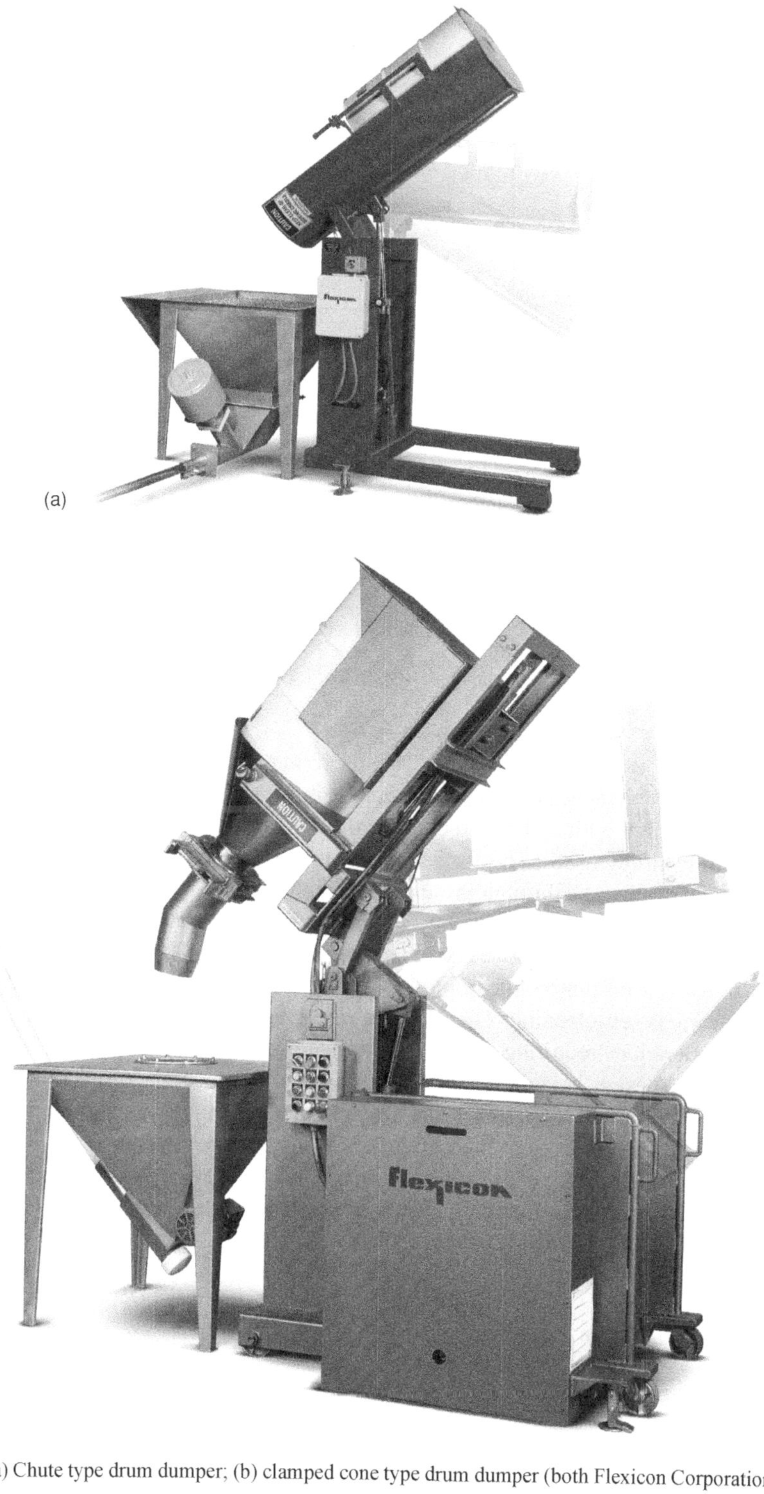

(a)

(b)

Figure 7.18 (a) Chute type drum dumper; (b) clamped cone type drum dumper (both Flexicon Corporation).

Figure 7.19 Typical flexible screw conveyor (Spiroflow Ltd).

contained within the transfer section. This receiving hopper can also be fitted with either a flexible screw conveyor or a pneumatic conveying system.

Another design with a clamped cone involves lifting the drum above hopper height and then rotating through 180°. The discharge cone is then aligned with the inlet to the hopper and the material discharged through these aligned pipe sections. As with all process equipment, it is important that the cone of the dumper is designed such that the material will not 'hang-up' in this section. With the small quantities of material being discharged it is unlikely that a mass flow design will be required; however, the slope needs to be adequate to ensure that the cone is self-cleaning. These devices are available in either carbon steel or stainless steel for operations requiring a hygienic surface. Importantly, for the small operation both variations of design have an aspect of portability. Some designs are fitted with casters, whereas others have a sack trolley like design that allows the drum and tipping device to be moved around the site. All of the designs are manually operated.

A variation on this type of device is a drum rotator. The device works on the same principle, and can be electrically or manually operated, but requires a separate hopper and/or delivery chute (S.T.S. Safety Trolley Systems 2006).

7.2.6 *Flexible screw conveyors*

This section is devoted entirely to the specification and use of flexible screw conveyors (see Figure 7.19). More information on the design of screw conveyors can be found in Chapter 5. The key advantage of a flexible screw conveyor is that within the flexibility of the conveying tube, the material can be delivered through a flexible routing. This same advantage is enjoyed by a pneumatic conveying system, but screw conveyors are much more suitable for routes of less than 10 m and can also convey at any angle. Another advantage of a flexible screw conveyor is that the screw itself is the only moving part. There are no internal bearings, and the material is fed directly into the screw from the feed hopper. The screw, constructed from a spring steel or stainless steel, is of the ribbon type; that is, it has no shaft.

There are a number of materials that a flexible screw conveyor is capable of conveying. Bourton and Clague (2000) list 141 materials, including food products such as frozen blueberries and rice flour and hulls, minerals and sands such as alumina and limestone, chemical powders such as ammonium chloride and potassium bicarbonate, plastic powders such as PVC and acrylic powders, rubber products and sticky materials such as china clay and gelatine. Fitzpatrick (1985) lists 85 materials divided into three areas: chemical–pharmaceutical, foods and plastics. Examples in the chemical–pharmaceutical area are soda ash, aspirin granulation and acetylsalicylic acid; in the foods area are powdered sugar, instant coffee and sesame seed; and in the plastics area are pellets, colour concentrates and polyethylene powder. Despite this versatility, care should be taken to ensure that the properties of the material are taken fully into account before attempting to convey the material. The following list gives an indication of some of the key material considerations:

- Bulk density (packed, loose and average)
- Particle size
- Particle shape
- Angle of repose (flowability)
- Temperature
- Moisture content
- Fat/oil content
- Abrasive/corrosive
- Degradable (affecting process/final product)
- Compressible/rubbery
- Aerates/fluidises
- Solidifies (if left to sit or vibrate) agglomerates
- Packs, cakes or smears under pressure
- Bridges when hoppered (or ratholes)
- Mechanically interlocks
- Hygroscopic
- Hazardous

The majority of these items would be covered by a standard material characterisation exercise (for more details see Chapter 1); however, it is worth discussing here how each of the criteria affects the conveying performance. The bulk density of the material is used to approximate the throughput of the conveyor. That is, while the volumetric throughout is reasonably consistent, most processes rely upon a mass throughput and thus the 'conveyed' bulk density is required. While a precise value of mass throughput is not usually possible, due to the nature of the conveying process, the loose bulk density normally gives a good estimate. A typical range of bulk densities that can be conveyed is 100–3000 kg/m^3. The particle size and shape determine both the size of the conveyor tube and the clearance

Table 7.2 Maximum particle size for various tube sizes (Bourton & Clague 2000).

Diameter of the outer tube (mm)	57	67	80	90	114	168	220
Maximum particle size (mm)	3	5	10	15	20	25	30

between the screw and the tube. While tube size is not critical for particle sizes below, say, 1 mm, once the particle size is a significant percentage of the tube diameter, care should be taken to ensure that the material will not jam the screw. For example, a typical size of the tube is normally in the range of 50–220 mm and the maximum particle size is approximately 25 mm. It is clearly not practical to convey this largest particle in the smallest tube size. Indeed, Fitzpatrick (1985) gives a maximum particle size of 1/4″ (6.35 mm), but as no tube diameter is given to relate this too, this figure may be misleading. For particles where degradation and/or abrasion are an issue, the particle size and shape are used to determine an appropriate screw/tube clearance. Particle size is critical as it should be ensured that it is not possible for a particle to be trapped between the screw and the tube, and particle shape is critical as particle size is usually some average value and non-spherical shapes may result in jamming when the mean size is less than the gap. Table 7.2 gives indicative maximum particle sizes based upon the diameter of the outer tube.

The flowability of the bulk material is also important. While Bourton and Clague (2000) associate only angle of repose with flowability, several of the other criteria listed above could also be grouped under flowability – compressible/rubbery; aerates/fluidises; solidifies (if left to sit or vibrate) agglomerates; packs, cakes or smears under pressure; bridges when hoppered (or ratholed); and, mechanically interlocks. So, while the angle of repose is important in determining the angle of the screw flights, many other aspects of the material's flowability should be assessed when considering the design of the flexible screw conveyor. Temperature, moisture, abrasion and corrosion are other important material characteristics in the specification of a flexible screw conveyor.

The influence of particle size on degradation has been discussed, but screw speed should also be selected to minimise the damage caused. The two final criteria, hygroscopic and hazardous affect the environment in which the material is conveyed. Both require that the system be as closed as possible, in order that for the hygroscopic material, excess moisture is not allowed to ingress, and for the hazardous material, no material is allowed to escape to the atmosphere. The importance of a number of the selection criteria outlined above is minimised by the fact that the screw is self-centring. There are two forces applied upon the conveyed material within the screw area: an axial force that acts to convey the material along the tube and a radial force that presses the material against the conveyor tube. This radial force provides a barrier between the tube and the screw. Thus, providing the tube is not excessively curved, the screw will remain in the centre of the tube (Bourton & Clague 2000). This is the phenomenon that allows flexible screws to be 'self-centring'. Some designs also have CIP capability. This is made possible by the fact that the screw is the only moving part, and further assisted by the fitting of an end cap that allows direct access to the screw inlet.

The feed section of the screw is generally sited in a U-shaped trough, situated directly below the feed hopper. This section of the screw picks up and feeds the material into the tubular section of the screw conveyor and the material is conveyed to the outlet point.

The motor drive for the screw is generally mounted on the discharge chute. This not only means it is away from the 'dirtier' feed end, but it will generally be well isolated from the material itself, which has obvious operating advantages in terms of dust protection, etc. The discharge chute simply provides a gravity fed slide for the material to be directed away from the screw conveyor body and into the receiving hopper. For extended routes or elevations, it is possible to connect the discharge chute to the inlet of a subsequent conveyor.

While the tube is generally made of a flexible plastic material, it can also be constructed of a rigid plastic or steel. These designs retain the advantages of the ribbon screw while providing a rigid conveying route where the flexible plastic is not necessarily required. Another advantage of the plastic tube material is that the weight of the overall assembly is reduced over a rigid steel screw conveyor. This lends the assembly to adaptation for a portable unit. Fitzpatrick (1985) gives two additional advantages of using polyethylene tubes: (i) the slick, smooth surface greatly assists with cleaning, that assists with batch integrity and multiple use; (ii) the surface is also largely resistant to chemical attack from most materials. The feed hopper is either mounted on a frame with casters, with the screw conveyor attached to the hopper, or the screw conveyor can be supported separately if necessary. This portability is further enhanced by the ability to use interchangeable screws. Thus, if the one screw conveyor is required to satisfy a number of duties on the one site, the only component that needs to be changed to handle the different material is the screw itself. Indeed, it may even be possible to convey different materials with the same conveyor and screw if the materials are suitably similar and/or the throughput rate is not critical.

Looking at the screw in more detail we find that for the majority of applications a circular cross section is suitable (see Figure 7.20). This design is used for two main reasons; a large number of materials can be conveyed using this design, and it also produces a greater radial force, thus providing a more resilient bearing face of conveyed material between the screw and the tube. A heavier gauge of screw material may be used for heavy bulk materials (bulk density >1500 kg/m^3). While this design is suitable for free-flowing powders and granular materials such as sugar, flour, soda ash, plastic powder, plastic pellets, rice, corn, regrind, etc. (Bourton & Clague 2000), a flat profile of one of two types is used for more difficult powders. For powders that aerate/fluidise easily a flat profile screw is used, as this provides a greater ratio of axial force to radial force compared to the circular profile. For the same size screw this design will give a higher conveying capacity than the circular cross section. This assertion is confirmed in Fitzpatrick (1985). For materials with particularly poor handlability, a tight toleranced flat screw and tube are required to minimise the radial force exerted. This results in the maximum forward motion of material for input power, and minimises the impact with the wall for materials that suffer from degradation and/or caking or packing if overloaded. Figure 7.21 demonstrates the different conveying characteristics of the three different screw profiles.

However, the division between the two types of cross section is not clear cut, and comprehensive testing of different screw configurations is recommended. Furthermore, the screw cross section is not the only variable that needs to be considered. The conveyor tube must be large enough for the desired throughput, with the screw diameter some significant percentage of this value, and the screw pitch must also be specified accordingly. More details on the specification of screw conveyors can be found in Chapter 5. However, due to the specialised nature of these screws with the flexible screw routing and changes in elevation, these figures will most likely only provide a design estimate.

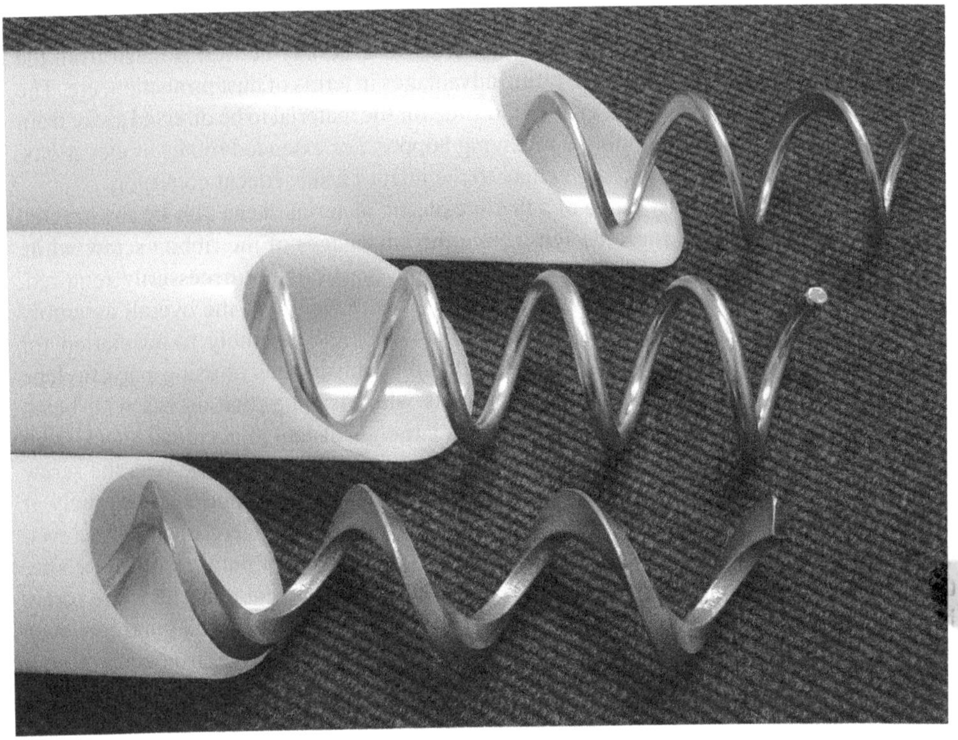

Figure 7.20 Typical flexible screw construction (Spiroflow Ltd).

Typical maximum lengths for flexible screw conveyors range from 4.5 to 12 m dependent on the conveyor diameter and the duty required from the conveyor. The maximum length tends to decrease with increasing conveyor diameter and increasing level of duty, where the duty of the conveyor is considered more strenuous with increasing elevation and material bulk density. Another key characteristic for flexible screw conveyor specification is the minimum required bend radius to avoid fouling of the screw on the conveyor tube. The three variables to be considered in this situation are the conveyor size, the degree of bend

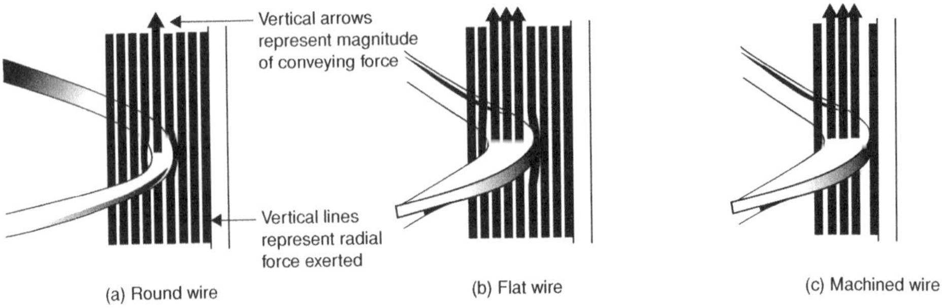

Figure 7.21 Comparative loading of screw cross section types. (Adapted from Bourton and Clague 2000.)

and the type of screw, whether it is flat or round. This minimum radius that ranges from 1.5 to 6 m tends to increase with increasing conveyor size and degree of bend. Furthermore, flat screws require a larger minimum bend radius than do round ones.

The capacity of the flexible screw conveyor, typically given as a volumetric flow rate, ranges from as low as 0.12 m^3/h to 40.0 m^3/h. It can be seen from Figure 7.20 that for the same size screw a flat or machined profile screw will have a greater capacity than a round screw, and from simple geometry a larger screw diameter corresponds to a larger capacity. For example, a typical maximum with a 67 mm diameter conveyor is approximately 1 m^3/h and to achieve the 40 m^3/h capacity a 220 mm diameter screw would be required. The volumetric throughput is also dependent on the rotational speed of the screw, with approximately double the capacity for a doubling in screw speed – however, this relationship becomes less and less reliable with increasing screw size.

7.2.7 *Sack fillers*

For smaller operations where the output is smaller than that catered for by a bulk bag or FIBC, this type of equipment is ideal. In its simplest form this piece of equipment consists of a storage hopper with a feed chute mounted on the side that is used to fill the sack. It will also have some sort of supporting platform for the sack that may or may not include a load cell arrangement to assist with weigh filling (see Figure 7.22) (Accrapak Ltd 2006).

Four other types of machine that are available on the market are the turbo impeller packer, the force flow packer, the screw packer and the twin head packer. The impeller packer allows continuous filling while still operating in a dust-free mode. This is ideal for high throughput operations where the bag size needs to be smaller for ease of handling, etc. The force flow packer is useful for products that are poor flowing, yet have good aeration characteristics. By a combination of an upward airstream to provide fluidisation and a pressurised delivery hopper, these poor-flowing materials are fed into the sack. The screw packer is for those materials that are poor flowing and have poor aeration characteristics, this making them unsuitable for any of the previous options. Basically a screw conveyor/feeder with a sack filling attachment, this sack filler is to be used with the most cohesive of products. It has a ribbon type screw as found in a flexible screw conveyor. The twin head arrangement allows greater throughput without the need for the impeller operation described above (see Figure 7.23). Most configurations also come with variable sack attachment devices to handle a range of different sack sizes and types.

Typical materials that can be used in these devices are abrasive grit, agri-chemicals, cat litter, coarse powders, grain, industrial minerals, industrial sand, pet foods, plastic granules, rubber crumb, salt, seeds and sugar (A.T. Sack Fillers Ltd 2006).

7.2.8 *Small volume batch conveyors*

While negative pressure conveying systems have been covered in some detail in Chapter 4, there is a small sub-set that will be considered here. Small volume batch conveyors, or simply 'vacuum conveyors', generally consist of a small tank and a relatively short conveying line in which a vacuum is generated, and this is connected to a material feed hopper. Once the required negative pressure has been reached using the attached vacuum pump, a valve is opened on the feed hopper and the material is conveyed into the vacuum tank. The system

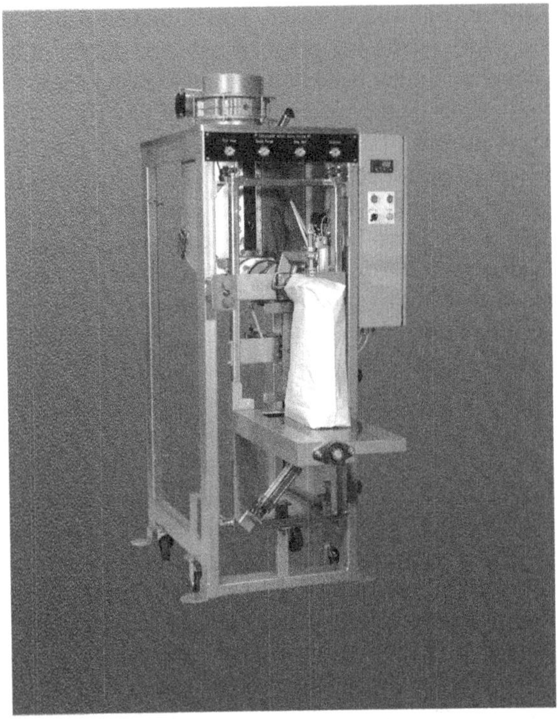

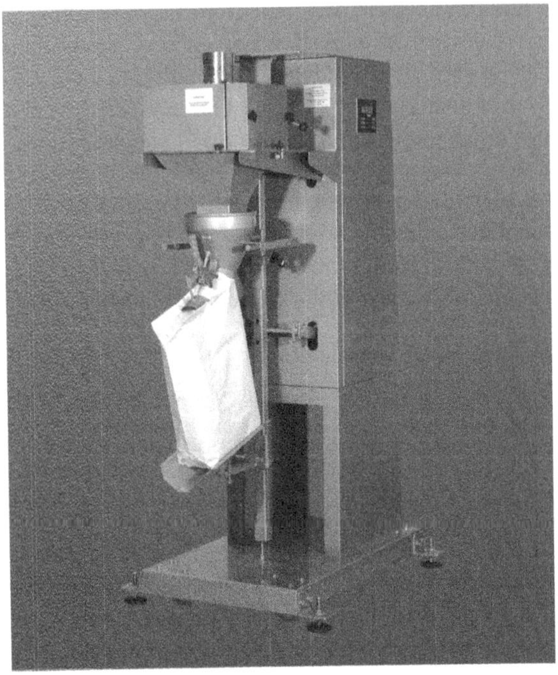

Figure 7.22 Typical sack filling devices (A.T. Sack Fillers Ltd).

Figure 7.22 (*continued*).

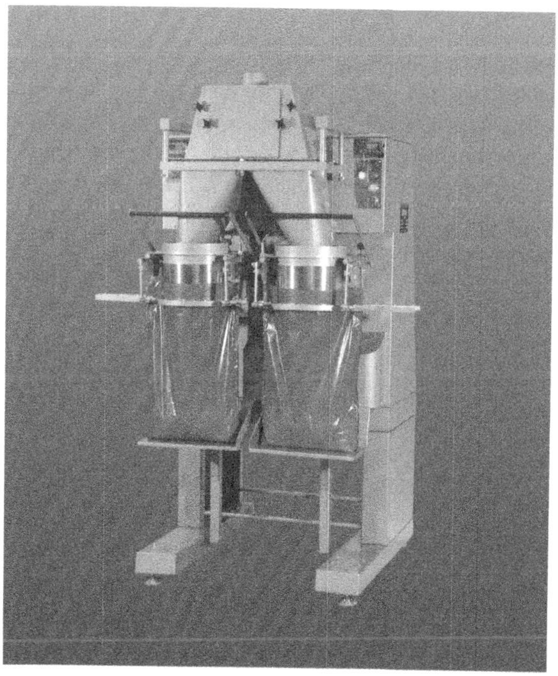

Figure 7.23 Twinhead sack filler (A.T. Sack Fillers Ltd).

is generally controlled either pneumatically or electronically, so once the vacuum tank is full, or has reached a pre-set level, a valve at the bottom of the vacuum tank opens and the material is discharged into the down-stream process. The valves then all return to their original settings and the next cycle begins resulting in a subsequent batch being delivered to the vacuum tank. The vacuum tank will generally include an integral dust filter. Investigations of two suppliers of this type of equipment (K-Tron 2006; PIAB 2006) indicate that a modular approach can be taken to the selection of a vacuum conveyor for a particular duty. That is, the size of tank, the size of vacuum pump and the bottom valve unit can all be combined, if required, to produce the best conveyor for a duty. These two suppliers have been chosen as examples of this type of equipment due to the availability of a large amount of readily available equipment specification and selection information. However, there are a number of alternative suppliers that also need be mentioned, such as Clyde Materials Handling (2006), Vac-U-Max (Vibrair 2006), Pneuvac Conveyors (Spiroflow Ltd 2006), Portasilo Bulk Handling Systems (2006) (for slightly larger systems) and Colortronic UK Ltd (2006). Some further aspects of the selection process are now discussed.

PIAB (2006) outlines procedures for the selection of both standard and customised conveyors on its website. These procedures based on recommended values and should only be used as a guide for system specification. It is recommended that the material to be conveyed is tested by the supplier or independently for its pneumatic conveying capability before committing to a conveyor system specification.

Both the standard and customised conveyor selection processes have the same skeleton, but there are more details given for the customised process. Step 1 of the standard procedure identifies a power requirement equation that is used to identify one of a number of models:

$$\mathrm{Pm} = L \times Q \tag{7.2}$$

where Pm is the power requirement (m tonne/h); L is the total conveying distance (sum of the vertical and horizontal lengths) (m); Q is the capacity of system (tonne/h).

Thus, those conveyors with a larger Pm value require a larger pump unit, with values in the range of 3–5 being at the low end of the scale and 60–80 at the high end.

PIAB have also restricted the use of Equation (7.2) to conveying lengths of 4–30 m, materials of bulk density in the range of 500–1800 kg/m^3 and for particles sizes of less than 5 mm. This bulk density range covers cement, dry sand, flour, nylon pellets, salt and sugar, to name a few common materials (McGlinchey 2005; Powder and Bulk Dot Com 2006).

An example of using Equation (7.2) to select a conveying system is given below for a flour application.

Conveying length: 20 m; throughput: 2 tonne/h; bulk density: 670 kg/m^3; maximum particle size: 150 μm

$$\begin{aligned}\mathrm{Pm} &= L \times Q\\ &= 20 \times 2\\ &= 40 \text{ metric tonne/h}\end{aligned}$$

This gives a choice of at least three conveyor configurations (PIAB 2006). The decision from here becomes a balance of economics and versatility. The configurations with a larger pump unit have some excess capacity that could be used if required, but the smaller configurations both have the capacity required, at a cheaper cost. The bulk density of the material, the maximum particle size and the conveying length are all within the constraints required for the use of Equation (7.2).

Once the conveyor has been chosen, it is necessary to choose the conveying line diameter that will be used to convey the material from the loading hopper to the vacuum receiver. The selection of pipe diameter is based upon the conveyor model chosen and the material's bulk density. Typical pipe diameters are in the range of 32–102 mm, and generally lighter products require a larger pipe dimension. This is due to the fact that the velocity of the conveyed product will be the same for each conveyor selected from the previous process, so to achieve a greater mass throughput a greater cross-sectional area is required. Once again it is worth stating that these figures are recommendations and should be confirmed by product testing.

Therefore, for the three conveyors selected in the previous step, the pipeline diameters are 76 mm or 102 mm (PIAB 2006).

The customised selection process takes the customer through the individual units that go to make up a complete conveyor (PIAB 2006):

- Pump unit – the part of the conveyor that allows the whole to meet the capacity and conveying distance, as shown in the first part of the standard selection process.
- Filter unit – dependent on the type of material and its bulk characteristics.
- Connection unit – the connection between the receiving vessel and the pipeline, this is also dependent on the material characteristics.
- Bottom valve unit – available in a number of materials, and with or without fluidisation, the application and material characteristics decide the selection of this item.
- Control unit – connected to both the pump and the bottom valve, this unit 'runs' the conveying system.
- Nylon tubing unit – provides the connection between the control unit, pump and bottom valve, including fluidisation ports, if required.

Another supplier of this equipment is K-Tron (2006), where the main application appears to be the use of this type of equipment as a loading/re-filling mechanism for a range of feeders. Thus, the selection guide is geared more in this direction. K-Tron use three selection parameters: material type, material flow and conveying capability (conveying rate and distance); and as with the PIAB equipment, the selection process is only given as a guide, and it is recommended that the material to be conveyed is tested either by the equipment supplier or independently to assess its conveying characteristic. Material type is divided into ranges of particle size, and also takes into account particle shape. However, this appears slightly redundant as the selection table infers that any of their conveying systems are capable of handling any of the different shaped products. Furthermore, the only apparent consideration for the finer powders is to use a conveying system with an in-built filter. The material flow section is divided into five flow regimes: very free flowing; free flowing; average flowing; sluggish flowing; and non-free flowing. No further information is given regarding these sub-divisions, but there are a number of different methods where a similar characterisation can be achieved, such as Carr's compressibility index or the angle of repose (see Chapter 1 and McGlinchey 2005). Conveying rate is divided into four sections, up to 500 kg/h, up to 1000 kg/h, up to 2500 kg/h and up to 5000 kg/h. To achieve a short (5 m), medium (5–25 m) or long (>25 m) conveying distance, different conveyors are then specified for each conveying rate. This information can then be used to select a conveyor configuration for the flour application given above (K-Tron 2006).

Apart from the feeder application mentioned above another application of small volume batch pneumatic conveyors is for a receiving vessel above a mixer inlet (Burgess 2006). The vacuum receiver vessel is mounted on load cells, and the control unit is set up to convey each of the ingredients into the receiving vessel. Once all the ingredients have been loaded and the mixer is emptied from the previous cycle, the vacuum receiver can be emptied into the mixer and while the mixing is taking place, the next load can be conveyed. This has the advantage of reduced down-time for the mixer as it is not waiting for material to be loaded whilst not in mixing mode. Two other applications noted by Burgess (2006) are a check screen and a metal separation screen. Because of their wide use in the food industry (being good for small batch weighing and with a sanitary finish such as 316 stainless steel) both of these applications are quite important. Occasionally, the bag slitters that have been discussed previously in this section can leave scrap pieces of bag in the process material. A check screen mounted in the vacuum conveying line can help reduce the amount of this 'tramp' material, with the air passing through only serving to increase the collecting efficiency. Metal separation is achieved by mounting a 'bullet' magnet in-line in the vacuum conveying line. A design advantage of the 'bullet' magnet is that it minimises the obstruction to flow and with sufficient air velocity no residual product remains attached to it, thus reducing the hygiene risk. Metal pieces attached to the magnet are cleaned from the 'bullet' by access through a trap door.

For shorter distance conveying, remembering that vacuum systems are generally limited to a maximum of 0.5 bar gauge line pressure drop, it is also possible to convey materials in a low velocity dense phase mode (Burgess 2006). Some examples of products conveyed in this mode are roasted coffee beans, puffed rice, oats, spray dried milk powder, peanuts and breadcrumbs. Whole almonds were conveyed along a 65 mm flexible tube at a rate of 1500 kg/h into a hopper in another application (Walker 1994). Indeed, the range of materials that can be conveyed in a vacuum conveying system stretch well beyond these examples given for the food industry. Examples of other industry sectors where this technology is applied are plastics and rubber, e.g. polyester pellets, polythene powder, rubber granules; chemicals/detergents, e.g. aluminium oxide, detergents, soap flakes, magnesium oxide, titanium dioxide; and pharmaceuticals, e.g. penicillin powder (Jenco Engineered Solutions 2006).

7.3 Summary

This section has provided an introduction to the equipment that is available to smaller operators in order to make their operations more efficient. While one piece of equipment may provide some benefits, it may be necessary to consider the combination of a number of pieces of equipment as described above. It is recommended that some knowledge of characterisation be obtained to assist with the use of all this equipment. Labour-saving equipment is not going to be of use if it has not been correctly specified for the bulk solid it is handling. There are also a number of good references on the subject of particle and bulk solid characterisation for those who wish further information in this area (McGlinchey 2005; Woodcock & Mason 1987).

References

Accrapak Ltd (2006) *Sack Fillers*. http://www.accrapak.co.uk/. Accessed 21 November 2006.

Angus, W. (1999) The aero-mechanical conveyor. In: *Presentation to the Solidex Symposium 99*, Harrowgate, England.

Aptech (Powder Systems) Limited (2006) *Bulk Bag Dischargers, Bulk Bag Fillers, Sack Tipping Stations*. http://www.aptech.uk.com/index.php. Accessed 21 November 2006.

Arodo Total Bag Handling (2006) *Bulk Bag Fillers, Sack Fillers*. http://www.arodo.com/en/index.html. Accessed 21 November 2006.

A.T. Sack Fillers Ltd (2006) *Sack Fillers*. http://www.simplafillsystems.co.uk/. Accessed 21 November 2006.

Bourton, K. & Clague, K. (2000) Understanding flexible screw conveyors. In: *From Powder to Bulk: International Conference on Powder and Bulk Solids Handling, IMechE Conference Transactions*. London, UK, pp. 419–431.

Burgess, R. (October 2006) Vacuum conveying. *Solids & Bulk Handling*, **32**(8), 8–9.

Clyde Materials Handling (2006) *Vacuum Conveyors*. http://www.clydematerials.co.uk/products/vacuum/ index.cfm?Content_ID=9C334158-2AE7-41DD-B6BB2FCD42340CC8. Accessed 20 November 2006.

Colortronic UK Ltd (2006) *Vacuum Conveyors*. http://www.colortronic.co.uk/products/conveying/conveying_central.asp. Accessed 20 November 2006.

Dynamic Air Conveying Systems (2006) *Bag Breakers, Big Bag Unloaders*. http://www.dynamicair.co.uk/products.html. Accessed 21 November 2006.

Entecon UK Ltd (2006) *Aeromechanical Conveyors, Automatic Sack Opener, Bulk Bag Dischargers, Bulk Bag Fillers, Sack Tipping Stations*. http://www.entecon.co.uk/page.php?page=_products. Accessed 21 November 2006.

European Agency for Health and Safety at Work (EU-OSHA) (2006) http://osha.europa.eu/OSHA. Accessed 5 December 2006.

Fairport Process Materials Handling Division (2006) http://www.fairport.co.uk/pmh/index.htm. Accessed 5 December 2006.

Fitzpatrick, D.T. (1985) Flexible high-speed screw conveyors applications and benefits. In: *Proceedings of the Technical Program – Powder & Bulk Solids Conference/Exhibition*, Rosemont, IL, pp. 445–451.

Flexicon Corporation (2006) *Bag Dump Stations, Bulk Bag Dischargers, Bulk Bag Fillers, Drum Dumpers, Flexible Screw Conveyors*. http://www.flexicon.com/us/index.asp. Accessed 21 November 2006.

Flomat Bagfilla International (2006) *Bulk Bag Dischargers, Bulk Bag Fillers*. http://www.fairport.co.uk/pmh/flomat/index.htm. Accessed 21 November 2006.

Health and Safety Executive (HSE) (2006) http://www.hse.gov.uk/index.htm. Accessed 21 November 2006.

Hosokawa Containment (2006) *Bulk Bag Dischargers, Bulk Bag Fillers*. http://www.hosokawa.co.uk/fibc.php. Accessed 21 November 2006.

ISL Cone Valve Technology (2006) *Cone Valves for Bulk Bag Discharge*. http://www.conevalve.com/. Accessed 21 November 2006.

Jenco Engineered Solutions (2006) *Vacuum Conveyors* http://www.jenco.co.uk/products.asp?step=2&id=2&pstring=2. Accessed 20 November 2006.

K-Tron Process Group (2006) *Small Volume Batch Conveyors*. http://www.ktron.com/Products/Pneumatic_Conveyors/loaders_overview.cfm?x loop=1&CFID=8052470&CFTOKEN=a40d8ba4a9ca8c27-5CDC52E0-3048-422D-483029DD67B572A7. Accessed 21 November 2006.

Matcon (2006) *Bulk Bag Dischargers, Bulk Bag Fillers*. http://www.matcon-cone.com/. Accessed 5 December 2006.

McGlinchey, D. (ed) (2005) *Characterisation of Bulk Solids*. Blackwell, Oxford, UK.

Morray Beltyne (2006) *Bag Slitting Machines, Bulk Bag Dischargers*. http://www.beltyne.com/Products.htm. Accessed 21 November 2006.

Palamatic Materials Handling (2006) *Bag Openers, Sack Tipping Stations, Vacuum Bag Lifters*. http://www.palamatic.com/materials/. Accessed 21 November 2006.

PIAB AB (2006) *Small Volume Batch Conveyors*. http://www.piab.com/Templates/Page.aspx?id=10676. Accessed 21 November 2006.

Portasilo Bulk Handling Systems (2006) *Vacuum Conveyors*. http://www.portasilo.co.uk/vacuum.asp. Accessed 20 November 2006.

Powder and Bulk Dot Com (2006) *Bulk Density Chart*. http://www.powderandbulk.com/resources/bulk_density/material_bulk_density_chart_a.htm. Accessed 21 November 2006.

Spiroflow Ltd (2006) *Aeromechanical Conveyors, Bin, Sack and Drum Emptiers, Bulk Bag Dischargers, Bulk Bag Fillers, Flexible Screw Conveyors, Sack Fillers, Vacuum Conveyors*. http://www.spiroflow.com/index.html. Accessed 21 November 2006.

S.T.S. Safety Trolley Systems (2006) *Drum Rotator*. http://www.sts-trolleys.co.uk/index.htm. Accessed 21 November 2006.
Unitrak Corporation Ltd (2006) *Aeromechanical Conveyors, Bulk Bag Dischargers, Flexible Screw Conveyors*. http://www.unitrak.com/. Accessed 21 November 2006.
U.S. Department of Labor Occupational Safety and Health Administration (OSHA) (2006) www.osha.gov. Accessed 5 December 2006.
Vibrair (2006) *Bag Tip Stations, Bulk Bag Dischargers, Flexible Screw Conveyors*. http://www.vibrair.com/products.html. Accessed 21 November 2006.
Walker, K.A. (January/March 1994) Vacuum conveying to improve productivity and cut costs. *Powder Handling Process.*, **6**(1), 106.
Webster Griffin Ltd (2006) *Bulk Bag Dischargers, Bulk Bag Fillers*. http://www.webstergriffin.com/. Accessed 21 November 2006.
WJ Morray Engineering Ltd (2006) *Bag Slitters*. http://www.morray.com/. Accessed 21 November 2006.
Woodcock, C.R. & Mason, J.S. (1987) *Bulk Solids Handling: An Introduction to the Practice and Technology*. Chapman and Hall, New York.
Wypych, P.W. (1999) Pneumatic conveying of powders over long distances and at large capacities. *Powder Technol.*, **104**(3), 278–286.

Index

Printed and bound by CPI Group (UK) Ltd, Croydon, CR0 4YY

15/09/2026

14962393-0003